About Island Press

Island Press is the only nonprofit organization in the United States whose principal purpose is the publication of books on environmental issues and natural resource management. We provide solutions-oriented information to professionals, public officials, business and community leaders, and concerned citizens who are shaping responses to environmental problems.

Since 1984, Island Press has been the leading provider of timely and practical books that take a multidisciplinary approach to critical environmental concerns. Our growing list of titles reflects our commitment to bringing the best of an expanding body of literature to the environmental community throughout North America and the world.

Support for Island Press is provided by the Agua Fund, The Geraldine R. Dodge Foundation, Doris Duke Charitable Foundation, The Ford Foundation, The William and Flora Hewlett Foundation, The Joyce Foundation, Kendeda Sustainability Fund of the Tides Foundation, The Forrest & Frances Lattner Foundation, The Henry Luce Foundation, The John D. and Catherine T. MacArthur Foundation, The Marisla Foundation, The Andrew W. Mellon Foundation, Gordon and Betty Moore Foundation, The Curtis and Edith Munson Foundation, National Fish and Wildlife Foundation, Oak Foundation, The Overbrook Foundation, The David and Lucile Packard Foundation, Wallace Global Fund, The Winslow Foundation, and other generous donors.

The opinions expressed in this book are those of the author(s) and do not necessarily reflect the views of these foundations.

About the Pacific Institute for Studies in Development, Environment, and Security

The Pacific Institute for Studies in Development, Environment, and Security, in Oakland, California, is an independent, nonprofit organization created in 1987 to conduct research and policy analysis in the areas of environmental protection, sustainable development, and international security. Underlying all of the Institute's work is the recognition that the urgent problems of environmental degradation, regional and global poverty, and political tension and conflict are fundamentally interrelated, and that long-term solutions dictate an interdisciplinary approach. Since 1987, we have produced more than sixty research studies, organized roundtable discussions, and held widespread briefings for policymakers and the public. The Institute has formulated a new vision for long-term water planning in California and internationally, developed a new approach for valuing well-being in local communities, worked on transborder environment and trade issues in North America, analyzed ISO 14000's role in global environmental protection, clarified key concepts and criteria for sustainable water use in the lower Colorado basin, offered recommendations for reducing conflicts over water in the Middle East and elsewhere, assessed the impacts of global warming on freshwater resources, and created programs to address environmental justice concerns in poor communities and communities of color.

For detailed information about the Institute's activities, visit www.pacinst.org, www.worldwater.org, and www.globalchange.org.

THE WORLD'S WATER
2006-2007

THE WORLD'S WATER

2006-2007

The Biennial Report on Freshwater Resources

Peter H. Gleick

with Heather Cooley, David Katz, Emily Lee, Jason Morrison, Meena Palaniappan, Andrea Samulon, and Gary H. Wolff

Pacific Institute for Studies in Development,
Environment, and Security
Oakland, CA

ISLANDPRESS

Washington • Covelo • London

ISLAND PRESS is a trademark of The Center for Resource Economics.

Library of Congress Card Catalog Number 98–24877
ISBN 10: 1-59726-105-X (cloth)
ISBN 13: 978-1-59726-105-0 (cloth)

ISBN 10: 1-59726-106-8 (paper)
ISBN 13: 978-1-59726-106-7 (paper)

ISSN 1528-7165

Printed on recycled, acid-free paper ✪

Manufactured in the United States of America.
10 9 8 7 6 5 4 3 2 1

To the past and present staff of the Pacific Institute, who have made this a job I still look forward to every morning.

Peter Gleick
2006

Contents

Foreword   xiii

Introduction   xv

ONE   Water and Terrorism   1
          by Peter H. Gleick

          Introduction   1
          The Worry   2
          Defining Terrorism   3
          History of Water-Related Conflict   5
          Vulnerability of Water and Water Systems   15
          Responding to the Threat of Water-Related Terrorism   22
          Water Security Policy in the United States   25
          Conclusion   25

TWO   Going with the Flow: Preserving and Restoring Instream
          Water Allocations   29
          by David Katz

          Environmental Flow: Concepts and Applications   30
          Legal Frameworks for Securing Environmental Flow   34
          The Science of Determining Environmental
              Flow Allocations   38
          The Economics and Finance of Environmental
              Flow Allocations   40
          Making It Work: Policy Implementation   43
          Conclusion   45

THREE   With a Grain of Salt: An Update on Seawater Desalination   51
          by Peter H. Gleick, Heather Cooley, Gary Wolff

          Introduction   51
          Background to Desalination   52
          History of Desalination   54
          Desalination Technologies   54
          Current Status of Desalination   55
          Advantages and Disadvantages of Desalination   66
          Environmental Effects of Desalination   76
          Desalination and Climate Change   80

Public Transparency   81
Summary   82
Desalination Conclusions and Recommendations   83

FOUR   Floods and Droughts   91
          by Heather Cooley

          Introduction   91
          Droughts   92
          Floods   104
          The Future of Droughts and Floods   112
          Conclusion   113

FIVE   Environmental Justice and Water   117
          by Meena Palaniappan, Emily Lee, Andrea Samulon

          Introduction   117
          A Brief History of Environmental Justice in the
             United States   119
          Environmental Justice in the International
             Water Context   123
          Environmental Justice and International Water Issues   124
          Recommendations   137
          Conclusion   141

SIX   Water Risks that Face Business and Industry   145
          by Peter H. Gleick, Jason Morrison

          Introduction   145
          Water Risks for Business   146
          Some New Water Trends: Looking Ahead   150
          Managing Water Risks   153
          An Overview of the "Water Industry"   158
          Conclusion   163

WATER BRIEFS
One   Bottled Water: An Update   169
          by Peter H. Gleick

Two   Water on Mars   175
          by Peter H. Gleick

          Introduction   175
          Background   175
          Martian History of Water   178
          Future Mars Missions   180
Three   Time to Rethink Large International Water Meetings   182
          by Peter H. Gleick

          Introduction   182
          Background and History   182

Outcomes of International Water Meetings   183

Ministerial Statements   184

Conclusions and Recommendations   185

4th World Water Forum
  Ministerial Declaration   186

Four   Environment and Security: Water Conflict Chronology
       Version, 2006–2007   189
       by Peter H. Gleick

Five   The Soft Path in Verse   219
       by Gary Wolff

DATA SECTION

Data Table 1   Total Renewable Freshwater Supply by Country   221

Data Table 2   Freshwater Withdrawal by Country and Sector   228

Data Table 3   Access to Safe Drinking Water by Country,
               1970 to 2002   237

Data Table 4   Access to Sanitation by Country, 1970 to 2002   247

Data Table 5   Access to Water Supply and Sanitation by Region,
               1990 and 2002   256

Data Table 6   Annual Average ODA for Water, by Country, 1990 to 2004
               (Total and Per Capita)   262

Data Table 7   Twenty Largest Recipients of ODA for Water,
               1990 to 2004   268

Data Table 8   Twenty Largest Per Capita Recipients of ODA for Water,
               1990 to 2004   270

Data Table 9   Investment in Water and Sewerage Projects with Private
               Participation, by Region, in Middle- and Low-Income
               Countries, 1990 to 2004   273

Data Table 10   Bottled Water Consumption by Country,
                1997 to 2004   276

Data Table 11   Global Bottled Water Consumption, by Region,
                1997 to 2004   280

Data Table 12   Per Capita Bottled Water Consumption by Region,
                1997 to 2004   282

Data Table 13   Per Capita Bottled Water Consumption, by Country,
                1999 to 2004   284

Data Table 14   Global Cholera Cases and Deaths Reported to the World
                Health Organization, 1970 to 2004   287

Data Table 15   Reported Cases of Dracunculiasis by Country, 1972 to
                2005   293

Data Table 16   Irrigated Area, by Region, 1961 to 2003   298

Data Table 17   Irrigated Area, Developed and Developing Countries,
                1961 to 2003   301

Data Table 18   The U.S. Water Industry Revenue (2003) and Growth
                (2004–2006)   303

Data Table 19   Pesticide Occurrence in Streams, Groundwater, Fish, and
                Sediment in the United States   305

Data Table 20   Global Desalination Capacity and Plants—
                January 1, 2005   308

Data Table 21 100 Largest Desalination Plants Planned, in Construction,
              or in Operation—January 1, 2005   310
Data Table 22 Installed Desalination Capacity by Year, Number of
              Plants, and Total Capacity, 1945 to 2004   314

WATER UNITS, DATA CONVERSIONS, AND CONSTANTS 319

COMPREHENSIVE TABLE OF CONTENTS 329

Volume 1: The World's Water 1998–1999: The Biennial Report on
           Freshwater Resources   329
Volume 2: The World's Water 2000–2001: The Biennial Report on
           Freshwater Resources   332
Volume 3: The World's Water 2002–2003: The Biennial Report on
           Freshwater Resources   335
Volume 4: The World's Water 2004–2005: The Biennial Report on
           Freshwater Resources   338

COMPREHENSIVE INDEX 343

Foreword

A huge roadside billboard near the venue of the Fourth World Water Forum in Mexico in March 2006 proclaimed that "the next world war will be fought over water." Right or wrong, it certainly caught the delegates' attention and worried, irritated, and confronted them.

Inside that forum and around the world, sector specialists have increasingly been addressing the links between water and security. Meanwhile, the people working on water resources and on drinking water and sanitation, and those managing water for agriculture and for ecosystems, are starting to better understand each other. Dams and private sector participation remain contentious, but people are, at least, listening to each other's views on them. Most importantly, water professionals are now explaining their problems and solutions to political decision makers, after years of talking among themselves.

The concept of *peak oil*, that the world is reaching a maximum level of production of the oil on which much of the global economy depends, has become well-known and seriously debated recently. An equally important concept could be termed *peak water*, that the world is approaching a maximum level of exploitation of the water on which all of human life depends. In many parts of the world, water is already used faster than natural hydrological processes can replenish it. Almost everywhere, water resources are under huge stress from mankind. For example, economists rightly hail the large number of people in China and India who have climbed out of poverty in recent years. But many of them are living on borrowed water, as the economic growth and agricultural productivity of the North China Plain and the Indo-Gangetic Basin are accompanied by dropping groundwater levels and drying rivers. Sustaining those people's improved lifestyles will involve using water resources elsewhere to grow their food and manufacture their goods. Although water is usually categorized as a local resource, the water-related problems of some very important and populous parts of the world will be solved only by political, economic, and agricultural links with other places with more water. Meanwhile, the Millennium Development Goals will continue to command political attention—and almost every one of them is linked to water.

These water-related subjects are important to many groups of people, including politicians, researchers, media professionals, and water sector practitioners. They are the very people for whom *The World's Water* is published. Combining information and commentary, the book bridges the gap between science and policy. This is its fifth biennial volume. Unlike many periodicals, the successive volumes do not supersede the previous ones but complement each other to form one cumulative body of knowledge.

The lead author, Peter Gleick, modestly avoids the limelight and prefers to influence others through the thoroughness of his research and the intellectual quality of his arguments. He and his colleagues at the Pacific Institute are respected for their scholarship, impartiality, inter-disciplinary approach, and commitment to the beneficial application of science. Perhaps the strongest areas of their work are on water and security and on water and climate. They also contribute relevant material on water and poverty, agricultural and ecosystem uses of water, water governance, and many other subjects. All of these topics benefit from clear, objective data and analysis, and *The World's Water* provides just that.

Since Peter Gleick started this series in 1998-99, a similar publication has come on the scene: *The World Water Development Report,* produced every three years by a collaborative effort of no fewer than twenty-seven UN agencies. Meanwhile, Web sites on water have multiplied, and vast amounts of data and opinions are available to people with Internet access. So why do we still need *The World's Water?* I believe its strength lies in illuminating the complex world of water from one viewpoint. Where *The World Water Development Report* is encyclopedic, *The World's Water* is analytical and concentrates on particular topics that the authors have identified as especially relevant and important. The two publications complement rather than compete with each other; both should have a place on the bookshelves of people seriously interested in grasping global water issues.

Every person uses water in many different ways every day. So water permeates every social, economic, and political area of human activity; it cannot be treated simply as a technical subject. But if we look deeper and examine our attitudes to and uses of water, I believe we can see three behavior patterns that give particular hope for the future. The first is the rights-based and democratic approach that encourages ordinary people to challenge and question political leaders about their decisions. The second is the application of human intelligence to solving scientific problems and harnessing natural forces for the benefit of mankind. The third is our sense of interdependence and the need to work together for the planet, the environment, and ourselves.

Democratic accountability, human ingenuity, and collaborative working can solve the world's water problems and prove the Mexican billboard wrong. This book will contribute to that process. I commend it to you.

Jon Lane
Blantyre, Malawi
May 2006

Introduction

Welcome to the fifth volume of *The World's Water: The Biennial Report on Freshwater Resources*. It is difficult to believe that ten years have gone by since I first began producing these global assessments, and more than twenty years have passed since this book's predecessor, *Water in Crisis*, was published. I'd like to say that all water problems have now been solved, but, of course, I cannot. Indeed, it is easy to wonder whether the world will ever truly commit the resources, effort, and money required to meet basic human needs for water for all, to protect environmental needs for water, and to resolve water conflicts with discussion, not violence. But I still hope.

As with each of the preceding volumes, this book offers information on issues of topical importance and data and insights into water challenges facing the public, policy makers, and scientists. In this volume, with new authors and new data, we hope to add a little more to the vast information available on the world's water problems and solutions, and to draw lessons and insight from that information. Each of the volumes of *The World's Water* builds and adds on to the previous ones. The chapters are an evolving mix of new and updated discussions, information, and raw data. In the last volume, we included a complete Table of Contents, and we continue that here, with an integrated index that permits readers to find information across all five books.

The importance of water continues to grow. Concerns about availability and quality of freshwater resource are at the top of the world's environmental agenda. In December 2003, the United Nations General Assembly officially declared the years 2005 to 2015 to be the "Water for Life" International Decade for Action. In March 2006, nearly twenty thousand water activists, scientists, experts, and policy makers met at the Fourth World Water Forum in Mexico City to discuss water issues. Among the top priorities reaffirmed at that forum was meeting the Millennium Development Goals (MDGs) set by the United Nations to address the woes of the world's poorest populations. Two of these MDGs are aimed at reducing unmet basic water supply and sanitation needs—a regular theme in previous volumes of *The World's Water* and continued here.

Progress is being made, but as I seem to note every two years, it seems excruciatingly—and unnecessarily—slow. The MDGs for water are again discussed in several chapters in this volume, and new data are presented in Data Tables 3, 4, and 5 that suggest we are going to fail to meet those goals. Inadequate resources are being devoted to meeting them at all levels and in all regions.

In the post 9/11 world, concerns over security have spilled over directly into concerns over water and terrorism. The link between water and security has long been a theme in *The World's Water*, with several chapters addressing water and conflicts and international distribution of freshwater and the regular update of the Water Conflict Chronology. My opening chapter in the current volume explicitly tackles the question of how vulnerable our water systems may be to terrorism of different kinds, explores the history of past terrorism events related to water, and offers approaches that we might pursue to reduce those vulnerabilities.

This volume also addresses, in a chapter by David Katz, the question of the critical needs of the environment for water and approaches and tools for providing ecosystem flows. Many of the most serious environmental problems we face globally are associated with how and when humans take water for their own use. New efforts and new tools are now needed to protect aquatic ecosystems while still meeting human needs, and Katz brings clarity to this complex challenge.

We update and expand our previous work on desalination with a chapter by Heather Cooley, me, and Gary Wolff. Based in part on a new detailed assessment of the risks and benefit of desalination, this chapter explores the status of new projects around the world, describes the reliability, environmental, and economical advantages and disadvantages of seawater desalination, and offers suggestions for policy makers considering this innovative, albeit still costly, approach to water supply. New desalination data are also provided in the Data Tables section of the book.

Cooley also delves into the twin challenges of floods and droughts. These extreme events are the most damaging and lethal of all natural disasters and are the result of both natural hydrologic variability as well as the way we manage our land, developments, and expanding populations. Cooley reviews the history and definitions of floods and droughts and policies that may help reduce human exposures and risks.

A new issue addressed in this volume is the environmental justice implications of water developments, allocations, economics, and impacts. Meena Palaniappan, Emily Lee, and Andrea Samulon explore the history of the environmental justice movement in the United States and worldwide, with definitions and discussions of the special challenges that water poses for poorer, often politically powerless, communities. Among the issues reviewed are inequitable access to basic human needs for water, severe impacts that the construction of major dams and water infrastructure often has on rural communities with little or no say in water decision making, and the differential effects of water-related diseases, climate change, and other environmental challenges on rich and poor, North and South, and women and minorities everywhere.

The last chapter, by me and Jason Morrison, provides an overview of the close connections between water use and industrial development and production. Water is a key element in industrial production, and a more than $400-billion-per-year industry has developed to provide water technology, service, consulting, and commercial goods. At the same time, inadequate water quality or quantity, or poor management and planning, may expose businesses to new and unanticipated risks to operations, corporate reputation, and profit and viability. Gleick and Morrison offer data and insight into both the risks facing the business sector from water challenges and strategies to reduce those risks.

As always, *The World's Water* includes shorter Water Briefs on some fun, interesting, or developing topics around water. The 2004-2005 volume included a chapter on bottled water that was timely and popular, given the serious increase in consumption and the growing concerns about bottled water costs and consequences. As a result, we provide an update to the bottled water industry here as a Water Brief, with new data on use and new information on some of the challenges that this sector poses and faces. A new set of Data Tables on bottled water use is also presented.

Continuing a trend, we also provide a Water Brief on the status of the search for water outside of our own planet—in this case, on the planet in our solar system most like our own: Mars. In the past two years, remarkable progress has been made in landing instruments on Mars and in collecting and understanding data on current and past Martian hydrology. It now appears unambiguous that significant amounts of water were present, in liquid form, during Mars' past, and that water still exists in frozen and vapor forms, with possible groundwater in some places. There is no doubt our understanding of Martian hydrology and paleohydrology will continue to change and improve over time.

While we regularly discuss and expand upon the concept of the Soft Path for water in the biennial reports, this volume offers the first paean in verse form to the concept, with a poem by Gary Wolff. This poem was presented at the Fourth World Water Forum in Mexico City and touches on both the human and technological aspects of the Soft Path. The Mexico City conference is also addressed in the following Water Brief, which offers a commentary on the value of large water meetings, in general, based on an analysis done by Peter Gleick and Jon Lane, who has also kindly provided the Foreword to this volume.

The last Water Brief is our regular update of the enormously popular Water Conflict Chronology—published in each of the volumes of *The World's Water*. Researchers and readers from around the world have consistently commented on its value as an analytical and educational tool and on its historical interest, and we are delighted that many readers send new entries to be considered and added. This version is a substantial revision from the past versions, with recategorized entries. We will continue to maintain this as a regular feature, also available at http://www.worldwater.org.

Finally, as always, *The World's Water* offers an extensive section of water-related data. Data Tables 1 and 2 on water availability and use by country remain consistently the most sought-after data by researchers, media, and the public, and we update them again here. Other data are also updated from previous volumes, including Data Tables 3, 4, and 5 on progress toward meeting basic human needs for water and sanitation, tables on desalination capacity and bottled water use, and the disturbing slowing trend in irrigated acreage, especially in developed countries. We also provide new data this year on the industrial water sector, overseas development assistance trends, public/private investment experience in water projects, and pesticide occurrence in U.S. surface and groundwater. All five volumes now offer unique data sets as well as some consistently updated data (take a look at the complete Table of Contents to get a listing of the data tables from all the volumes).

I must offer a special note of thanks to the many researchers, activists, and policy makers who refuse to despair about the state of the world's water and who provide information, feedback, and encouragement for the effort required to produce these books. Thanks are also owed to Todd Baldwin, my editor at Island Press, who continues to be enthusiastic and encouraging about the series, as well as patient. As the world of communications, electronic information exchange, and data on-demand expands, I think it is likely that the format and structure of *The World's Water* will also change, and that possibility is exciting and challenging. Stay tuned.

Peter H. Gleick
Oakland, California
May 2006

C H A P T E R   1

Water and Terrorism

Peter H. Gleick

Introduction

As the twenty-first century unfolds, concerns over the real and imagined risks of terrorism have risen much higher on the global list of worries and priorities. These concerns have spilled over into the area of water resources, causing experts and policy makers to review possible terrorist threats to developed water systems and, ultimately, human health and politics. Modern society depends on a complex, interconnected set of water infrastructure designed to provide reliable safe water supplies and to remove and treat wastewater. This infrastructure is both vital for civilization and vulnerable to intentional disruption from war, intrastate violence, or terrorism.

Violence of all kinds against, or in the name of, the environment is not new, but it does appear to be on the increase. Because water is such a fundamental resource for human and economic welfare, threats to water systems must be viewed with alarm, and care must be taken to both understand and reduce those risks. Past chapters of *The World's Water* have looked in detail at the issue of water and conflict and the role that water plays in the realm of international security (Gleick 1998, chap 4; Gleick 2000, chap 2; and http://www.worldwater.org/conflict.html). The focus of this chapter is the connections between water and terrorism.

The chance that terrorists will strike at water systems is real. Water has been used as a political or military target or tool for over four thousand years. Water resources and systems are attractive to terrorists because there is no substitute for water—it is a vitally necessary resource. Whether its lack is due to natural scarcity, a physical supply interruption, or contamination, a community of any size that lacks fresh water will suffer greatly. Furthermore, a community does not have to *lack* water to suffer. Too much water at the wrong time can also lead to death and great damage.

Water resources are vulnerable to terrorist attacks in the form of explosives used against delivery or treatment systems or with the introduction of poison or disease-causing agents. The damage is done by hurting people, rendering the water unusable, or destroying the purification and supply infrastructure. Some important water facilities, such as dams, reservoirs, and pipelines, are easily accessible to the public at various points. Many large dams are tourist attractions and offer tours to the public, and many reservoirs are open to the public for recreational boating and swimming. Pipelines are often exposed for long distances. Water and wastewater treatment plants dot our urban and rural landscape.

What is less clear, however, is how significant such threats are today, compared with other targets that may be subject to terrorist attack, or how effective such attacks would actually be. Analysis and historical evidence suggest that massive casualties from attacking water systems are difficult to produce, although there may be some significant exceptions. At the same time, the risk of societal disruptions, disarray, and even over-reaction on the part of governments and the public from any attack may be high.

As an example of the economic and human chaos that even moderate disruption or contamination might cause, an outbreak of *Cryptosporidium* in Milwaukee in 1993 killed over 100 people, affected the health of over 400,000 more (MacKenzie et al. 1994), and cost millions in lost wages and productivity. The outbreak, unrelated to terrorism, was thought to be due to a combination of an improperly functioning water treatment plant and pollution discharges upstream. But a similar undetected outbreak in a larger city might cost billions, kill many more people, and, if caused intentionally, lead to panic.

This chapter will not offer any new information for those hoping to harm water systems. Rather, I hope it will be useful in identifying where productive and protective efforts to reduce risks would be most useful on the part of water managers and planners and in helping to reduce unnecessary fear and worry. Proper and appropriate safeguards can significantly reduce the risks and the consequences, if an event occurs.

The Worry

A popular scenario for a terrorist attack on domestic water supplies involves putting a chemical or biological agent in a water-supply system, such as a reservoir, or using conventional explosives to damage basic infrastructures, such as pipelines, dams, and treatment plants. This is not as straightforward as it sounds. The number of casualties that would result from such an attack depends on the system for water treatment already in place, the type and dosage of poison ingested, individual resistance, the timing of an attack, and the speed and scope of discovery and response by local authorities.

Most biological pathogens cannot survive in water, and most chemicals require very large volumes to contaminate a water system to any significant degree. Many pathogens and chemicals are vulnerable to the kinds of water treatment used to make it potable for human use. Indeed, the whole purpose of municipal water systems is to destroy biological pathogens and reduce chemical concentrations through chlorination, filtration, ultraviolet radiation, ozonation, and many other common treatment approaches. Many contaminants are also broken down over time by sunlight and other natural processes. Most infrastructure has built-in redundancy that reduces vulnerability to physical attacks.

Because of these safeguards, one commentator noted:

> It is a myth that one can accomplish [mass destruction] by tossing a small quantity of a 'supertoxin' into the water supply . . . it would be virtually impossible to poison a large water supply: hydrolysis, chlorination, and the required quantity of the toxin are the inhibiting factors (Kupperman and Trent 1979).

Perhaps more important than the actual casualties, however, would be the public perception and response. Society reacts differently to natural and human-caused disasters, with a level of acceptance to natural disasters not matched by reaction to intentional acts of violence. As we have seen in the past several years, overreaction to terrorism is a common governmental and public response, when the impacts are compared to impacts from natural disasters or accidents. As a result, the adverse reactions that result from an effort to contaminate or damage public water systems may be both significant and underestimated. The solution to this must include efforts to both prevent such attacks and educate the public and media about actual risks and consequences. In all cases, these efforts must be commensurate with the freedoms we enjoy in an open, democratic society.

Defining Terrorism

There are many definitions of terrorism, and more than 100 have been reported (Schmid 1997). One of the earliest known definitions appeared in the Oxford English Dictionary in 1795 as "a government policy intended to strike with terror those against whom it is adopted." This early definition, however, has been substantially modified in typical modern usage. The term now usually excludes official government actions and refers to violence perpetrated by subnational groups or individuals for political or social ends.

The problem of defining terrorism has now become part of some actual definitions. The Oxford Concise Dictionary of Politics (2nd edition, 2003) begins its definition:

> Term with no agreement amongst government or academic analysts, but almost invariably used in a pejorative sense, most frequently to describe life-threatening actions perpetrated by politically motivated self-appointed sub-state groups. But if such actions are carried out on behalf of a widely approved cause . . . then the term 'terrorism' is avoided and something more friendly is substituted. In short, one person's terrorist is another person's freedom fighter.

The U.S. Federal Bureau of Investigation (FBI) and the U.S. *Code of Federal Regulations* (28 *CFR* Section 0.85) defines terrorism as "the unlawful use of force or violence against persons or property to intimidate or coerce a government, the civilian population, or any segment thereof, in furtherance of political or social objectives." Title 22, Section 2656 of the U.S. *Code* states:

> Terrorism means premeditated, politically motivated violence perpetrated against noncombatant targets by subnational groups or clandestine agents.

Both of these definitions focus on motive rather than target. Such motives can be religious, cultural, political, economic, or psychological. In the traditional definitions of terrorism, targets are usually governments, political figures, objects of economic or social significance, or random civilians. But both motives and targets can include environmental and ecological resources.

Environmental Terrorism, Ecoterrorism, and Environmental Warfare

Important distinctions should be made between three different categories of environmental violence: environmental terrorism, ecoterrorism, and more traditional forms of war that may intentionally or unintentionally affect the environment. The focus of this chapter is on the first of these categories, but I discuss the others as well to provide some context.

In recent years, U.S. law enforcement agencies have had to deal with a range of concerns and activities increasingly defined as *terrorism*. For example, in 2006, the FBI announced arrests in several cases of property destruction thought to have been caused by extreme animal rights or environmental groups. Indeed, FBI Director Mueller said one of the bureau's "highest domestic terrorism priorities" is prosecuting people who commit crimes "in the name of animal rights or the environment" (Janofsky 2006). This kind of activity, however, should be considered *ecoterrorism* not *environmental terrorism* (Schwartz 1998; Schofield 1999).

An important distinction exists between the two. Environmental terrorism is the unlawful use of force against environmental resources or systems, with the intent to harm individuals or deprive populations of environmental benefit(s) in the name of a political or social objective. This distinguishes it from ecoterrorism, which is the unlawful use of force against people or property, with the intent of saving the environment from further human encroachment and destruction. The professed aim of ecoterrorists is to slow or halt exploitation of natural resources and to bring public attention to environmental issues (Lee 1995; Chalecki 2001). In 2002, Jarboe, Domestic Terrorism Section Chief of the FBI, offered the following definition of ecoterrorism:

> the use or threatened use of violence of a criminal nature against innocent victims or property by an environmentally oriented, subnational group for environmental-political reasons, or aimed at an audience beyond the target, often of a symbolic nature.

Simply put, environmental terrorism involves targeting natural resources and the environment for a political, social, or economic objective. Ecoterrorism involves targeting social, political, or economic resources for an environmental objective.

A difference also exists between environmental warfare and environmental terrorism. The easiest distinction is that although both target environmental assets and natural resources, warfare is usually conducted by the state and terrorism by nonstate actors.

Warfare is sometimes governed by two complementary criteria: *jus ad bellum* ("just cause") (war must be declared for a good reason) and *jus in bello* (war must be conducted in a just fashion). The first criterion states that the cause for war must be right and that legal, economic, diplomatic, and all other recourses must have been attempted. However, the government of a state fighting a civil war might see rebel forces as threats to the existence of the state, whereas the same rebel forces may see clear justifications for fighting a government that it considers oppressive or illegitimate. Accordingly, evaluating "just cause" is problematic as a way to define terrorism. Hence, the popular expression "one person's terrorist is another person's freedom fighter" was adopted in the Oxford definition, as shown earlier.

The second criterion, *jus in bello*, implies behavioral constraints on the part of the combatants, chiefly among them the principle that noncombatants are not to be targeted in the conflict (Beres 1995; Stern 1999; Chalecki 2001; Maiese 2003). The *jus in bello* criterion, the guiding force behind the Geneva Conventions and the Environmental Modification Convention, acknowledges that although collateral environmental damage may occur, environmental resources are not to be *intentionally* targeted during war unless there is a direct military advantage to doing so, and, even then, effort must be made to avoid or reduce incidental damage. Terrorism violates the *jus in bello* criterion, because targeting noncombatants often lies at the core of its strategy. The idea that a target is environmental and not human does not blur the distinction between warfare and terrorism. Environmental warfare operates within the larger objective of war: to defeat the enemy's military forces or capacity (Centner 1996). Because of the larger risks of collateral casualties, however, the international community has highlighted and imposed special constraints on environmental warfare, as noted earlier.

History of Water-Related Conflict

A long history exists of the use of water resources as both a target and tool of war and terrorism (Gleick 1993; see also Water Brief 4). Water resources or systems can be used as delivery vehicles to cause violence to a human population. Water supplies can be poisoned, and dams can be destroyed to harm downstream populations. Both of these tactics have been used in traditional warfare and intrastate conflicts. The Water Conflict Chronology (see pages 189–218) presents the history of water-related conflict, including terrorism, from ancient times to the present. An early example, from 1700 B.C., is the intentional damming of the Tigris River by Abi-Eshuh, a grandson of Hammurabi, in an effort to prevent the retreat of rebels seeking the independence of Babylon (Hatami and Gleick 1994). In 1573, at the beginning of the eighty-years war against Spain, the Dutch flooded the land to break the siege of Spanish troops on the town Alkmaar. The same defense was used to protect Leiden in 1574. This strategy became known as the Dutch Water Line (2002) and was used frequently for defense in later years. In 1938, Kai-shek ordered the destruction of flood-control dikes along the Huang He (Yellow) River to flood areas threatened by the Japanese army (Yang Lang 1989, 1994). The German army flooded land in 1944 in an effort to stop the liberation of Europe (Kirschner 1949), and the allies bombed dams during the same conflict.

Conversely, the environment or resources themselves may be targeted for destruction or compromise, with the collateral damages being felt by civilian populations. As noted previously, a subset of water-related violence can be described as environmental terrorism. Table 1.1 lists examples from the Water Conflict Chronology that can be described as terrorism. Even popular culture reflects public interest and concern over these issues. Box 1.1 lists some popular novels and films with plots touching on environmental terrorism or ecoterrorism.

The recorded history of attacks on water systems goes back 4,500 years ago, when Urlama, King of Lagash, from 2450 to 2400 B.C., diverted water from this region to boundary canals, drying up boundary ditches to deprive the neighboring city-state of Umma of water. His son Il later cut off the water supply to Girsu, a city in Umma. In an early example of biowarfare (or bioterrorism, depending on one's understanding of "states" and "governments" at the time), Solon of Athens besieged Cirrha around 600 B.C., for a wrong done to the temple of Apollo, and put the poison hellebore roots (or rye ergot; reports differ) into the local water supply. This reportedly caused the Cirrhaeans to become violently ill, which eased the seizure of the city (Eitzen and Takafuji 1997).

Many of the recorded instances of violence by individuals and nonstate groups around water focus on perceived inequities associated with water development

TABLE 1.1 Water and Terrorism Chronology[1]

Date	Parties Involved	Violent Conflict or In the Context of Violence?	Description	Sources
1748	United States	Yes	Mar 28: Ferry house on Brooklyn shore of East River burns down. New Yorkers accuse Brooklynites of having set the fire as revenge for unfair East River water rights.	Museum of the City of New York (MCNY n. d.)
1841	Canada	Yes	A reservoir in Ops Township, Upper Canada (now Ontario) was destroyed by neighbors who considered it as a health hazard.	Forkey 1998
1844	United States	Yes	A reservoir in Mercer County, Ohio, was destroyed by a mob that considered it as a health hazard.	Scheiber 1969
1840s	Canada	Yes	Partly successful attempt to destroy a lock on the Welland Canal in Ontario, Canada, either by Fenians protesting English policy in Ireland or by agents of Buffalo, NY, grain handlers unhappy at the diversion of trade through the canal.	Styran and Taylor 2001
1850s	United States	Yes	Attack on a New Hampshire dam that impounded water for factories downstream by local residents unhappy over its effect on water levels.	Steinberg 1990
1853–1861	United States	Yes	Repeated destruction of the banks and reservoirs of the Wabash and Erie Canal in Southern Indiana by mobs regarding it as a health hazard.	Fatout 1972, Fickle 1983
1887	United States	Yes	Dynamiting of a canal reservoir in Paulding County, Ohio, by a mob regarding it as a health hazard. State militia called out to restore order.	Walters 1948
1907–1913	Owens Valley, Los Angeles, California	Yes	The Los Angeles Valley aqueduct/pipeline suffers repeated bombings in an effort to prevent diversions of water from the Owens Valley to Los Angeles.	Reisner 1986, 1993
1965	Israel, Palestinians	Yes	First attack ever by the Palestinian National Liberation Movement Al-Fatah is on the diversion pumps for the Israeli National Water Carrier. Attack fails.	Naff and Matson 1984, Dolatyar 1995

Year	Country	Description		Sources
1970	United States	The Weathermen, a group opposed to American imperialism and the Vietnam war, allegedly attempts to obtain biological agents to contaminate the water supply systems of U.S. urban centers.	Threat	Kupperman and Trent 1979, Eitzen and Takafuji 1997, Purver 1995
1972	United States	Two members of the right-wing group called *Order of the Rising Sun* are arrested in Chicago with 30–40 kg of typhoid cultures that are allegedly to be used to poison the water supply in Chicago, St. Louis, and other cities. It was felt that the plan would have been unlikely to cause serious health problems due to chlorination of the water supplies.	Threat	Eitzen and Takafuji 1997
1972	United States	Reported threat to contaminate water supply of New York City with nerve gas.	Threat	Purver 1995
1973	Germany	Threat by a biologist in Germany to contaminate water supplies with bacilli of anthrax and botulinum, unless he was paid $8.5 million.	Threat	Jenkins and Rubin 1978, Kupperman and Trent 1979
1977	United States	Contamination of a North Carolina reservoir with unknown materials. According to Clark: "Safety caps and valves were removed, and poison chemicals were sent into the reservoir. . . . Water had to be brought in."	Yes	Clark 1980, Purver 1995
1978–1984	Sudan	Demonstrations in Juba, Sudan, in 1978 opposing the construction of the Jonglei Canal led to the deaths of two students. Construction of the Jonglei Canal in the Sudan was forcibly suspended in 1984 following a series of attacks on the construction site.	Yes	Suliman 1998, Keluel-Jang 1997
1980s	Mozambique, Rhodesia/Zimbabwe, South Africa	Regular destruction of power lines from Cahora Bassa Dam during fight for independence in the region. Dam targeted by RENAMO.	Yes	Chenje 2001
1982	United States	Los Angeles police and the FBI arrest a man who was preparing to poison the city's water supply with a biological agent.	Threat	Livingstone 1982, Eitzen and Takafuji 1997
1983	Israel	The Israeli government reports that it had uncovered a plot by Israeli Arabs to poison the water in Galilee with "an unidentified powder."	No	Douglass and Livingstone 1987
1984	United States	Members of the Rajneeshee religious cult contaminate a city water supply tank in The Dalles, Oregon, using *Salmonella*. A community outbreak of over 750 cases occurred in a county that normally reports fewer than five cases per year.	Yes	Clark and Deininger 2000

continues

TABLE 1.1 *continued*

Date	Parties Involved	Violent Conflict or In the Context of Violence?	Description	Sources
1985	United States	No	Law enforcement authorities discovered that a small survivalist group in the Ozark Mountains of Arkansas, known as *The Covenant, the Sword, and the Arm of the Lord (CSA)* had acquired a drum containing 30 gallons of potassium cyanide, with the apparent intent to poison water supplies in New York, Chicago, and Washington, D.C. CSA members devised the scheme in the belief that such attacks would make the Messiah return more quickly by punishing unrepentant sinners. The objective appeared to be mass murder in the name of a divine mission rather than to change government policy. The amount of poison possessed by the group is believed to have been insufficient to contaminate the water supply of even one city.	Tucker 2000, NTI 2005
1991	Canada	Threat	A threat is made via an anonymous letter to contaminate the water supply of the city of Kelowna, British Columbia, with "biological contaminants." The motive was apparently "associated with the Gulf War." The security of the water supply was increased in response and no group was identified as the perpetrator.	Purver 1995
1992	Turkey	Yes	Lethal concentrations of potassium cyanide are reported discovered in the water tanks of a Turkish Air Force compound in Istanbul. The Kurdish Workers' Party (PKK) claimed credit.	Chelyshev 1992
1993	Iran	No	A report suggests that proposals were made at a meeting of fundamentalist groups in Tehran, under the auspices of the Iranian Foreign Ministry, to poison water supplies of major cities in the West "as a possible response to Western offensives against Islamic organizations and states."	Haeri 1993
1994	Moldavia	Threat	Reported threat by Moldavian General Nikolay Matveyev to contaminate the water supply of the Russian 14th Army in Tiraspol, Moldova, with mercury.	Purver 1995

Year	Location	Casualties	Description	Source
1998	Tajikistan	Threat	On November 6, a guerrilla commander threatened to blow up a dam on the Kairakkhum channel if political demands were not met. Col. Makhmud Khudoberdyev made the threat, reported by the ITAR-Tass News Agency.	WRR 1998
1998/1994	United States	No	*The Washington Post* reports that a 12-year-old computer hacker broke into the SCADA computer system that runs Arizona's Roosevelt Dam, giving him complete control of the dam's massive floodgates. The cities of Mesa, Tempe, and Phoenix, Arizona, are downstream of this dam. This report turns out to be incorrect. A hacker did break into the computers of an Arizona water facility, the Salt River Project, in the Phoenix area. But he was 27, not 12, and the incident occurred in 1994, not 1998. And although clearly trespassing in critical areas, the hacker could have never had control of any dams—leading investigators to conclude that no lives or property were ever threatened.	Gellman 2002, Lemos 2002
1998	Democratic Republic of Congo	Yes	Attacks on Inga Dam during efforts to topple President Kabila. Disruption of electricity supplies from Inga Dam and water supplies to Kinshasa.	Chenje 2001, Human Rights Watch 1998
1999	Lusaka, Zambia	Yes	Bomb blast destroyed the main water pipeline, cutting off water for the city of Lusaka; population 3 million.	FTGWR 1999
1999	South Africa	Yes	A homemade bomb was discovered at a water reservoir at Wallmansthal near Pretoria. It was thought to have been meant to sabotage water supplies to farmers.	Pretoria Dispatch 1999
1999	Angola	Yes	100 bodies were found in four drinking water wells in central Angola.	*International Herald Tribune* 1999
1999	East Timor	Yes	Militia opposing East Timor independence kill pro-independence supporters and throw bodies in water well.	BBC 1999
1998–1999	Kosovo	Yes	Contamination of water supplies/wells by Serbs disposing of bodies of Kosovar Albanians in local wells. Other reports of Yugoslav federal forces poisoning wells with carcasses and hazardous materials.	CNN 1999, Hickman 1999.

continues

TABLE 1.1 *continued*

Date	Parties Involved	Violent Conflict or In the Context of Violence?	Description	Sources
2000	Belgium	Yes	In July, workers at the Cellatex chemical plant in Northern France dumped 5,000 liters of sulfuric acid into a tributary of the Meuse River when they were denied workers' benefits. A French analyst pointed out that this was the first time "the environment and public health were made hostage in order to exert pressure, an unheard-of situation until now."	*Christian Science Monitor* 2000
2000	Australia	Yes	In Queensland, Australia, on April 23, 2000, police arrested a man for using a computer and radio transmitter to take control of the Maroochy Shire wastewater system and release sewage into parks, rivers, and property.	Gellman 2002
2001	Israel, Palestine	Yes	Palestinians destroy water supply pipelines to West Bank settlement of Yitzhar and to Kibbutz Kisufim. Agbat Jabar refugee camp near Jericho disconnected from its water supply after Palestinians looted and damaged local water pumps. Palestinians accuse Israel of destroying a water cistern, blocking water tanker deliveries, and attacking materials for a wastewater treatment project.	Israel Line 2001a,b; ENS 2001
2001	Pakistan	Yes	Civil unrest over severe water shortages caused by the long-term drought. Protests began in March and April and continued into summer. Riots, four bombs in Karachi (June 13), 1 death, 12 injuries, 30 arrests. Ethnic conflicts as some groups "accuse the government of favoring the populous Punjab province (over Sindh province) in water distribution."	Nadeem 2001, Soloman 2001
2001	Macedonia	Yes	Water flow to Kumanovo (population 100,000) cut off for 12 days in conflict between ethnic Albanians and Macedonian forces. Valves of Glaznja and Lipkovo Lakes damaged.	AFP 2001, Macedonia Information Agency 2001

Year	Country		Description	Reference
2001	Philippines	No	Philippine authorities shut off water to six remote southern villages after residents complained of a foul smell from their taps, raising fears that Muslim guerrillas had contaminated the supplies. Abu Sayyaf guerrillas, accused of links with Saudi-born militant Osami bin Laden, had threatened to poison the water supply in the mainly Christian town of Isabela on Basilan island if the military did not stop an offensive against them.	World Environment News 2001
2002	Nepal	Yes	The Khumbuwan Liberation Front (KLF) blew up a hydroelectric powerhouse of 250 kilowatts in Bhojpur District on January 26. The power supply to Bhojpur and adjoining areas was cut off. Estimated repair time was 6 months; repair costs were estimated at 10 million Rs. By June 2002, Maoist rebels had destroyed more than seven microhydro projects as well as an intake of a drinking water project and pipelines supplying water to Khalanga in Western Nepal.	*Kathmandu Post* 2002, FTGWR 2002a
2002	Rome, Italy	Threat	Italian police arrest four Moroccans allegedly planning to contaminate the water supply system in Rome with a cyanide-based chemical, targeting buildings that included the United States embassy. Ties to Al-Qaida were suggested.	BBC 2002
2002	United States	Threat	Papers seized during the arrest of a Lebanese national who moved to the United States and became an Imam at an Islamist mosque in Seattle included "instructions on poisoning water sources" from a London-based Al-Qaida recruiter. The FBI issued a bulletin to computer security experts around the country indicating that Al-Qaida terrorists may have been studying American dams and water-supply systems in preparation for new attacks. "U.S. law enforcement and intelligence agencies have received indications that Al-Qaida members have sought information on Supervisory Control And Data Acquisition (SCADA) systems available on multiple SCADA-related Web sites," reads the bulletin, according to SecurityFocus. "They specifically sought information on water supply and wastewater management practices in the U.S. and abroad."	McDonnell and Meyer 2002, MSNBC 2002

continues

TABLE 1.1 *continued*

Date	Parties Involved	Violent Conflict or In the Context of Violence?	Description	Sources
2002	Colombia	Yes	Colombian rebels in January damaged a gate valve in the dam that supplies most of Bogota's drinking water. Revolutionary Armed Forces of Colombia (FARC) detonated an explosive device planted on a German-made gate valve located inside a tunnel in the Chingaza Dam.	Waterweek 2002
2002	United States	Threat	Earth Liberation Front threatens the water supply for the town of Winter Park. Previously, this group claimed responsibility for the destruction of a ski lodge in Vail, Colorado, that threatened lynx habitat.	Crecente 2002, *Associated Press* 2002
2003	United States	Threat	Al-Qaida threatens U.S. water systems via call to Saudi Arabian magazine. Al-Qaida does not "rule out . . . the poisoning of drinking water in American and Western cities."	*Associated Press* 2003a, Waterman 2003, NewsMax 2003, *U.S. Water News* 2003
2003	United States	Yes	Four incendiary devices were found in the pumping station of a Michigan water-bottling plant. The Earth Liberation Front (ELF) claimed responsibility, accusing Ice Mountain Water Company of "stealing" water for profit. Ice Mountain is a subsidiary of Nestle Waters.	*Associated Press* 2003b
2003	Colombia	Yes	A bomb blast at the Cali Drinking Water Treatment Plant killed 3 workers on May 8. The workers were members of a trade union involved in intense negotiations over privatization of the water system.	PSI 2003
2003	Jordan	Threat	Jordanian authorities arrested Iraqi agents in connection with a botched plot to poison the water supply that serves American troops in the eastern Jordanian desert near the border of Iraq. The scheme involved poisoning a water tank that supplies American soldiers at a military base in Khao, which lies in an arid region of the eastern frontier near the industrial town of Zarqa.	MJS 2003

Year	Location		Description	Source
2003	Iraq	Yes	Sabotage/bombing of main water pipeline in Baghdad. The sabotage of the water pipeline was the first such strike against Baghdad's water system, city water engineers said, and involved an explosive fired at the six-foot-wide water main in the northern part of Baghdad.	Tierney and Worth 2003
2003–2004	Sudan	Yes	The ongoing civil war in the Sudan has included violence against water resources. In 2003, villagers from around Tina said that bombings had destroyed water wells. In Khasan Basao they alleged that water wells were poisoned. In 2004, wells in Darfur were intentionally contaminated as part of a strategy of harassment against displaced populations.	*Toronto Daily* 2004, Reuters Foundation 2004
2004	Pakistan	Yes	In military action aimed at Islamic terrorists, including Al Qaida and the Islamic Movement of Uzbekistan, homes, schools, and water wells were damaged and destroyed.	Reuters 2004a
2004	India, Kashmir	Yes	Twelve Indian security forces were killed by an IED planted in an underground water pipe during "counter-insurgency operation in Khanabal area in Anantnag district."	TNN 2004
2004	Gaza Strip	Yes	The United States halts two water development projects as punishment to the Palestinian authority for their failure to find those responsible for a deadly attack on a U.S. diplomatic convoy in October 2003.	*Associated Press* 2004

Notes:

1. This table is a subset of all water-related conflicts reported in the Pacific Institute's Water Conflict Chronology (http://www.worldwater.org). Included are those incidents that fall only under the broad definition of environmental terrorism, defined here as "the unlawful use of force against environmental resources or systems with the intent to harm individuals or deprive populations of environmental benefit(s) in the name of a political or social objective." Please remember the caution, described in the text, that one person's "terrorist" is another person's "freedom fighter." As a result, I have no doubt that the description of some of these events as "terrorism" will be controversial to some of the parties involved. My objective is not to offend. Also, because of the evolution of the concept of nations and states, I've excluded from this list all water and conflict events before the mid-1700s. I've excluded development disputes in which individuals or subnational groups take violent action as a result of water disputes, shortages, or allocation controversies, that is, where people fight over water for the sake of water. I note, however, the difficulty in defining "terrorism" (as opposed to military target, tool, or goal or other category) and caution users to use care with applying these categories.

Sources: see *Water Brief 4*, page 189, for all sources.

Box 1.1   Environmental Terrorism, Ecoterrorism, Water, and Popular Culture

Popular culture often portrays terrorism in dramatic ways that either influence perceptions of threats (Jenkins 2000) or reflect public fears and concerns. Environmental and ecoterrorism around water have long been included among those threats. Kurt Vonnegut's classic book *Cat's Cradle* describes an amoral genius who creates "ice-nine"—a chemical that freezes water at room temperature and ends up destroying the world. Edward Abbey's (1975) novel *The Monkey Wrench Gang* and Johnson and Bent's film *Christie Malry's Own Double Entry* featured environmental terrorism or ecoterrorism, such as blowing up dams, poisoning water supplies, and attacking resources for political or environmental purposes. Wilson and Leeson's 2002 movie *The Tuxedo*, starring Jackie Chan, features a power-hungry, bottled-water mogul trying to destroy the world's natural water supply to force everyone to drink his bottled water. The movie *Batman Begins*, released in 2005, portrayed a terrorist attempt to destroy Gotham by introducing a vapor-borne hallucinogen into the water system. The 2005 film *Syriana*, by Baer and Gaghan, includes an attack on energy resources in the form of a liquefied natural gas tanker. In early 2006, an independent feature film, *Waterborne*, was released, which follows the fictional aftermath of a bioterrorist attack on the water supply of Los Angeles. And *V for Vendetta* (2006) features corrupt government leaders contaminating London's water supply to kill people, spread fear, and consolidate power.

If readers have other examples, please send them to info@pacinst.org.

projects or controversial decisions about allocations of water. Often, marginalized groups, faced with the construction of water systems that appropriate local water resources, have responded by threatening or attacking those systems. In one of the earliest reported acts, an angry mob in New York, in 1748, burned down a ferry house on the Brooklyn shore of the East River, reportedly as revenge for unfair allocation of East River water rights (MCNY n.d.). In the 1840s and 1850s, groups attacked small dams and reservoirs in the eastern and central United States because of concerns about threats to health and to local water supplies (Table 1.1). In a now-famous case, between 1907 and 1913, farmers in the Owens Valley of California repeatedly dynamited the aqueduct system being built to divert their water to the growing city of Los Angeles (Reisner 1986, 1993).

The first reported attack of the Palestinian National Liberation Movement Al-Fatah was in 1965 on the diversion pumps for the Israeli National Water Carrier (Naff and Matson 1984), and the region has seen many more examples. In 2001, Palestinians attacked and vandalized water pipes leading to the Israeli settlement of Yitzhar to try to force the Israelis out of the settlement. Around the same time, Palestinians accused

Israel of destroying a water cistern, blocking water tanker deliveries, and attacking materials for a wastewater treatment project (Israel Line 2001a, 2001b; ENS 2001).

Rivers and water-supply infrastructure, such as reservoirs, are particularly vulnerable to this type of environmental terrorism, because they are publicly accessible in many places. In July 1999, engineers discovered a bomb in a water reservoir near Pretoria, South Africa. The security personnel felt that the bomb, which had malfunctioned, would have been powerful enough to damage the system, thus depriving farmers, a nearby military base, and a hydrologic research facility of water (Pretoria Dispatch 1999). In 2000, a simulated terrorist attack on the Lake Nacimiento Dam caused some local panic in central California, until the media was notified that the situation was merely a disaster preparedness drill (Gaura 2000).

Motives for such attacks can be economic as well as political. In July 2000, workers at the Cellatex chemical plant in Northern France dumped 5,000 liters of sulfuric acid into a tributary of the Meuse River when they were denied workers' benefits. Whether they were trying to kill wildlife, people, both, or neither is unclear, but a French analyst pointed out that this was the first time "the environment and public health were made hostage in order to exert pressure, an unheard-of situation until now" (Christian Science Monitor 2000).

More recently, a series of events in India, Pakistan, the Persian Gulf, and the Middle East have reaffirmed the attractiveness of water and water systems as targets of terrorists in a wide range of unrelated conflicts and disputes. The major water pipeline to Baghdad was attacked in 2003. In the same year, Al-Qaida threatened U.S. water systems in a call published in a Saudi Arabian magazine: "Al-Qaida does not 'rule out . . . the poisoning of drinking water in American and Western cities'" (Associated Press 2003; Waterman 2003). In 2004, twelve Indian security forces were killed by an explosive device planted in an underground water pipe during a "counter-insurgency operation in Khanabal area in Anantnag district" (TNN 2004). In an unusual twist to this problem, the United States responded to a terrorist attack on U.S. diplomatic personnel in the Middle East by canceling plans for a water-development project in the Gaza Strip (Associated Press 2004).

Vulnerability of Water and Water Systems

Infrastructure Attacks

The most traditional form of environmental terrorism has involved physical attacks on water infrastructure—specifically water-supply dams and pipelines. One such attack might target a large hydroelectric dam on a major river or a major water-supply system for a city. Terrorists equipped with a relatively small conventional explosive might not be able to cause serious structural damage to a massive dam, which is, after all, usually a giant block of rock, earth, or concrete. But the adverse consequences of a major dam failure make the risk worth both assessing and reducing. A major failure can kill thousands of people, and even more modest damage might interrupt power generation or affect some other important water-system operation.

Some disasters with water infrastructure offer insights into the risks of water-related terrorism. In 1975, the Banqiao and Shimantan dams on tributaries of the Huang He

(Yellow) River in China failed in sequence, contributing to the ultimate failure of dozens of lower dams and the deaths of 85,000 people (Yi 1998). The famous Johnston Flood of 1889 killed more than 2,200 people when the collapse of a poorly built dam sent a massive wall of water through the poor steel town of Johnston, Pennsylvania. At least 400 people died in California in 1928, when the St. Francis Dam failed in San Francisquito Canyon. Worldwide, millions of people live in the floodplains below large dams and reservoirs, vulnerable to failure. In addition to the potential loss of life, secondary impacts also exist, including water-quality problems, loss of freshwater supply and hydroelectric power, damage to property and commercial fisheries, and recreation losses.

Although many municipal water systems are built with redundancy and backup systems, others have particularly vulnerable points, such as single large pipelines, pumping plants, or treatment systems. The bombing of the major water pipeline entering Baghdad, in 2003, is an example of such a vulnerability (Tierney and Worth 2003), and the Water Conflict Chronology offers several more historical examples from Angola, Eritrea, Zambia, the West Bank, Nepal, and the United States (http://www.worldwater.org/chronology.html).

A more modern concern is the use of remote computers to attack valves, pumps, and chemical-processing equipment though computer-based controls. These control systems were typically developed with no attention to security. As a result, many of the Supervisory Control and Data Acquisition (SCADA) networks used by water agencies to collect data from sensors and control equipment "may be susceptible to attacks and misuse" (Heilprin 2005). In Queensland, Australia, on April 23, 2000, police arrested a man for using a computer and radio transmitter to take control of the Maroochy Shire wastewater system and to release sewage into parks, rivers, and property. This is one of the first documented cases of cyberterrorism in the water industry (Gellman 2002).

Chemical and Biological Attacks

Of growing concern and greater media attention is the risk of chemical and biological attacks. This type of attack on water is often portrayed as follows: A terrorist hikes to a publicly accessible city reservoir and drops a certain amount of concentrated water-soluble contaminant (chemical or biological) near the intake pipe. In the best-case scenario, the contaminant is detected as it enters the water treatment plant, and the plant is shut down while the contaminant is neutralized. This can result in interruption of potable water service to the city and a "boil-water" alert for city residents. In the worst-case scenario, the contaminant is undetected and people begin to get sick, panic ensues, and health and economic damages soar.

This type of attack may not be as easy as often portrayed. To be effective as a tool of water-related terrorism, a chemical or biological weapon must be

- Weaponized: It must be produced and disseminated in quantities sufficient to have the intended effect.

- Appropriate for water dissemination: It must be viable, dissolvable, stable, and transportable in water.

- Effective over time and treatment: It must maintain its effectiveness in water long enough to reach and affect humans, and it must not be negated by standard water treatment systems likely to be in place.

- Infectious, virulent, or toxic: It must be effective at causing illness or death, with no widespread immunity in the target population.

According to easily available open literature, a wide range of chemical and biological agents could be used in water. Table 1.2, described in a recent U.S. EPA review of water-related threats, should be considered illustrative of the relevant contaminant classes. As noted, some of these substances are likely to be found in military stockpiles only; others may be produced by subnational terrorist groups; others may have more mundane industrial or even household applications (U.S. NRC 1995; Hickman 1999; U.S. EPA 2003a). All of the listed agents have strengths and weaknesses, especially in their usefulness as weapons for use in water. These details will not be described here.

TABLE 1.2 Chemical and Biological Contaminants of Water: *Classes, Availabilities, and Restrictions*

Class	Examples (Not Exhaustive)	Limited Sources[1]	Access
Microbiological Contaminants			
Bacteria	*Bacillus anthracis, Brucella* spp., *Burkholderia* spp., *Campylobacter* spp., *Clostridium perfringens,* E. coli O157:H7, *Francisella tularensis, Salmonella typhi,* *Shigella* spp., *Vibrio cholerae,* *Yersinia pestis, Yersinia* *enterocolitica*	Naturally occurring, microbiological laboratories, state-sponsored programs	Yes for select agents
Viruses	Caliciviruses, enteroviruses, hepatitis A/E, variola, VEE virus	Naturally occurring, microbiological laboratories,[1] state-sponsored programs	Yes for select agents
Parasites	*Cryptosporidium parvum, Entamoeba histolytica, Toxoplasma gondii*	Naturally occurring, microbiological laboratories[1]	No
Inorganic Chemicals			
Corrosives and caustics	Hydrochloric acid (toilet bowl cleaners), sulfuric acid (tree-root dissolver), sodium hydroxide (drain cleaner)	Retail, industry	No
Cyanide salts or cyanogenics	Sodium cyanide, potassium cyanide, amygdalin, cyanogen chloride, ferricyanide salts	Supplier, industry (esp. electroplating)	Yes
Metals	Mercury, lead, osmium, their salts, organic compounds, and complexes (even those of iron, cobalt, copper are toxic at high doses)	Industry, supplier, laboratory	Yes[2]

continues

TABLE 1.2 *continued*

Class	Examples (Not Exhaustive)	Limited Sources[1]	Access
Nonmetal oxyanions, organononmetals	Arsenate, arsenite, selenite salts, organoarsenic, organoselenium compounds	Some retail, industry, supplier, laboratory	Yes[3]
Organic Chemicals			
Fluorinated organics	Sodium trifluoroacetate (a rat poison), fluoroalcohols, fluorinated surfactants	Supplier, industry, laboratory	Yes
Hydrocarbons and their oxygenated and/ or halogenated derivatives	Paint thinners, gasoline, kerosene, ketones (e.g., methyl isobutyl ketone), alcohols (e.g., methanol), ethers (e.g., methyl tert-butyl ether or MTBE), halohydrocarbons (e.g., dichloromethane, tetrachloroethane)	Retail, industry, laboratory, supplier	No
Insecticides	Organophosphates (e.g., Malathion), chlorinated organics (e.g., DDT), carbamates (e.g., Aldicarb), some alkaloids (e.g., nicotine)	Retail, industry, supplier (varies with compound)	Yes
Malodorous, noxious, foul-tasting, and/or lachrymatory chemicals[4]	Thiols (e.g., mercaptoacetic acid, mercaptoethanol), amines (e.g., cadaverine, putrescine), inorganic esters (e.g., trimethylphosphite, dimethylsulfate, acrolein)	Laboratory, supplier, police supply, military depot	Yes
Organics, Water-miscible	Acetone, methanol, ethylene glycol (antifreeze), phenols, detergents	Retail, industry, supplier, laboratory	No
Pesticides other than insecticides	Herbicides (e.g., chlorophenoxy or atrazine derivatives), rodenticides (e.g., superwarfarins, zinc phosphide, α-naphthyl thiourea)	Retail, industry, agriculture, laboratory	Yes
Pharmaceuticals	Cardiac glycosides, some alkaloids (e.g., vincristine), antineoplastic chemotherapies (e.g., aminopterin), anticoagulants (e.g., warfarin). Includes illicit drugs, such as LSD, PCP, and heroin.	Laboratory, supplier, pharmacy, some from a natural source	Yes

Class	Examples (Not Exhaustive)	Limited Sources[1]	Access
Chemical Warfare Agents			
Chemical weapons	Organophosphate nerve agents (e.g., sarin, tabun, VX), vesicants, [nitrogen and sulfur mustards (chlorinated alkyl amines and thioethers, respectively)], Lewisite	Suppliers, military depots, some laboratories	Yes
Biotoxins			
Biologically produced toxins	Biotoxins from bacteria, plants, fungi, protists, defensive poisons in some marine or terrestrial animals. Examples include ricin, saxitoxin, botulinum toxins, T-2 mycotoxins, microcystins.	Laboratory, supplier, pharmacy, natural source,[5] state-sponsored military programs	Yes
Radiological Contaminants			
Radionuclides	Does not refer to nuclear, thermonuclear, or neutron bombs. Radionuclides may be used in medical devices and industrial irradiators (Cesium-137 Iridium-192, Cobalt-60, Strontium-90). Class includes both metals and salts.	Laboratory, state sources, waste facilities	Yes[2]

Notes:
1. The quantity of bacteria, viruses, or parasites needed for widespread contamination of a water system is not typically available in a typical clinical laboratory, although the seed cultures could be available. For viruses, vaccine production-grade volumes would be needed, requiring special equipment and facilities, perhaps with state sponsorship.
2. Availability may be commercially limited for the more toxic materials, especially the heavy metals, which can be quite expensive. Iron and copper are readily available but not usually in soluble (bioavailable) forms.
3. Availability of arsenicals and selenium compounds in the retail sector has been reduced owing to environmental regulations, but such products can occasionally be found as part of older inventories of merchandise, especially in small-town hardware stores. Supplies of such materials may generally be too small to cause concern.
4. This grouping includes riot-control agents and other mucous membrane irritants.
5. The quantity available from laboratories, suppliers, and pharmacies needed for widespread contamination of a water system is typically not available from these sources. Many biotoxins that occur naturally would need to be purified or prepared to be of significant concern to water, which could make production beyond the capabilities of most terrorists.

Source: Modified from U.S. EPA 2003a and U.S. NRC 1995.

Although some of the biological and chemical contaminants listed in Table 1.2 have been produced for military use, military-grade chemical weapons are far more difficult to produce, handle, and disseminate. Commercial chemicals that are commonly produced, distributed, and used throughout the world are more likely to be used by terrorists to contaminate water supplies. Of particular concern are pesticides and related chemicals that are used to kill insects, rodents, and plants. They include organophosphate pesticides, chlorinated pesticides, and rodenticides. Organophosphates affect the nervous system, as do organochlorine pesticides. Rodenticides, such as sodium fluoroacetate, strychnine, and thallium sulfate, are all capable of incapacitating or killing humans in appropriate doses (Hickman 1999).

Several inorganic chemicals are also widely available and potential threats to water systems, including various forms of arsenic and cyanide. Both are soluble in water and can be lethal. Hickman (1999) discusses the challenge posed by a material, such as sodium cyanide (NaCN), for a small water system. Sodium cyanide is relatively plentiful and accessible because of its use in the mining and metals industry. It is an odorless white salt, which is stable and highly soluble in water.

A conference, "Early Warning Monitoring to Detect Hazardous Events in Water Supplies," held in May 1999 in Reston, Virginia, concluded that terrorist use of bioweapons can, under some circumstances, pose a significant threat to drinking water. Although most biological warfare agents were developed for the purpose of aerial dissemination, some can be effective if digested, and some are stable and soluble in water. The two main types of biological threats are pathogens and toxins. Pathogens are live organisms, including bacteria, viruses, and protozoa. Toxins are chemicals that derive from biological processes (Valcik 1998). Table 1.3 shows a subset of known biological threats to water supplies, including at least four reported in the public literature to have been produced as biological weapons. Table 1.4 shows known biological toxins that pose a water threat, including three known to have been turned into weapons. Both tables also describe tolerance to chlorine. As noted earlier, some of these pathogens or toxins are neutralized by chlorine levels commonly used in municipal

TABLE 1.3 Biological Pathogens Considered Water Threats

Pathogen	Type	Weaponized	Stable in Water	Chlorine Tolerance
Anthrax	B	Yes	2 years spores	Spores resistant
Brucellosis	B	Yes	20-72 days	Unknown
Clostridium perfringens	B	Probable	Common in sewage	Resistant
Tularemia	B	Yes	<90 days	Inactivated, 1 ppm, 5 min.
Shigellosis	B	Unknown	2-3 days	Inactivated, 0.05 ppm, 10 min.
Cholera	B	Unknown	Yes	"Easily killed"
Plague	B	Probable	16 days	Unknown
Q fever	R	Yes	Unknown	Unknown
Hepatitis A	V	Unknown	Unknown	Inactivated, 0.4 ppm, 30 min.

Notes: B, bacteria; R, rickettsia; V, virus.
Source: Modified from Valcik 1998.

TABLE 1.4 Biological Pathogens Considered Water Threats

Toxin	Weaponized	Stable in Water	Chlorine Tolerance
Botulinum	Yes	Yes	Inactivated at 6 ppm, 20 min.
T-2 mycotoxin	Probable	Yes	Resistant
Aflatoxin	Yes	Probable	Resistant
Ricin	Yes	Unknown	Resistant at 10 ppm
Staphylococcus enterotoxins	Probable	Probable	Unknown
Microcystins	Possible	Probable	Very resistant at 100 ppm
Anatoxin A	Unknown	Inactivated in days	Unknown
Tetrodotoxin	Possible	Unknown	Inactivated, 50 ppm
Saxitoxin	Possible	Yes	Resistant at 10 ppm

Source: Modified from Valcik 1998.

water systems. Less information is available on how these threats may be affected or neutralized by some of the newer, nonchlorine-based water-treatment systems, including advanced filtration, ultraviolet disinfection, and ozonation.

In 1970 (sometimes dated "early 1970s"), the U.S. radical group Weather Underground reportedly attempted to blackmail a homosexual officer at the U.S. Army's bacteriological warfare facility in Fort Detrick, Maryland, into supplying organisms that could then be used to contaminate urban water supplies (Mullins 1992; Berkowitz et al. 1972, citing *The New York Times* 21 November 1970). According to one source, the terrorists apparently succeeded in gaining the cooperation of the officer in question, but "This plot was discovered when the officer requested issue of several items unrelated to his work" (Purver 1995). Another reported incident was the arrest by Los Angeles police and FBI agents of a man "who was preparing to poison the city's water system with a biological poison" (Livingstone 1982).

Individuals and groups have been known to plan and carry out attacks on water systems in the belief that they can be effective. A few cases of actual contamination of water supplies, or confirmed plans to conduct such attacks, have been reported in the open literature. In 1972, a right-wing, neo-Nazi group known as the Order of the Rising Sun, "dedicated to creating a new master race," acquired 30 to 40 kilograms of typhoid bacteria cultures to use against water supplies in Chicago, St. Louis, and other Midwestern cities (Kupperman and Trent 1979; Purver 1995). According to Ponte (1980), those arrested had "in their possession detailed plans for dumping the deadly germs into the water supplies." It is likely that typhoid bacteria, even if introduced into an urban water supply, would have been destroyed by normal chlorination (U.S. OTA 1991).

In a case of criminal extortion, in 1973, a German biologist threatened to contaminate water supplies with bacilli of anthrax and botulinum, unless he was paid a financial ransom of $8.5 million (Jenkins and Rubin 1978:228; Kupperman and Trent 1979:46). The Israeli government reported in 1983 that it had uncovered a plot by Israeli Arabs to poison the water supply of the city of Galilee with "an unidentified powder" (Douglass and Livingstone 1987). In 1985, federal law enforcement authorities discovered that The Covenant, the Sword, and the Arm of the Lord (CSA)—a survivalist group in the Ozark

Mountains of Arkansas—had acquired a drum containing 30 gallons of potassium cyanide. Its goal was to poison water supplies in New York, Chicago, and Washington D.C. in the belief that this would make the Messiah return more quickly by punishing unrepentant sinners (http://www.nti.org/h_learnmore/cwtutorial/chapter02_02.html).

A chemical poisoning attempt was reported in March 1992, when lethal concentrations of potassium cyanide were found in the water tanks at a Turkish air force base in Istanbul. The Kurdish Workers' Party (PKK) claimed credit (Chelyshev 1992). The media reported that proposals were made at an early February 1993 meeting of fundamentalist groups in Tehran, under the auspices of the Iranian Foreign Ministry, to poison the water supplies of major cities in the West "as a possible response to Western offensives against Islamic organizations and states" (Haeri 1993).

Responding to the Threat of Water-Related Terrorism

No easy estimate of the true risk of water-related terrorism is possible. The numerous examples of actual and planned attacks on water systems in the past suggest that the risk is real. What is more challenging is evaluating both the probability of future attacks and their consequences—the separate components of calculating risk. In the absence of any definitive assessment of risk, however, it is vital to understand vulnerabilities and to put in place measures to reduce those vulnerabilities and, ultimately, the overall risk. This can be done by addressing both the probability of water-related terrorism and the consequences of an attack, if one occurs.

Addressing the probability requires a wide range of actions, from reducing the fundamental motivation for terrorist attacks to limiting the vulnerability of water resources and systems through selective and focused efforts of protection and detection. Addressing the consequences of attacks requires putting in place an array of suitable responses for different kinds of events. They can include rapid repair teams to fix infrastructure to the development of redundant delivery and treatment systems to prepare the health system to detect and treat water-related illnesses.

Denying Physical Access

Perhaps the most fundamental action that can be taken to protect water systems is to limit or to deny physical access to vulnerable points. Sometimes this may be as easy as locking gates or buildings, or reducing public access to sensitive locations. Often, however, this approach is not possible, given the vast exposed length of pipelines or aqueducts, or the public uses of lakes, reservoirs, rivers, and land. As examples of new activities put in place since September 11, 2001, the Coast Guard increased patrols in the area of Chicago's water intakes from Lake Michigan. New York City increased the number of daily water samples it takes. California has limited access to some dams and pumping plants and blocked off some roads close to water reservoirs. Many water agencies have stationed guards at "critical sites" (Center for Defense Information 2002).

Among the recommendations for reducing the physical risk to infrastructure are

- Facilities (treatment plants, reservoirs, dams, storage facilities, pumping plants, intake facilities, and control systems) should be identified and inventoried. Access to those most critical to operations, or vulnerable to attack, should be limited with physical barriers.

- Access to water distribution maps and facility plans should be limited when there is a clear security risk.

- Lighting, surveillance cameras, and motion detectors should be installed in appropriate places.

- To prevent hacking, supervisory control and data acquisition systems for monitoring and controlling water parameters should not be connected to the Internet or should be connected with appropriate electronic security, firewalls, and passwords.

- Water treatment chemicals should be kept in secure facilities and inventoried on a regular basis.

Detection and Protection Challenges

Unlike more traditional weapons used by terrorists, water-related threats pose some special challenges in the areas of detection and response. As noted previously, an attack on a water system may be done surreptitiously through the introduction of a chemical or biological agent. In this case, unless immediate publicity is an objective of the attack, the first evidence may be increased incidences of sickness and death. Identifying the nature of the illness, the source of the contamination, and then identifying and quantifying the specific threat could take a substantial amount of time.

New security measures, such as more extensive monitoring of pipelines, water supplies, or more guards at power plants, will be expensive and mean higher costs for consumers. Nevertheless, it seems clear that some such measures will be required. In 2002, the U.S. Congress passed the Public Health, Security, and Bioterrorism Preparedness and Response Act (Bioterrorism Act), which President Bush signed into law on June 12, 2002. Among other things, the Bioterrorism Act established requirements that community water systems serving more than 3,300 individuals perform a system-specific vulnerability assessment for potential terrorist threats, including intentional contamination (http://cfpub.epa.gov/safewater/watersecurity/bioterrorism.cfm). This sort of assessment, if properly done, can provide valuable information for planning and protection.

Early Warning Systems (EWS)

"Early warning" monitoring systems can help identify contamination events early enough to permit an effective response. An early warning system (EWS) must be reliable: it should minimize the potential for significant numbers of both false negatives (missing a true event) and false positives (reporting a false event). It must be easy to install and operate, provide continuous monitoring, and result in rapid notification of an event. Continuous monitoring reduces the likelihood that contamination events will be missed. The development of standard systems would reduce cost, permit sharing among users, and facilitate repair and replacement (Foran and Brosnan 2000).

New technologies are being developed to rapidly detect pathogens in real time, both in source water and water distribution systems (U.S. EPA 2005). Included among the technologies are DNA microchip arrays (Betts 1999a), immunologic techniques (Betts 1999b), microrobots (Hewish 2000), and various optical tools, molecular probes, and other techniques (Pelley 1999; Sobsey 1999).

Such technology would be useful for a wide range of purposes, including regular water-quality monitoring at municipal systems, but wide development and dissemination of such systems are moving forward slowly. Most of these technologies are not yet commercially available and have not been tested in large drinking water systems.

Some organizations are now working to improve both available technology and knowledge about useful tools for detection and response. The American Water Works Association (AWWA), for example, offers seminars on these topics for water managers. In 2003 and 2004, the U.S. Environmental Protection Agency (EPA) (2003b) published a series of guides for water utilities to help them identify and respond to contamination attacks. Similarly, in 2003, the World Health Organization updated and released a comparable planning document.

Public and Governmental Responses

Adequate physical barriers, EWS, and other preventive measures are unlikely to prevent all attacks. Threats, alone, can possibly trigger reactions. A threat to a drinking water system, whether real or a hoax, may cause as much of a problem as an actual terrorist act. Developing tools for responding to both real and threatened events is thus vital.

Responses may include public advisories, shutdown of the system, identification and use of alternative water supplies, chemical and biological treatment and disinfection, additional data gathering or monitoring, epidemiologic studies, health interventions, or some combination of these actions. Responses to actual events will depend on the nature of the attack, the population affected, and characteristics of the water system itself.

A key component to the success of any response will be the advance preparation of a process or plan that provides guidelines to all appropriate stakeholders. Such a plan should be considered part of comprehensive emergency planning for various threats to public health, both waterborne and nonwaterborne. Major stakeholders include

- Individuals served by a water system
- Individuals with specific expertise (e.g., microbiologists, toxicologists)
- Community leaders and elected officials
- Health departments, hospitals, and other health care professionals
- Water utility representatives and employees
- Water regulatory agency representatives
- Emergency responders (police, fire)
- Representatives of high-risk groups
- Law enforcement agencies
- Local and regional media.

Emergency response plans are already developed in different communities, though recent experience with Hurricane Katrina has revealed gaping holes in those plans. The U.S. EPA (2003a, 2003b), AWWA (2006), American Society of Civil Engineers, U.S. Federal Emergency Management Agency, the National Infrastructure Protection Center of the Federal Bureau of Investigation, and the Emergency Management and Emergency Preparedness Office of the U.S. Health and Human Services all offer some guidelines for water plans, and some effort has been made to develop post-event responses (Macintyre et al. 2000; Simon 1997; Waeckerle 2000). The state of local, regional, and national planning, however, is still inadequate.

Water Security Policy in the United States

Even before September 11, 2001, many analyses were prepared and risks and threats of terrorism were evaluated (Gilmore Commission 1999, 2000). The focus of U.S. security policy, however, underwent a fundamental shift in 2001, toward domestic security and challenges. In June 2002 the President signed PL 107-188 (http://www.fda.gov/oc/bioterrorism/bioact.html), the Bioterrorism Act (US FDA 2002). Title IV of the act pertains to drinking water security and safety. Specifically, the bill requires community water systems that serve more than 3,300 people to assess their vulnerability to terrorist attack. Systems that serve more than 100,000 people must have completed the assessment by March 31, 2003; those that serve 50,000 to 100,000 must have completed the assessment by December 31, 2003; and those that serve 3,300 to 49,999 must have completed the assessment by June 30, 2004. In addition, community water systems must prepare or revise emergency response plans that incorporate the results of the vulnerability assessment. Water systems must certify to the Administrator of EPA within six months of the completion of the vulnerability assessment that they have completed an emergency response plan. According to the EPA in February 2006, all large- and medium-size systems had completed their assessments; 97% of small systems had completed assessments (Johnson, personal communication, February 6, 2006). No separate information is available on the adequacy or comprehensiveness of the assessments, or whether actual response plans have been put in place.

In early 2006, the U.S. EPA announced a new effort called the Water Sentinel Initiative to design, deploy, and evaluate a water contamination warning system. This program was called for by the Homeland Security Presidential Directive 9 (HSPD-9), which charges EPA to develop surveillance and monitoring systems to provide early detection and awareness of water contamination events. HSPD-9 also directs EPA to develop a network of integrated federal and state water testing laboratories.

Conclusion

A long history exists of water-related violence and conflicts, including what must be categorized as environmental terrorism that targets water resources and infrastructure. The threat of future attacks is real. But the actual risks of serious human health consequences is less clear, given the complex nature of our developed water systems, protections already put in place to identify and eliminate biological and chemical contaminants, and the attractiveness and vulnerability of other targets. It is vital that sensitive water systems be protected through a combination of physical barriers, real-time chemical and biological monitoring and treatment, and the development of smart and integrated response strategies at all levels. However, it is equally important that the risks not be exaggerated, so that limited financial resources can be spent efficiently and effectively, and so that the public is not made fearful of risks that are low or manageable.

The best approaches will require careful assessment of both the probability and the consequences of attacks. By evaluating both, it will be easier to identify vulnerabilities and the appropriate and measured responses to those vulnerabilities. We must maintain a safe, affordable, and reliable water system, as well as maintain our open and democratic government and institutions.

REFERENCES

Abbey, E. 1975. *The monkey wrench gang.* New York: Harper Collins Publishers.

American Water Works Association. 2006. Water Infrastructure Security Enhancement (WISE) program. http://www.awwa.org/science/wise/.

Associated Press. 2003a. Water targeted, magazine reports. *AP*, May 29, 2003.

Associated Press. 2004a. US dumps water projects in Gaza over convoy bomb. *AP*, May 6, 2004.

Beres, L. R. 1995. The legal meaning of terrorism for the military commander. *Connecticut Journal of International Law* 11:1–27.

Berkowitz, B. J., and others. 1972. *Superviolence: The civil threat of mass destruction weapons.* Santa Barbara, CA: ADCON (Advanced Concepts Research) Corporation, Report A72-034-10, 29 September.

Betts, K. S. 1999a. DNA chip technology could revolutionize water testing. *Environmental Science and Technology* 33(15):300A–301A.

Betts, K. S. 1999b. Testing the waters for new beach technology. *Environmental Science and Technology* 33(16):353A–354A.

Center for Defense Information (CDI). 2002. Securing U.S. water supplies. Center for Defense Information, Washington, D.C. July 19, 2002. http://www.cdi.org/terrorism/water-pr.cfm.

Centner, C. M. 1996. Environmental warfare: Implications for policymakers and war planners. *Strategic Review* 24:71–76.

Chalecki, E. 2001. A new vigilance: Identifying and reducing the risks of environmental terrorism. A report of the Pacific Institute for Studies in Development, Environment, and Security. Oakland, CA. http://www.pacinst.org/reports/environment_and_terrorism/.

Chelyshev, A. 1992. Terrorists poison water in Turkish Army Cantonment. Telegraph Agency of the Soviet Union (TASS), 29 March. http://www.jewishvirtuallibrary.org/jsource/Terrorism/chemterror.html.

Christian Science Monitor. 2000. Ecoterrorism as negotiating tactic. July 21, 2000, 8.

Douglass, J. D., Jr., and Livingstone, N. C. 1987. *America the vulnerable: The threat of chemical and biological warfare.* Lexington, MA: Lexington Books.

Dutch Water Line. 2002. Information on the historical use of water in defense of Holland. http://www.xs4all.nl/~pho/Dutchwaterline/dutchwaterl.htm.

Eitzen, E. M., and Takafuji, E. T. 1997. Historical overview of biological warfare. In *Textbook of military medicine, medical aspects of chemical and biological warfare.* Washington, DC: Office of The Surgeon General, Department of the Army, 415–424.

Environment News Service (ENS). 2001. Environment as a weapon in the Israeli-Palestinian conflict. February 5, 2001. http://www.ens-newswire.com/ens/feb2001/2001-02-05-01.asp.

Foran, J. A., and Brosnan, T. M. 2000. Early warning systems for hazardous biological agents in potable water. *Environmental Health Perspectives* 108(10), October. http://ehp.niehs.nih.gov/realfiles/docs/2000/108p993-995foran/foran-full.html.

Gaura, M. A. 2000. Disaster simulation too realistic media fooled by 'news' of terrorist attack. *San Francisco Chronicle*, October 27, 2000, A1, A23. http://sfgate.com/cgi-bin/article.cgi?f=/c/a/2000/10/27/MN1002CH.DTL&hw=dam&sn=001&sc=1000.

Gellman, B. 2002. Cyber-attacks by Al Qaeda feared. *Washington Post,* June 27, 2002, A1.

Gilmore Commission. 1999. First annual report to the President and the Congress of the Advisory Panel to assess domestic response capabilities for terrorism involving weapons of mass destruction. I. Assessing the threat. Santa Monica, CA: RAND, December 15, 1999.

Gilmore Commission. 2000. Second annual report to the President and the Congress of the Advisory Panel to assess domestic response capabilities for terrorism involving weapons of mass destruction. II. Toward a national strategy for combating terrorism. Santa Monica, CA: RAND, December 15, 2000.

Gleick, P. H. (Summer 1993). Water and conflict. *International Security* 18(1):79–112.

Gleick, P. H. 1998. *The world's water 1998–1999: The biennial report on freshwater resources.* Covelo, CA: Island Press.

Gleick, P. H. 2000. *The world's water 2000–2001: The biennial report on freshwater resources.* Covelo, CA: Island Press, 315.

Haeri, S. 1993. Iran: Vehement reaction. *Middle East International* 19 March, 8.

Hatami, H., and Gleick, P. 1994. Chronology of conflict over water in the legends, myths, and history of the ancient Middle East. In *Water, war, and peace in the Middle East.* 6. Washington, DC: Heldref Publishers; *Environment* 36(3):6.

Heilprin, J. 2005. EPA watchdog finds security lapses in remote controls for water systems. *Associated Press.* January 10. http://www.sfgate.com/cgi-bin/article.cgi?file=/news/archive/2005/01/10/national1827EST0682.DTL.

Hewish, M. 2000. Mini-robots sniff out chemical agents. Jane's IDR, Features.

Hickman, D. C. 1999. Chemical and biological warfare threat: USAF water systems at risk. 36. Future Warfare Series No. 3. Air University, US Air Force Counterproliferation Center, Maxwell AFB, AL. http://www.au.af.mil/au/awc/awcgate/cpc-pubs/hickman.htm.

Israel Line. 2001a. Palestinians loot water pumping center, cutting off supply to refugee camp. *Israel Line.* http://www.israel.org/mfa/go.asp?MFAH0dmp0, http://www.mfa.gov.il/mfa/go.asp?MFAH0iy50.

Israel Line. 2001b. Palestinians vandalize Yitzhar water pipe. *Israel Line,* January 9, 2001. http://www.mfa.gov.il/mfa/go.asp?MFAH0izu0.

Janofsky, M. 2006. Feds accuse 11 of ecoterrorism: Targeted meatpacker, ski resort, timber firm. *New York Times News Service,* January 21, 2006. http://www.chicagotribune.com/news/nationworld/chi-0601210075jan21,1,7234189.story?coll=chi-newsnationworld-hed.

Jarboe, J. F. 2002. The threat of eco-terrorism. Domestic Terrorism Section Chief, Counterterrorism Division, FBI Testimony Before the House Resources Committee, Subcommittee on Forests and Forest Health, February 12. http://www.fbi.gov/congress/congress02/jarboe021202.htm.

Jenkins, B. M., and Rubin, A. P. 1978. New vulnerabilities and the acquisition of new weapons by nongovernment groups. In *Legal aspects of international terrorism.* 221-276. Evans, S., and Murphy, J. F. (editors). Lexington, MA: Lexington Books.

Jenkins, B. M. 2000. *RAND Review.* 5. Fall 2000.

Kirschner, O. 1949. Destruction and protection of dams and levees. Military Hydrology, Research and Development Branch, U.S. Corps of Engineers, Department of the Army, Washington District. From Schweizerische Bauzeitung 14 March 1949, Translated by H.E. Schwarz, Washington, DC.

Kupperman, R. H., and Trent, D. M. 1979. *Terrorism: Threat, reality, response.* Stanford, CA: Hoover Institution Press.

Lee, M. F. 1995. Violence and the environment: The case of "earth first!" *Terrorism and Political Violence* 7(3):109–127.

Livingstone, N. C. 1982. *The war against terrorism.* Lexington and Toronto, Canada: Lexington Books.

Macintyre, A. J., Christopher, G. W., Eitzen, E., Gum, R., Weir, S., DeAtley, C., Tonat, K., and Barbera, J. A. 2000. Weapons of mass destruction events with contaminated casualties. *Journal of the American Medical Association* 283(2):242–249.

MacKenzie, W. R., Hoxie, N. J., Proctor, M. E., Gradus, M. S., Blair, K. A., Peterson, D. E., Kazmierczak, J. J., Addiss, D. G., Fox, K. R., Rose, J. B., and Davis, J. P. 1994. A massive outbreak in Milwaukee of Cryptosporidium infection transmitted through the public water supply. *New England Journal of Medicine* 331(3):161–167.

Maiese, M. 2003. Jus ad Bellum. Beyond intractability. http://www.beyondintractability.org/essay/jus_ad_bellum/.

McLean, I., McMillan A. 2003. *The Oxford Concise Dictionary of Politics.* 2nd edition. United Kingdom: Oxford University Press.

Mullins, W. C. 1992. An overview and analysis of nuclear, biological, and chemical terrorism: The weapons, strategies and solutions to a growing problem. *American Journal of Criminal Justice* 16(2)95–119.

Museum of the City of New York (MCNY). n.d. The greater New York consolidation timeline. http://www.mcny.org/Exhibitions/GNY/timeline.htm.

Naff, T., and Matson, R. C. (editors). 1984. *Water in the Middle East: Conflict or cooperation?* Boulder, CO: Westview Press.

Pelley, J. 1999. Rapid, genetic-based test can identify human viruses in beach water. *Environmental Science and Technology* 33(18):399A.

Ponte, L. 1980. The dawning age of technoterrorism. *Next* (July-August), 49–54.

Pretoria Dispatch Online. 1999. Dam bomb may be "aimed at farmers." July http://www.dispatch.co.za/1999/07/21/southafrica/RESEVOIR.HTM.

Purver, R. 1995. Chemical and biological terrorism: The threat according to the open literature.

Canadian Security Intelligence Service (CSIS). http://www.csis-scrs.gc.ca/en/publications/other/c_b_terrorism01.asp.

Reisner, M. 1986, 1993. *Cadillac desert: The American West and its disappearing water.* New York: Penguin Books.

Schmid, A. 1997. The problems of defining terrorism. In *Encyclopedia of world terrorism.* Vol. 1, 12–22. Crenshaw, M., and Pimlott, J. (editors). Armonk, NY: M. E. Sharpe.

Schofield, T. 1999. The environment as an ideological weapon: A proposal to criminalize environmental terrorism. *Boston College Environmental Law Review* 26:619–647.

Schwartz, D. 1998. Environmental terrorism: Analyzing the concept. *Journal of Peace Research* 35(4)483–496.

Simon, J. D. 1997. Biological terrorism: Preparing to meet the threat. *Journal of the American Medical Association* 278(5):428–430.

Simpson, J., and E. Weiner (editors). 1989. *Oxford English Dictionary.* Second edition, Clarendon Press, United Kingdom.

Smith, V. (Summer 1994). Disaster in Milwaukee: Complacency was the root cause. *EPA Journal, Vol. 20, Issue 1-2, pp. 16-18. Washington. D.C.*

Sobsey, M. 1999. Methods to identify and detect microbial contaminants in drinking water. In *Identifying future drinking water contaminants.* 177–203. Washington, DC: U.S. National Academy of Science, National Academy Press.

Stern, J. 1999. *The ultimate terrorists.* 214. Cambridge, MA: Harvard University Press.

Tierney, J., and Worth, R. F. 2003. Attacks in Iraq may be signals of new tactics. *The New York Times,* August 18, 2003, 1. http://www.nytimes.com/2003/08/18/international/worldspecial/18IRAQ.html?hp.

Times News Network (TNN). 2004. IED was planted in underground pipe. December 5, 2004. http://timesofindia.indiatimes.com/articleshow/947432.cms.

U.S. Environmental Protection Agency (U.S. EPA). 2003a. Response protocol toolbox: Planning for and responding to drinking water contamination, threats, and incidents. Washington, DC. http://www.waterisac.org/epa/Guidance/guide_response_module1.pdf.

U.S. Environmental Protection Agency (U.S. EPA). 2003b. Large Water System Emergency Response Plan outline: Guidance to assist community water systems in complying with the Public Health Security and Bioterrorism Preparedness and Response Act of 2002. United States Environmental Protection Agency, Office of Water, Office of Ground Water and Drinking Water, EPA 810-F-03-007. July. http://www.epa.gov/safewater/security, http://www.epa.gov/ogwdw/security/pdfs/erp-long-outline.pdf.

U.S. Environmental Protection Agency (U.S. EPA). 2005. The monitor: The newsletter of the ETV Advanced Monitoring Systems (AMS) Center. 8(6). http://www.epa.gov/etv/pdfs/newletters/monitor/01_mon_nov05.pdf.

U.S. Environmental Protection Agency (U.S. EPA). 2006. WaterSentinel fact sheet. http://www.epa.gov/safewater/watersecurity/pubs/water_sentinel_factsheet.pdf.

U.S. Food and Drug Administration (U.S. FDA). 2002. The Bioterrorism Act of 2002. http://www.fda.gov/oc/bioterrorism/bioact.html.

U.S. National Research Council (U.S. NRC). 1995. *Guidelines for chemical warfare agents in military field drinking water.* Commission on Life Sciences (CLS). Washington, DC: National Academies Press.

U.S. Office of Technology Assessment (U.S. OTA). 1991. *Technology against terrorism: The Federal effort.* OTA-ISC-487. United States Congress. Washington, DC: U.S. Government Printing Office.

Valcik, J. E. 1998. *Biological warfare agents as potable water threats.* Medical Issues Information Paper No. IP-31-017. U.S. Army Center for Health Promotion and Preventative Medicine. Aberdeen Proving Ground, MD.

Waeckerle, J. F. 2000. Domestic preparedness for events involving weapons of mass destruction. *Journal of the American Medical Association* 283(2):252–254.

Waterman, S. 2003. Al-Qaida threat to U.S. water supply. *United Press International (UPI),* May 28, 2003.

Yang Lang. 1989, 1994. High dam: The sword of Damocles. In *Yangtze! Yangtze!* 229–240. Dai Qing (editor). London, UK: Probe International, Earthscan Publications.

Yi Si. 1998. The world's most catastrophic dam failures: The August 1975 collapse of the Banqiao and Shimantan dams. In *The river dragon has come!* Qing D. (editor). New York: M. E. Sharpe.

Going with the Flow: Preserving and Restoring Instream Water Allocations

David Katz

Most water experts, and most water analyses, focus on challenges of water scarcity and poor water quality for humans. Far less attention has been paid, even in previous editions of this biennial series, to the critical dependence of natural ecosystems on water and on the role that such systems play in maintaining both the hydrological cycle and water quality. Until relatively recently, leaving water to flow in streams, aquifers, or wetlands was considered by many to be a waste of a precious resource. As a consequence, water management over the past century has largely been a story of widespread, large-scale diversions of water out of natural systems. More than 60 percent of the world's rivers have undergone major hydrological alteration (Revenga et al. 1998) and, by one estimate, humanity now appropriates over half the world's accessible runoff (Postel et al. 1996). The means of achieving this include over 800,000 dams, including nearly 50,000 dams over 15 meters high (WCD 2000), extensive channelization of river systems, massive pumping of groundwater aquifers, and long-distance water-conveyance systems. This infrastructure has traditionally been seen as an indication of progress, given the substantial benefits it provides, including stable water supplies, food security, hydropower, and flood control.

This style of water development, however, has also taken a heavy toll on the world's freshwater ecosystems. More than half the world's wetlands have been lost since 1900 (Finlayson and Davidson 1999). The abundance of global freshwater species has decreased by 50 percent since 1970, a rate much faster than it declines in either terrestrial or marine life (Loh and Wackernagel 2004). Over 20 percent of the world's freshwater fish are estimated to be at serious risk or already extinct (Moyle and Leidy 1992), and even higher rates of threat and extinction exist for other forms of aquatic wildlife. Moreover, much of the terrestrial, avian, and estuarine wildlife that are dependent on freshwater flows is also at risk (Allan and Flecker 1993).

Recently, water managers, policy makers, and the general public have begun to recognize the environmental, economic, and aesthetic benefits that freshwater ecosystems provide. Most efforts to protect aquatic ecosystems have targeted water quality and pollution prevention (Karr 1991; Bernhardt et al. 2005). With a growing understanding of the importance of water flow as a "master variable" in determining aquatic ecosystem functioning (Poff et al. 1997), however, decision makers are increasingly focusing attention on securing sufficient flows for ecological, recreational, and other instream objectives (Petts and Maddock 1996; Baron et al. 2002).

Around the world, new efforts are now under way to design, implement, and monitor environmental flows, as water left instream has come to be called (Tharme 2003; Moore 2004; TNC 2005). This chapter will introduce the concept of environmental flows, discuss their ecological and social importance, and present examples of efforts currently under way to preserve and restore them. It will then review legal frameworks that are currently in place for protecting environmental flows and survey existing scientific and economic methods for determining optimum flow levels. It will conclude with a discussion of some of the challenges facing policy implementation.

Environmental Flow: Concepts and Applications

Flow Characteristics

As global water withdrawals continue to rise at rates exceeding human population growth,[1] people are increasingly confronted with trade-offs when allocating water. Traditionally, decision makers have seen such trade-offs as those between urban, agricultural, and industrial consumption. Streams and lakes have been left with the water remaining after other sectors were satisfied, in many cases, leaving little or no water to maintain the natural ecosystem. The many services provided by instream water, such as maintaining water quality, fish stocks, and recreational opportunities, were largely overlooked. Currently, however, a widespread effort is under way to document these services, delineate their ecological, economic, and social value, and understand the types of flows necessary for their provision.

The World Commission on Dams (WCD) published a thematic report on methods for measuring environmental flow needs (King et al. 1999), bringing international attention to the issue. Although the WCD's recommendations are not binding, they emphasize balancing developmental and environmental needs, and represent a standard of best practices for dam design and operation that many countries have committed to apply.

In the past, to the extent that issues of environmental flow were addressed at all, concern was limited to providing the minimum flow necessary to maintain specific commercial or recreational fish populations. Today, a growing consensus has developed among water professionals that the key to maintaining healthy freshwater ecosystems and the services that they provide is preservation (or restoration) of some semblance of the natural flow regime around which the native flora and fauna developed (Poff et al. 1997; Bunn and Arthington 2002). This includes daily, seasonal,

---

1. Based on calculations of water withdrawals and population from the United Nations' Food and Agricultural Organization's AQUASTAT database, accessed online at: http://www.fao.org/ag/agl/aglw/aquastat/dbase/index.stm.

and interannual flow variations. Richter et al. (1996) outlined "five fundamental characteristics" of hydrological regimes: (1) magnitude of flow, (2) timing of flow, (3) frequency of various flow events, (4) duration of flow events, and (5) rate of change between types of flows. All of these have been significantly altered by dam construction and operation, surface and groundwater withdrawals, and channel alteration (see Figure 2.1).

Of the five characteristics, magnitude of flow is perhaps the most obvious concern. Low flow can mean reduced stream and wetland habitat and a disconnection of a river from its floodplain. This can lead to lower populations of fish and wetland wildlife, invasion of exotic species, diminished opportunities for recreation, such as swimming and boating, and reduced possibilities for pollution assimilation and protection of downstream water quality. For these reasons, securing minimum stream flows received much of the early attention among water professionals.

Too much water instream can also be harmful to ecosystems. Flow of unnaturally high magnitude can inundate shallow water habitat, such as stream riffles; alter geomorphological characteristics of streams, such as channel course and substrate; and increase flood risks. Such flows may result from dam releases, discharge of sewage waters, or interbasin water supply transfers.

The remaining four flow characteristics can be of equal importance to ecosystem functioning (Richter et al. 1996). The timing and frequency of different flow events often play a role in seed dispersal, signaling mating periods, preventing invasion of exotic species, flushing out sediments, and cycling nutrients. The duration and rate of change of flow patterns determine the type and length of connectivity between a stream and its floodplain, which, in turn, determines how long inundated riparian zones are available as breeding and nursery grounds by fish and which types of flora will be able to develop in wetlands. Many species of plants and aquatic animals have narrow windows of time in which to develop during or following flooding. For example, germination of several types of riparian and emergent wetland vegetation is dependent on a gradual recession of flood waters that permits seeds to establish and root.

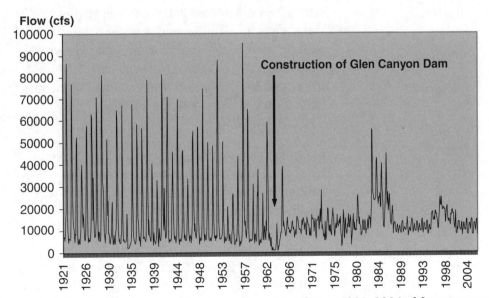

FIGURE 2.1 CHANGES IN FLOW LEVELS ON THE COLORADO RIVER, 1921–2004. Measurements taken at Lee's Ferry gauge station. Flow rate in cubic feet per second.
Source: USGS 2005.

Environmental Flow Projects in Practice

Although the science and practice of environmental flow protection is still in its infancy in much of the world, flow conservation and restoration projects are being undertaken in at least 70 nations (Moore 2004; TNC 2005), including both industrialized and developing nations, ranging from the chronically water-scarce (e.g., Israel and Tunisia) to the most water-rich (e.g., Canada and New Zealand) (see Box 2.1).

The goals of the environmental flow projects vary from establishing habitat to enhancing groundwater recharge and downstream water quality. The means for implementation are also diverse. They include restriction of groundwater withdrawals, redesign of dams and of dam-release schedules, dam removal, and purchase of stream withdrawal rights. Environmental flow projects range in scale and expense from a few thousand dollars to lease water rights in drought years for small stream stretches to $2 billion for flow restoration in China's longest interior river, the Tarim (Parish 2005), and more than $10 billion for restoration of the Kissimmee River and Florida Everglades (King 2005). A range of environmental flow projects from around the world is presented in Box 2.1.

Box 2.1   Selected Flow Preservation and Restoration Activities

Region	Flow Preservation/Restoration Activity
Africa (Sub-Saharan)	In Cameroon, experimental floods were released along the Logone River to restore floodplain flora and fauna.[1] In South Africa, experimental floods were released along the Kromme River to improve habitat and enhance sediment transport,[2] dam release schedules were altered in the Lesotho Highlands Water Project,[3,4] and several flow contingency plans have been drawn up for the Sabie River to support wildlife in Kruger National Park.[5] In Zambia, high flows are being restored to provide habitat for fish and the Kafue lechwe, a type of wetland antelope.[1,6]
Asia	In China, flows are being restored on the Tarim River to combat desertification and enhance habitat[7] and on the Suzhou River to improve water quality.[6] Experimental flushing flows are being implemented on the Yellow River to improve sediment transport.[6] In Sri Lanka, seasonal flushing flows were restored on the Mahaweli Ganga River to assist in channel flushing and sediment transport.[6]

Box 2.1

Region	Flow Preservation/Restoration Activity
Asia *continued*	In Thailand, the gates to the Pak Mun dam were opened in 2001-2003, allowing the return of over 150 species of fish. Authorities, however, have recently announced that the dam will be open only four months a year.[8]
Australia and New Zealand	In Australia, a cap was placed on total withdrawals from the Murray-Darling River system, and seasonal floods were restored to riparian forests and wetlands to improve habitat for migratory birds and other wildlife.[9,10] In New Zealand, seasonally adjusted minimum flows were restored to several rivers, including the Temuka and Waitaki, to enhance fish, avian, and invertebrate habitat, and to improve water quality and groundwater recharge.[6]
Europe	An international attempt to partially restore flows in the Danube River is under way.[1] Initial activities include establishing seasonal minimum flows in Austria[6] and releasing experimental floods on the River Spöl in Switzerland.[11] In Finland, dam releases were redesigned to improve habitat in several lakes, including Kostonjärvi and Oulujärvi.[12] In Spain, seasonal minimum flows were established on several rivers to enhance aquatic wildlife and habitat.[13]
Middle East and North Africa	In Israel, a portion of the Hula wetlands along the Jordan River was reflooded to enhance habitat for migratory birds and to improve water quality,[14] and a program has been developed to restore flows to its coastal streams.[15] In Jordan, the planned location of a dam along Wadi Mujib was moved downstream to preserve the desert oasis and nature reserve.[16] In Tunisia, flows into lakes and wetlands were increased to reduce salinity and to improve habitat conditions for fish, flora, and waterfowl.[6]

continues

Box 2.1 *continued*

Region	Flow Preservation/Restoration Activity
North America	Around 500 dams have been removed throughout the United States and Canada, many to restore flows for ecosystem services, fisheries, or recreational opportunities.[17] Examples include dams along the Sacramento and Kennebec rivers in the United States[18] and the Pamehac and Theodosia in Canada.[6]
	In the state of Maryland, high flows have been reduced and low flows restored along the Patauxent River and Sawmill Creek.[6]
	In Puerto Rico, water intake timing and infrastructure were altered to protect freshwater shrimp populations in rivers in the Caribbean National Forest.[5]
Central and South America	In Brazil[19] and in Chile,[20] minimum instream flows have been established for several rivers based on naturally occurring low flows.
	In Costa Rica, water user fees have been levied to finance flow protection.[21]

Sources: 1) Postel and Richter 2003; 2) King, et al. 1999; 3) Arthington et al. 2003b; 4) Hirji and Panella 2003; 5) King and Louw 1998; 6) TNC 2005; 7) Parish 2004; 8) IRN 2005; 9) MDRC 2005; 10) Arthington and Pusey 2003a; 11) Robinson et al. 2004; 12) Hellsten et al. 1996; 13) García de Jalón 2003; 14) NAS 1999; 15) NRA 2004; 16) EcoPeace/FoEME 2005; 17) Gleick 2000, The Heinz Center 2002; 18) Gleick 2000; 19) Benetti et al. 2004; 20) Davis and Riestra 2002; 21) Dyson et al. 2003.

Legal Frameworks for Securing Environmental Flow

International Conventions and Agreements

Over the past two decades, several international legal frameworks have recognized the need to protect aquatic ecosystems. Some focus specifically on this topic (e.g., Ramsar Convention for wetland protection), whereas others address it within a broader context of calls for sustainable development (e.g., Agenda 21 of United Nations Conference on Environment and Development [UNCED; Article 18]), biodiversity protection (e.g., Convention on Biological Diversity), or equitable water management (e.g., Dublin Statement on Water and Sustainable Development).

International law pertaining to transboundary water allocation also addresses freshwater ecosystem protection, in general, and environmental flow, in particular. The United Nations Convention on the Law of the Non-Navigational Uses of International Watercourses (1997) lists environmental conservation as a criterion to be considered in the allocation of transboundary waters. The Berlin Conference Report of 2004—a summary of international water law drafted by the International Law Association as an update to its Helsinki Rules of 1967—specifically acknowledges the legal legitimacy and importance of environmental flow for ecological and other instream purposes (Articles 22 and 24). Currently, some nations are attempting to apply these principles by negotiating environmental flows in specific international water basin agreements, such as the Mekong River Commission, in Southeast Asia.

The European Union's Water Framework Directive of 2000 obliges all member countries to monitor the quality of water bodies and to take actions to ensure their "good ecological, status." It specifically mentions the importance of the "quantity and dynamics of water flow" in rivers and other surface water (Annex V, 1.1.1) and the need to take into account "natural flow conditions" for purposes of environmental protection (Article 34), albeit falling short of setting specific minimum flow requirements.

Incorporating Environmental Flow in National Water Laws

Though international law provides broad guidance regarding environmental flow, national laws and policies often have the most direct impact on flow rates. Several countries have federal water laws that specifically address environmental flow. Japan amended its River Law in 1997 to declare conservation of riverine environments a vital part of water management (Tamai 2005). In 2004 Israel amended its Water Law of 1959 to officially recognize nature as a beneficial user of water and to require the National Water Commission to submit yearly reports to parliament, listing the amount and quality of water allocated for ecosystem purposes (Amendment 19). Switzerland goes even further, mandating specific minimum flow quantities in its Water Protection Act of 1991 (Article 31).

Australia and South Africa stand out in terms of the legal protection that they afford freshwater ecosystems. Australia developed a Water Reform Framework in 1994, recognizing the need to protect the aquatic environment and calling for each of the states to allocate water for this purpose (Postel and Richter 2003). All states in Australia have since enacted such legislation. In one application of the framework, a cap has been placed on total withdrawals from the Murray-Darling River Basin to accommodate ecosystem needs (MDBC 2005).

South Africa's National Water Law of 1998 established water *reserves*—uses for which allocations are nonnegotiable. The first reserve is for basic human needs, such as drinking, cooking, and sanitation. The second is to "maintain the ecological functions on which humans depend" to ensure "the long term sustainability of aquatic and associated ecosystems." Under this system, after the reserves are met, only water left over is available for allocation for other purposes, such as agriculture, industry, hydropower, and nonessential domestic consumption. This stands in stark contrast to the policy existing in most of the world, under which ecosystems tend to receive only the amount of water left over after all other uses have been satisfied. Interdisciplinary teams, including those under South Africa's Water Research Commission, are now attempting to calculate the size of these reserves and to decide how best to secure them. Among the wide variety of issues being addressed are education of rural communities as to the importance of environmental flow, integration of water quantity and quality concerns in forming a reserve, and prioritizing areas of high conservation need (WRC 2003).

A Grab-Bag Policy Approach—Laws for Environmental Flow in the United States

The primary responsibility for allocation of water in the United States lies with the states rather than with the federal government. Almost all of the more arid Western states have some legislation that recognizes and provides for instream water rights, and several of the more humid Eastern states have also initiated some form of legal protection of streamflow. Environmental purposes were also being explicitly considered as a criterion for allocation of waters in negotiations over an interstate compact (Apalachicola-Chattahoochee-Flint basin [Richter et al. 2003]), though a final agreement may be left for the courts.

The United States lacks comprehensive federal water laws but offers several other examples of federal legislation that can be used to secure environmental flows. The U.S. Clean Water Act (CWA) obliges states to take action to protect the "chemical, physical, and biological integrity" of the nation's waters and to strive to ensure that all are of fishable and swimmable quality. This mandate allows states to preserve instream flow of water to achieve quality standards, a principle upheld by the U.S. Supreme Court.[2] In practice, however, the CWA is rarely invoked for this purpose.

The Wild and Scenic Rivers Act protects "certain selected rivers" in "their free-flowing condition" (Sax et al. 2000). Streams flowing through national parks have implied water rights, and the federal government has, at times, specifically legislated environmental flow on federally managed lands or from federally operated water programs. For instance, the 1992 Central Valley Project Improvement Act (CVPIA) specifies that 1.2 to 1.35 million-acre feet (roughly 1.5 billion cubic meters) of water per year be set aside for instream use and rehabilitation of wetlands and wildlife preserves in California (CBO 1997). Perhaps the single most effective piece of federal legislation for securing environmental flow is the Endangered Species Act, which requires preservation of habitat of listed species.

Additional federal law mandates regulatory consideration of environmental factors, such as flow. The National Environmental Protection Act (NEPA) obliges the government to perform benefit-cost analyses on government projects, including water works. Under the Federal Power Act, the Federal Energy Regulatory Commission (FERC) can make adoption of environmental flow a criterion for dam licensing or relicensing (Gillilan and Brown 1997). According to the Electric Consumers Protection Act, FERC is required to give equal consideration to environmental and power issues (Loomis 1998). According to one source, for example, between 1980 and March of 1983, "59% of hydropower licenses contained special articles governing instream flows (Kerwin 1990)" and "instream flow conditions were included in 80% of the licenses issued during 1985 (Kerwin and Robinson 1985)" (cited in Lamb et al. 1998).

Critics of U.S. federal policy on this matter note that despite the varied array of legal options for provision of environmental flows, it is in essence a piecemeal approach

---

2. *PUD No. 1 v. Washington Dep't of Ecology,* 511 U.S. 700 (1994).

that limits opportunities for effective application. The Endangered Species Act, for instance, takes effect only when situations are already critical, often too late to be effective. Moreover, it is targeted at maintaining a single species, rather than the health of an ecosystem or other instream objectives (Postel and Richter 2003). The Wild and Scenic Rivers Act applies only to rivers that "possess outstandingly remarkable" scenic, environmental, cultural, or recreational values, and in essence protects a relatively small number of rivers that are still in a seminatural state.

Legal Principles for Flow Protection

Several legal principles support the protection of environmental flow. In many nations, water is the property of the state, and the state has an explicit obligation to manage it for the public good. This includes protecting freshwater environments. The Public Trust Doctrine, originating from Roman law, holds that water is to be treated as a common resource and managed by the state for the common benefit. This doctrine was adopted by many European countries. It has been incorporated into English, French, and Spanish legal codes (Smith and Hoar 1999; IFC 2002) and has become part of the legal framework of many formerly colonized nations (Postel and Richter 2003). The Public Trust doctrine served as the basis for the South African National Water Law and for a precedent-setting court ruling in the United States on the issue that granted environmental flow protection to Mono Lake in California.[3]

Under riparian water rights systems, in place in countries such as the United Kingdom and many of its former colonies, including Canada, India, and the Eastern United States, downstream users along a stream are entitled to its waters in a relatively undiminished and undamaged state. This principle, however, has been extremely difficult to enforce. In addition, many, if not most, legal systems, including international law, have found that strict interpretation of the undiminished principle is neither practical nor equitable, and that upstream users are entitled to some form of reasonable use (Wolf 1999). Furthermore, the lack of legal standing of ecosystems means that, although landowners along a stream may be entitled to water of sufficient quantity and quality, this need not apply to flora and fauna or to nonlandowners who may benefit from downstream flow.

Legal systems with private water rights, such as those in place in Chile, Mexico, and the in the Western United States, allow flows to be reserved for or acquired by downstream users. Under the Prior Appropriation Doctrine, private water rights are acquired specifically by offstream diversions and are prioritized by the date of diversion, the so-called *first in time, first in right*. Historically, instream uses have not fared well under such a *use it or lose it* system. More recently, instream users have been recognized as legitimate consumers of water, but such users often have low seniority, as newly assigned rights. Such junior rights may be of little use in drought situations on highly appropriated streams, exactly when they are needed most. To truly ensure environmental flows, senior rights need to be retired or transferred downstream. As will be discussed later, private water-rights systems may allow for market transactions as a means of attaining flows, which can be an effective means in situations in which highly valued environmental or recreational services are at stake.

---

3. *National Audubon Society v. Superior Court of Alpine County,* 189 Cal. Rptr. 346 (1983).

The Science of Determining Environmental Flow Allocations

Development of the Knowledge Base

Scientists began investigating the ecological functions of flows in the 1950s (Petts and Maddock 1996), although prior to the 1970s flow recommendations were "often based more on a biologist's or engineer's guess than on a quantified evaluation of the relationship between discharge and the ecology of the stream" (Fraser 1972; cited in Petts and Maddock 1996, 61). Only in the late 1960s and the 1970s did scientists—primarily stream ecologists—develop the first empirically based methods for flow prescriptions (IFC 2002). They generally did so, however, without addressing more fundamental questions regarding which values and objectives society was attempting to achieve with such flows. The results were, in large part, prescriptions for minimum flows designed to protect target fish species, mostly highly valued game species and some endangered species.

These recommendations were largely unsuccessful on several counts. By focusing solely on minimum flows, they often ignored the temporal and spatial flow variation necessary to maintain the aquatic ecosystem on which the target species depend. In addition, even in cases when target species did succeed, overall biodiversity and ecosystem health often continued to decline because ecosystem needs were not met for nontarget species.

Over the last three decades, dozens of more sophisticated techniques have been developed to address various environmental goals. Methodologies have been developed for cold and warm water streams, large river floodplains, ephemeral streams, wetlands, waterfowl, and even boating and instream recreation. Moreover, many of the methods now concentrate on overall ecosystem health rather than target species.

A Comparison of Methodologies

This section will review the most popular categories of environmental flow methodologies and their relative advantages and disadvantages (Table 2.1). Several comprehensive surveys of instream flow methods have been published recently (e.g., Jowett 1997; King et al. 1999; IFC 2002; and Tharme 2003). Jowett classifies instream flow assessment techniques into three categories: historic flow, hydraulic geometry, and habitat simulation. To this, Tharme adds the category of holistic methods.

Among the first techniques for determining environmental flows was the so-called Tennant Method. It attempted to outline a systematic correlation between percentage of average historic flow and fish habitat condition, based on extensive observations of numerous streams throughout the United States (Tennant 1976). Another commonly used historic flow method recommends minimum flows based on average annual seven-day low flows. Flow prescriptions based on such techniques are relatively easy and inexpensive to calculate. For this reason, they are among the most frequently used, especially in areas with little experience in flow protection or with limited budgets for such purposes (Reiser et al. 1989; Tharme 2003; Scatena 2004). Historic flow methods, however, tend to produce static minimum flow recommendations and have been criticized for not addressing the

TABLE 2.1   Environmental Flow Prescription Methodologies

Methodology	Pros and Cons
Historic Flow	*Pros:* Inexpensive. Quick. Low levels of expertise required. Uses flow data. *Cons:* Ignores physical and biological data. Simple models produce invariable flow recommendations.
Hydraulic Geometry	*Pros:* Inexpensive. Quick. Low levels of expertise required. Uses physical stream channel data. *Cons:* Ignores historic flow rates and biological data. Many models produce invariable flow recommendations.
Habitat Simulation	*Pros:* Specific to particular stream/wetland characteristics and to particular species' life-cycle needs. *Cons:* High levels of expertise required. Expensive. Time-consuming.
Holistic Approaches	*Pros:* Focuses on ecosystem health. Incorporates broad range of perspectives from across disciplines. *Cons:* High levels of expertise required. Can be expensive and time-consuming.

physical or biological characteristics of the streams. More advanced techniques, such as the Range of Variability Approach (Richter et al. 1997), are based not simply on average flow, but on a suite of flow features covering timing, frequency, duration, and rate of change, thus producing more dynamic flow prescriptions.

Hydraulic geometry methodologies take into consideration river channel attributes, such as the shape and width of a channel cross section. They may recommend, for instance, that a certain proportion of channel bed be submerged by water. These techniques are also relatively cheap and simple to calculate, but they, too, disregard biological information and natural flow variation. Thus, a prescription based on such a method might recommend flow allocations in a season in which a stream is naturally dry.

Habitat simulation techniques base recommendations on the flow needs of specific species at different life stages. Popular methodologies, such as the Instream Flow Incremental Methodology, use sophisticated computer models (e.g., PHABSIM) to determine flow needs for migration, spawning, and so on (Stalnaker et al. 1995; Bovee et al. 1998). Unlike the types of methods already mentioned, habitat simulation methods deal directly with the specific biological needs and thus are often considered scientifically more defendable (IFC 2002). But they often take a long time to complete, require a high level of expertise, and can be expensive. Furthermore, they tend to be species-specific, rather than addressing overall ecosystem health or other flow objectives.

Holistic methodologies begin with the natural flow regime of a river or stream and attempt to discern the various stages of environmental degradation associated with deviations from the natural state. They then seek consensus on desired development and environmental goals and on acceptable levels of degradation (Arthington et al. 2003a; King et al. 2003). These methodologies incorporate input from a range of disciplines that span the physical, biological, and even social sciences. They were developed in the 1990s by Australian and South African researchers within the context of their countries' relatively far-reaching legal commitments to securing environmental flows and are beginning to be applied elsewhere, such as India, Italy, Tanzania, and several countries in southern Africa (Postel and Richter 2003; Tharme 2003).

The preferred method for determining environmental flows is likely to depend on the specific environmental and social circumstances involved, as well as the desired objectives. Simple methods based on historic flow rates or channel structure are less likely to achieve broad ecological restoration goals but may be useful as convenient rules of thumb, when relatively little is at stake, or in situations in which budgets or time is limited. Habitat models, on the other hand, may be the most effective in efforts to preserve a particular endangered species, whereas holistic approaches may be the most appropriate, if more comprehensive ecosystem protection is desired or in situations with high environmental, economic, and social costs at stake. Many applications also make use of a combination of methods to fit specific contexts (Tharme 2003).

Linking Flow and Water Quality

The U.S. Environmental Protection Agency identified altered flow as the largest threat to water quality after agricultural pollution (EPA 2000). Appropriate flow levels can reduce pollutant concentrations, assist in the processing of nutrients, and maintain water temperature. Many of the benefits of restoring flow are achievable only with water of good quality; however, plans to use treated sewage to restore stream flows in water-scarce Israel, for instance, are currently being scrutinized by researchers attempting to determine what portion of the flows the reclaimed sewage should represent and what ecological value such flows would have (Gazith 2005). Though much of the focus of research and literature on environmental flows relates to water quantity, protection of water quality and flow regimes necessarily go hand-in-hand.

The Economics and Finance of Environmental Flow Allocations

The Economic Value of Instream Flows

One of the primary reasons for the growing shift in perceptions regarding the use of environmental flows is the growing understanding of the scale of their real and potential economic benefits. Numerous studies have looked at the economic value of ecosystem services provided by flows, including habitat creation, recreational opportunities, contribution to housing prices, groundwater recharge, contribution to water quality, and so on (NPS 2001; Emerton and Boss 2004). Both conventional and nonmarket valuation methods have been used to measure the public's willingness to pay for environmental flows, including travel cost, hedonic pricing, contingent valuation (stated preference surveys), and avoided expenditures (BoR 2001).

For policy making, the benefits of environmental flows must be measured against the opportunity costs of alternative uses of the water, such as foregone agricultural and hydropower production. Because many environmental flow benefits are difficult to quantify precisely, they have often been ignored in benefit-cost analyses and planning processes, resulting in inefficient and ecologically harmful water development projects. In some cases, however, even imprecise measures of flow benefits have proven sufficient to justify environmental flows on economic grounds. The value of recreation along the Missouri River has been estimated to be at least four times the revenue generated by river barge traffic (Abromovitz 1996). Ecological benefits of protecting Mono Lake were found to exceed the replacement cost of water from alternative sources by a factor of 50 (Loomis 1987, 1998). In the Hadejia Jama'are floodplain in Nigeria, ecological and social benefits from water in the floodplain were valued at $9,600 to $14,500 per cubic meter (in 1989-90 U.S. dollars), whereas upstream diversion of the water for irrigation was estimated to produce a return of merely $26 to $40 per cubic meter (Barbier and Thompson 1998).

Unlike most offstream applications of water, instream flow uses tend to be largely nonconsumptive. They can simultaneously provide benefits, such as habitat, recreation, and water quality improvement. Also, in many cases, water preserved for environmental flows at a certain point along a river may still be available for offstream use further downstream. Thus, opportunity costs may not be as large as they initially appear. A study on the Rio Grande River, for instance, found that reallocating water upstream, away from agriculture to provide minimum instream flow for an endangered minnow species, actually increased net economic benefits by making water available for higher value downstream uses (Ward and Booker 2003).

Many of the economic studies examining environmental flows often evaluated dichotomous choice situations, such as whether to build a dam or to provide minimum instream flows. Such all-or-nothing choices often miss economically optimal solutions. Minimum flows are seldom economically efficient flows. In many cases, economically optimum flows are orders of magnitude larger than minimum flows (Loomis 1998). Efficient flow rates are based on equating the marginal value of alternative uses of water. A great deal of evidence shows that the marginal value of water left instream often outweighs that of offstream options. One study of the Western United States found the marginal value of water for recreational fishing to be higher than that for irrigation (the primary consumer of water in the study region) in 52 out of 67 watersheds in which irrigation occurred (Hansen and Hallam 1991). In such situations, reallocation of at least some of the water for environmental flow makes both economic and environmental sense.

Mechanisms for Financing Environmental Flows

A restructuring of public-finance incentives is one of the most direct and economically efficient ways to preserve or restore flow. Subsidies for water-intensive agriculture, large flow-diversion projects, and regular dredging and channelization necessary for barge traffic are costly public expenses of questionable economic merit. International finance organizations are often also beginning to recognize the ecological and economic importance of environmental flows. The World Bank, for instance, has begun to incorporate the need for environmental flows in evaluation of dam and irrigation projects (Hirji and Panella 2003), although critics contend that actual implementation of such principles remains insufficient (e.g., Imhof 2005).

Aggressive water conservation programs can obviate the need for dams and other diversion infrastructure, as has happened in Bogotá, Colombia, in California, and in Boston, Massachusetts (Gleick et al. 2003; Postel 2005). Even in cases in which conservation is not the cheapest option, the public has often demonstrated a willingness to pay to avoid ecologically harmful diversions (e.g., Loomis 1987). User fees placed on water consumption have been used to finance environmental flow research in places like South Africa (Poff et al. 2003) and to finance actual flows and other watershed protection activities in Costa Rica (Dyson 2003). Under California's Central Valley Project Improvement Act, consumers pay a fee based on the volume of water consumption. Money generated by this user fee goes toward purchasing water from willing sellers and financing a fund specifically for stream and wetland restoration activities (CBO 1997). In Japan, experts have proposed financial levies on water or electricity use (or subsidies, if taxing proves to be politically unfeasible) as a means of financing flow releases from dams (Shirakawa and Tamai 2003).

Given the often relatively high marginal value of environmental flows, many observers have pointed out the potential role of market transactions in securing flow rights (e.g., Anderson and Johnson 1988; Colby 1990; Griffin and Hsu 1993). According to Landry (1998), between 1990 and 1997, over $61 million was spent to acquire over 2.1 million-acre feet (over 2.6 billion cubic meters) of water for environmental protection in the Western United States. In legal systems that prohibit the free transfer of market rights, the state can function as the sole purchaser. However, in systems allowing private transactions, individuals and nongovernmental organizations can play a potentially important role. Government purchases accounted for over 70% of the total quantity of acquired water cited by Landry, with the private sector accounting for the rest. Private-sector buyers include environmental organizations, community groups, and water trusts, organizations established with the specific objective of purchasing water for environmental purposes.

Options for market-based water acquisition include not only the purchase of water rights, but also short- and long-term leases, and dry-year options, by which right holders are paid to provide water in low rainfall years only. Leases have proved more popular in the United States, given a reluctance of many farmers and other rights holders to surrender their rights permanently. Another innovative use of the markets to secure environmental flows is a policy in the state of Oregon, where 25% of all water transferred between private users for any purpose is converted to instream use (McKinney and Taylor 1988). An additional finance mechanism is a tax on all water transfers that would go toward purchase of water rights for instream purposes (CWWM 1992).

In the Murray-Darling Basin in Australia, the market does not play a role in securing water for environmental flows; that is accomplished through a cap placed on total withdrawals. A pilot water-rights market is being used, however, as a means of ensuring that the water that is diverted offstream goes to the highest valued users (MDBC 2005). Such a *cap and trade* market also allows new water consumers to enter the market without coming at the expense of the aquatic ecosystems.

The potential role of market acquisitions of water for nature, although encouraging, should not be overemphasized. First, water markets can be a double-edged sword, because they can also transfer water away from instream uses. To protect against such occurrences, some legal systems with private water markets—Chile and much of the Western United States—have instituted protections for instream flow as a third-party right (Davis and Riestra 2002; Gillilan and Brown 1997).

Perhaps more importantly, significant problems of externalities, high transaction costs, inadequate and often asymmetric information, and problems associated with the public good nature of environmental flows, mean that the market is unlikely to bring about an economically—much less ecologically—optimum outcome (Colby 1990; Gillilan and Brown 1997). The benefits from environmental flows are often indirect and distributed across a wide range of beneficiaries. Also, much of their economic worth is in the form of indirect use or nonuse values, such as support of ecosystem functioning, and, in such cases, the potential for actual revenue generation is limited. Without centralized, funded advocates, environmental flows will often not be able to compete successfully in an open market for water. In sum, the market may be able to contribute significantly toward the provision of environmental flows but should not be viewed as an exclusive and, in most cases, not as a primary means of achieving this objective. Governments will continue to need to play a substantial and, most likely, leading role.

Making It Work: Policy Implementation

Challenges and Opportunities

In an analysis of the development of instream flow protection in the United States, Tarlock (1993) identified three distinct stages in the policy process: denial, recognition, and implementation (cited in Lamb 1995). Although recognition of the value of environmental flows is clearly on the increase, flow-related projects remain a relatively small share of total river restoration projects implemented (Bernhardt et al. 2005). Development of environmental flow policy is in its infancy in much of the world, especially in developing countries (Tharme 2003; Moore 2004), and successful implementation of policies remains a challenge for all nations due to uncertainties in both the science and economics involved and difficulties in enforcing legal instruments.

According to surveys of water professionals, the primary obstacles to implementation of environmental flows are a lack of policy guidance and management capacity, a lack of understanding of socioeconomic benefits and costs, legal and institutional constraints, and a lack of scientific expertise (Moore 2004; Scatena 2004). Insufficient public participation, funding, and hydrological data were also cited as impediments. Respondents who cited such difficulties included both governmental and nongovernmental representatives, from both industrialized and developing nations. In a sign that a change of perception among those involved in water management is perhaps already under way, 10% of those questioned in one survey gave water scarcity as a primary reason for lack of implementation, whereas 88% responded that "environmental flows were a necessary part of the efforts to solve problems related to water scarcity" (Moore 2004).

In countries in which environmental flow policies have been successfully implemented, the reasons most commonly offered for the establishment of the environmental flow concept included increasing public awareness, development of environmental flow expertise, implementation of pilot projects, and recognition of the importance of flows to local livelihoods (Moore 2004). In some cases, a triggering event, such as a proposed dam or strong advocacy and lobbying by interested parties, were also cited as driving factors.

Types of Flow Recommendations

In many cases, preservation of the natural flow regime is not practical or even possible. However, even provision of minimum base flow or seasonal flooding to flush sediments can often produce large ecological and environmental returns. A static minimum flow designed to achieve a particular ecological or social objective is still the most common form of environmental flow policy; however, new types of policies designed to achieve broader goals of overall ecosystem functioning are being developed and applied. In Southwest Florida, a percent of the natural flow at any given time is reserved as instream flow, which ensures a flow regime mimicking the timing and, at least, a portion of magnitude of natural flow variability. Several types of environmental flow allocation options are presented in Figure 2.2.

Some scholars have presented a different approach altogether. Instead of reserving the minimum amount of flow necessary for ecosystem survival and allocating the rest to development purposes, policy makers could set a specific development goal and then allow only the minimum amount of water necessary to achieve such a goal,

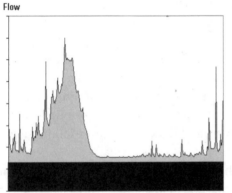

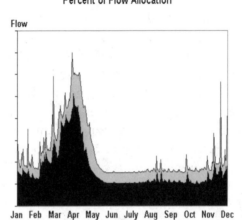

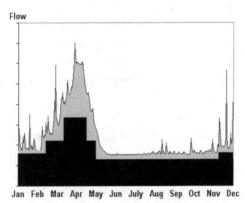

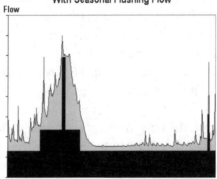

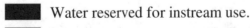

■ Water reserved for instream use.

▨ Water appropriated for offstream use.

FIGURE 2.2 Various environmental flow allocation scenarios.
Sources: Silk et al. 2000, Postel and Richter 2003.

leaving the remainder instream (Silk et al. 2000). Advocates claim that such an approach is more sustainable, given uncertainties in the science of determining ecosystem water demands.

Adaptive Management

Because policy for protection and restoration of environmental flows is still a relatively new endeavor, many restoration projects are being implemented in an experimental manner (Richter et al. 2006). In one of the most closely monitored projects, large floods were released along the Colorado River in 1996 to restore sandbars and improve habitat. After achieving only limited success, managers designed a new experimental flood, implemented in 2004, to distribute tributary sediments downstream, rather than sediments from the bottom of the main stem itself (USGS 2005).

Such an adaptive management approach to flow restoration not only allows more effective policy, but also helps reduce costs and political opposition. In the Roanoke River in North Carolina, for example, a plan was implemented according to which unnatural floods are to be decreased by half every five years until floodplain forests begin to regenerate. Under such an arrangement, sufficient water for ecological goals is guaranteed, and dam owners are not forced to commit to drastic reductions in diversions that go beyond what is necessary for forest health (Postel and Richter 2003).

Conclusion

Growing populations and lifestyle choices are putting greater pressure on the world's limited freshwater systems. Construction of dams and groundwater development projects will likely continue. However, a significant shift is underway in the way people perceive water in its natural setting. What was once considered a wasted resource is increasingly considered a valuable asset. Demand for freshwater ecosystems and the services that they supply, whether they be recreation, fisheries, or simply aesthetics, is growing, as is the scientific knowledge base regarding environmental flows. As a result, legal and policy frameworks for protecting and restoring them also continue to develop, and various governmental agencies, community-based organizations, private-sector actors, and individuals are becoming involved with implementing and monitoring these flows.

Although we have only limited experience with environmental flow restoration, early results are encouraging. Streams and wetlands are dynamic ecological systems, with impressive capacity to recover from even significant and sustained disturbances, once these disturbances are removed and natural conditions are restored. Despite the successes of restoration activities, however, as Baron et al. (2002) point out, preserving environmental flows is often much easier and inexpensive than restoring them.

Recent research highlights today's urgency of not only implementing policies to preserve environmental flow, but also designing policies that anticipate likely impacts of climate change on river flows and on aquatic biodiversity. One model has predicted that the changes in flow regimes due to both the combined impact of climate change and water withdrawals are likely to result in losses of 4 to 22 percent of local fish biodiversity in over 100 rivers that are expected to experience decreases in flow. Losses in individual rivers were projected to be as high as 75% (Xenopoulos et al. 2005). Furthermore, rivers expected to experience increased flow may also suffer unwanted ecological change due to invasion of nonnative species better acclimatized to higher flows.

Water managers and policy makers are increasingly being called upon to implement integrated management of water resources (OECD 1998; Gleick 1998). Initial steps in integrating environmental concerns into planning include education of the public and decision makers regarding the value of environmental flows, legal recognition of their beneficial uses, and establishment of physical and policy mechanisms to provide minimum flows. These measures, which have yet to be implemented in many areas, are necessary but insufficient to secure the potential benefits of environmental flow. Real progress entails creating formal channels for integrating the effect of flow variation on habitat, recreation, and water quality into project design and finance decision making. Promotion of flexible, forward-looking engineering, economic, and legal frameworks that can adapt as both demand for and scientific knowledge about environmental flows change, offers perhaps the best hope of achieving sustainable and efficient water management.

REFERENCES

Abromovitz, J. 1996. Imperiled waters, impoverished future: The decline of freshwater ecosystems. *Worldwatch Paper 128*, Worldwatch Institute, March.

Allan, J. D., and Flecker, A. 1993. Biodiversity conservation in running waters. *Bioscience* 43(1):32–43.

Anderson, T., and Johnson, R. 1988. The problem of instream flow. *Economic Inquiry* 24:535–554.

Arthington, A., and Pusey, B. 2003a. Flow restoration and protection in Australian rivers. *River Research and Applications* 19(5–6):377–395.

Arthington, A., Rall, J., Kennard, M., and Pusey, B. 2003b. Environmental flow requirements of fish in Lesotho rivers using the DRIFT methodology. *River Research and Applications* 19(5–6):641–666.

Barbier, E., and Thompson, J., 1998. The value of water: Floodplain versus large-scale irrigation benefits in Northern Nigeria. *Ambio* 27:434–440.

Baron, J., Poff, N. L., Angermeier, P., Dahm, C., Gleick, P., Nelson G. H., Jackson, R., Johnston, C., Richter, B., and Steinman, A. 2002. Meeting ecological and societal needs for freshwater. *Ecological Applications* 12(5):1247–1260.

Benetti, A. D., Lanna, A. E. and M. S. Cobalchini. 2004. Current practices for establishing environmental flows in Brazil. *River Research and Applications* 20(4):427–444.

Bernhardt, E. S., Palmer, M. A., Allan, J. D., Alexander, G., Barnas, K., Brooks, S., Carr, J., Clayton, S., Dahm, C., Follstad-Shah, J., Galat, D., Gloss, S., Goodwin, P., Hart, D., Hassett, B., Jenkinson, R., Katz, S., Kondolf, G., Lake, P., Lave, R., Meyer, J., Donnell, T., Pagano, L., Powell, B., and Sudduth, E. 2005. Synthesizing U.S. river restoration efforts. *Science* 308:636–637.

Bovee, K., Lamb, B., Bartholow, J., Stalnaker, C., Taylor, J., and Henriksen, J. 1998. *Stream habitat analysis using the Instream Flow Incremental Methodology.* Fort Collins, CO: U.S. Geological Survey, Biological Resources Division.

Bunn, S., and Arthington, A. 2002. Basic principles and ecological consequences of altered flow regimes for aquatic biodiversity. *Environmental Management* 30(4):492–507.

Bureau of Reclamation (BoR). 2001. *Economic nonmarket valuation of instream flows.* Denver, CO: Department of the Interior, Bureau of Reclamation, June.

Colby, B. 1990. Enhancing instream flows in an era of water marketing. *Water Resources Research* 26(6):1113–1120.

Committee on Western Water Management (CWWM). 1992. *Water transfers in the West: Efficiency, equity, and the environment.* Washington, DC: National Academy Press.

Congressional Budget Office (CBO). 1997. *Water use conflicts in the West: Implications of reforming the Bureau of Reclamation's water supply policies.* Washington, DC: Congressional Budget Office.

Davis, M., and Riestra, F. 2002. *Instream flow policies and procedures within integrated water management in Chile.* 4th Ecohydraulics Conference, Enviro-flows, 2002.

Dyson, M., Bergkamp, G., and Scanlon, J. (editors). 2003. *Flow: The essentials of environmental flows.* Gland, Switzerland, IUCN.

EcoPeace/Friends of the Earth Middle East (FoEME). 2005. *Crossing the Jordan—Concept document to rehabilitate, promote prosperity and help bring peace to the lower Jordan River.* EcoPeace/Friends of the Earth Middle East (FoEME), March.

Emerton, L., and Boss, E. 2004. *Value: Counting ecosystems as water infrastructure.* Gland, Switzerland, and Cambridge, UK, IUCN.

Food and Agriculture Organization (FAO) of the United Nations. 2005. AQUASTAT online database. http://www.fao.org/ag/agl/aglw/aquastat/dbase/index.stm.

Finlayson, C., and Davidson, N. 1999. *Global review of wetland resources and priorities for wetland inventory—Summary report.* Australia: Wetlands International and the Environmental Research Institute of the Supervising Scientist.

Fraser, J. 1972. Regulated discharge and the stream environment. In *River ecology and man.* Oglesby, R. T., Carlson, C. A., and McCann, J. A., editors. New York: Academic Press.

García de Jalón, D. 2003. The Spanish experience in determining minimum flow regimes in regulated streams. *Canadian Water Resources Journal* 28(2).

Gazith, A. 2005. Conference presentation. *Water crisis and environmental pollution.* Tel Aviv University: Porter School for Environmental Studies, 12 April.

Gillilan, D., and Brown, T. 1997. *Instream flow protection: seeking a balance in Western water use.* Washington, DC: Island Press.

Gleick, P. 1998. *The world's water 1998–1999.* Washington, DC: Island Press.

Gleick, P. 2000. *The world's water 2000–2001.* Washington, DC: Island Press.

Gleick, P. H., Haasz, D., Henges-Jeck, C., Srinivasan, V., Wolff, G., Cushing, K. K., and Mann, A. 2003. *Waste not, want not: The potential for urban water conservation in California.* A report of the Pacific Institute for Studies in Development, Environment, and Security. Oakland, CA.

Griffin, R., and Hsu, S. H. 1993. The potential for water market efficiency when instream flows have value. *American Journal of Agricultural Economics* 75:292–303.

Hansen, L., and Hallam, A. 1991. National estimates of the recreational value of stream flow. *Water Resources Research* 27(2):167–175.

The Heinz Center. 2002. *Dam removal: Science and decision making.* Washington, DC: The H. John Heinz Center for Science, Economics and the Environment.

Hellsten, S., Marttunen, M., Palomäki, R., Riihimäki, J., and Alasaarela, E. 1996. Towards an ecologically based regulation practice in Finnish hydroelectric lakes. *Regulated Rivers: Research and Management* 12(4–5):535–545.

Hirji, R., and Panella, T. 2003. Evolving policy reforms and experiences for addressing downstream impacts in World Bank water resources projects. *River Research and Applications* 19(5–6):667–681.

Imhof, A. 2005. 20 Years of success—Happy birthday IRN. *World Rivers Review* 20(4):1.

International Rivers Network (IRN). 2005. IRN's Pak Mun campaign. Accessed August 21, 2006. http://www.irn.org/programs/pakmun/.

Instream Flow Council (IFC). 2002. *Instream flows for riverine resource stewardship.* Lansing, MI: Instream Flow Council.

Jowett, I. 1997. Instream flow methods: A comparison of approaches. *Regulated Rivers: Research and Management* 13:115–127.

Karr, J. 1991. Biological integrity: A long-neglected aspect of water resource management. *Ecological Applications* 1:66–84.

King, J., Brown, C., and Sabet, H. 2003. A scenario-based holistic approach to environmental flow assessments for rivers. *River Research and Applications* 19(5–6):619–639.

King, J., and Louw, D. 1998. Instream flow assessments for regulated rivers in South Africa using the Building Block Methodology. *Aquatic Ecosystem Health and Management* 1:109–124.

King, J., Tharme, R., and Brown, C. 1999. *Definition and implementation of instream flows.* Cape Town, South Africa: World Commission on Dams, Thematic Review II.1: Dams, ecosystem functions and environmental restoration.

King, R. 2005. Everglades' restorations cost jumps $2.1 billion. *The Palm Beach Post.* 6 October.

Lamb, B. 1995. Criteria for evaluating state instream-flow programs: Deciding what works. *Journal of Water Resources Planning and Management* 121(3):270–274.

Lamb, B., Burkardt, N., and Taylor, J. G. 1998. Negotiation and decision-making. In *Hydroecological modelling: Research, practice, legislation and decision-making.* Blazkova, S., Stalnaker, C., Novicky, O., editors. 63–64. Prague, Czech Republic: T. G. Masaryk Water Research Institute.

Landry, C. 1998. Market transfers of water for environmental protection in the Western United States. *Water Policy* 1(5):457–469.

Loh, J., and Wackernagel, M. 2004. *Living Planet Report 2004*. Gland, Switzerland: World Wide Fund for Nature (WWF).

Loomis, J. 1987. Balancing public trust resources of Mono Lake and Los Angeles' Water Right: An economic approach. *Water Resources Research* 23(8):1449–1456.

Loomis, J. 1998. Estimating the public's values for instream flow: Economic techniques and dollar values. *Journal of the American Water Resources Association* 34(5):1007–1014.

McKinney, M., and Taylor, J. 1988. *Western state instream flow programs: A comparative assessment*. Fort Collins, CO: National Ecology Research Center, Fish and Wildlife Service, U.S. Department of Interior.

Moore, M. 2004. *Perceptions and interpretations of environmental flows and implications for future water resource management—A survey study*. Linköping University, Sweden: Masters Thesis, Department of Water and Environmental Studies.

Moyle, P., and Leidy, R. 1992. Loss of biodiversity in aquatic ecosystems: Evidence from fish faunas. In *Conservation biology: The theory and practice of nature conservation, preservation, and management*. Fiedler, P., and Jain, S. K., editors. 127–169. New York: Chapman and Hall.

Murray-Darling Basin Commission (MDBC). 2005. http://www.mdbc.gov.au/naturalresources/the_cap/the_cap.htm.

Naiman, R., Bunn, S., Nilsson, C., Petts, G., Pinay, G., and Thompson, L. 2002. Legitimizing fluvial ecosystems as users of water: An overview. *Environmental Management* 30(4):455–467.

National Academy of Sciences (NAS). 1999. *Water for the future: The West Bank and Gaza Strip, Israel, and Jordan*. Washington, DC: National Academy Press.

National Park Service (NPS). 2001. *Economic benefits of conserved rivers: An annotated bibliography*. Trails, Rivers, and Conservation Assistance Program, National Park Service, Department of the Interior, June.

Nature and Parks Authority (NRA) and Ministry of Environment (Israel). 2004. *Nature's right to water: A policy document*. Jerusalem (in Hebrew).

Organisation for Economic Co-Operation and Development (OECD). 1998. *Water consumption and sustainable water resources management*. OECD proceedings.

Parish, F. 2005. A review of river restoration practices in Asia. In *River restoration in East Asia*. Parish, F., Mokhtar, M., Abdullah, A. R., and May, C. O., editors. Kuala Lumpur, Malaysia: Global Environment Centre and Department of Irrigation and Drainage.

Petts, G. 1996. Water allocation to protect river ecosystems. *Regulated Rivers—Research and Management* 12(4–5):353–365.

Petts, G., and Maddock, I. 1996. Flow allocation for in-river needs. In *River Restoration*. Petts, G., and Callow, P., editors. Oxford, UK: Blackwell Science.

Poff, N. L., Allan, J. D., Bain, M., Karr, J., Prestegaard, K., Richter, B., Sparks, R., and Stromberg, J. 1997. The natural flow regime. *Bioscience* 47(11):769–784.

Poff, N. L., Allan, J. D., Palmer, M., Hart, D., Richter, B., Arthington, A., Rogers, K., Meyers, J., and Stanford, J. 2003. River flows and water wars: Emerging science for environmental decision making. *Frontiers in Ecology and the Environment* 1(6):298–306.

Postel, S. 2005. Liquid assets: The critical need to safeguard freshwater ecosystems. *Worldwatch Paper 170*. Worldwatch Institute, July.

Postel, S., Daily, G., and Ehrlich, P. 1996. Human appropriation of renewable fresh water. *Science* 271(5250):785–788.

Postel, S., and Richter, B. D. 2003. *Rivers for life: Managing water for people and nature*. Washington, DC: Island Press.

Reiser, D., Wesche, T., and Estes, C. 1989. Status of instream flow legislation and practices in North America. *Fisheries* 14(2):22–29.

Revenga, C., S. Murray, J. Abramovitz, and A. Hammond. 1998. *Watersheds of the world: Ecological value and vulnerability*. Washington, DC: World Resources Institute and Worldwatch Institute.

Richter, B., Baumgartner, J., Powell, J., and Braun, D. 1996. A method for assessing hydrologic alteration within ecosystems. *Conservation Biology* 10(4):1163–1174.

Richter, B., Baumgartner, J., Wigington, R., and Braun, D. 1997. How much water does a river need? *Freshwater Biology* 37(1):231–249.

Richter, B., Mathews, R., and Wigington, R. 2003. Ecologically sustainable water management: Managing river flows for ecological integrity. *Ecological Applications* 13(1):206–224.

Richter, B., Warner, A., Meyer, J., Lutz, K. 2006. A collaborative and adaptive process for developing environmental flow recommendations. *River Research and Applications*. Vol. 22: pp. 297–318.

Robinson, C., Uehlinger, U., and Monaghan, M. 2004. Stream ecosystem response to multiple experimental floods from a reservoir. *River Research and Applications* 20(4):359–377.

Sax, J., Abrams, R., Thompson, B., and Leshy, J. 2000. *Legal control of water resources: Cases and materials.* New York: West Wadsworth.

Scatena, F. 2004. A survey of methods for setting minimum instream flow standards in the Caribbean basin. *River Research and Applications* 20(2):127–135.

Shirakawa, N., and Tamai, N. 2003. Use of economic measures for establishing environmental flow in upstream river basins. *International Journal of River Basin Management* 1(1):15–19.

Silk, N., McDonald, J., and Wigington, R. 2000. Turning instream flow water rights upside down. *Rivers* 7:298–313.

Smith, G., and Hoar, A. (editors). 1999. *The Public Trust Doctrine and its application to protecting instream flows.* Anchorage, AK: National Instream Flow Program Assessment.

Stalnaker, C., Lamb, B., Henriksen, J., Bovee, K., and Bartholow, J. 1995. *The Instream Flow Incremental Methodology: A primer for IFIM.* Fort Collins, CO: National Biological Service.

Tamai, N. 2005. Principles and examples of river restoration. In *River restoration in East Asia.* Parish, F., Mokhtar, M., Abdullah, A. R., and May, C. O., editors. Kuala Lumpur, Malaysia: Global Environment Centre and Department of Irrigation and Drainage.

Tarlock, D. 1993. Future issues in instream flow protection in the West. In *Instream flow protection in the West.* Boulder, CO: University of Colorado, Natural Resources Law Center.

Tennant, D. 1976. Instream flow regimes for fish, wildlife, recreation and related environmental resources. *Fisheries* 1(4):6–10.

Tharme, R. 2003. A global perspective on environmental flow assessment: Emerging trends in the development and application of environmental flow methodologies for rivers. *River Research and Applications* 19(5–6):397–441.

The Nature Conservancy (TNC). 2005. *Flow restoration database.* http://www.freshwaters.org/tools/pdf/flowdb_6_05.pdf.

U.S. Environmental Protection Agency (EPA). 2000. *The Quality of Our Nation's Waters—A summary of the National Water Quality Inventory: 1998 Report to Congress.* Washington, DC: Office of Water, Environmental Protection Agency (EPA), June.

United States Geological Survey (USGS). 2004. *Interior scientists to evaluate effects of high flow test at Glen Canyon Dam.* Press release, 19 November. http://www.usgs.gov/newsroom/article.asp?ID=135.

United States Geological Survey (USGS). 2005. *Surface water data for the nation.* http://nwis.waterdata.usgs.gov/usa/nwis/discharge.

Ward, F., and Booker, J. 2003. Economic costs and benefits of instream flow protection for endangered species in an international basin. *Journal of the American Water Resources Association* 39(2):427–440.

Water Research Commission (WRC) (South Africa). 2003. *KSA-2 Water-linked ecosystems.* http://www.wrc.org.za/downloads/knowledgereview/2003/KSA2.pdf.

Wolf, A. 1999. Criteria for equitable allocations: The heart of international water conflict. *Natural Resources Forum* 23(1):3–30.

World Commission on Dams (WCD). 2000. *Dams and development: A new framework for decision-making. The report of the World Commission on Dams.* London and Sterling, VA: Earthscan Publications Ltd.

Xenopoulos, M., Lodge, D., Alcamo, J., Märker, M., Schulz, K., and van Vuuren, D. 2005. Scenarios of freshwater fish extinctions from climate change and water withdrawal. *Global Change Biology* 11:1–8.

With a Grain of Salt: An Update on Seawater Desalination

Peter H. Gleick, Heather Cooley, Gary Wolff

Introduction

The history of seawater desalination offers examples of both technological successes and economic failures. Long considered the "holy grail" of water supply, promoters, policy makers, and the public continue to hope that cost-effective and environmentally safe desalination will come to the rescue of water-short regions. That day may come; but it does not appear to be here yet, with some limited exceptions. In this chapter, we provide a comprehensive update on the history, benefits, and risks of ocean desalination, and the barriers that hinder its more widespread use, especially in the context of recent proposals for a massive increase in development around the world.

The potential benefits of desalination are great, but the barriers to wide commercialization are high. In most parts of the world, other alternatives can provide the same benefits of desalination at far lower economic and environmental costs. These alternatives include cleaning up local water sources, encouraging regional water transfers, improving conservation and efficiency, accelerating wastewater recycling and reuse, and implementing smart land-use planning. At present, the only significant seawater desalination capacity is in the Persian Gulf, islands with limited local supplies, and selected other locations, where water options are limited and the public is willing to pay high prices. In the United States, almost all seawater desalination facilities are small systems used for high-valued industrial and commercial needs.

This may be changing. Despite the major barriers to desalination, recent interest has mushroomed as technology has improved, demands on water have grown, and prices have dropped. Interest has been especially high in regions with both water scarcity and political concerns about water reliability and independence, such as Israel, Singapore, and, recently, California, where rapidly growing populations, poorly regulated land use, and ecosystem degradation has forced a rethinking of water policies and management. In the past five years, very large reverse osmosis (RO) plants have been built at Ashkelon, Israel, and Tuas, Singapore, and more than twenty proposals for large desalination facilities along the California coast have been put forward. Extensive development is also continuing in the Persian Gulf.

If the facilities proposed for California alone are built, they would increase the state's seawater desalination capacity by a factor of 100. Project proponents point to statewide water-supply constraints, the reliability advantages of supposedly "drought-proof" supply, the water-quality improvements offered by desalinated water, and the benefits of local control. Along with the proposals, however, has come a growing public debate about high economic and energy costs, environmental and social impacts, and consequences for coastal development policies.

A recent comprehensive assessment (Cooley et al. 2006) concluded that most of the recent desalination proposals in California and elsewhere appear to be premature because of continuing high costs and unresolved environmental and social concerns. Among the exceptions may be desalination proposals in regions where alternative water-management options have been substantially developed, explicit ecosystem benefits are guaranteed, environmental and siting problems have been identified and mitigated, and the construction and development impacts will be minimized. Even under these conditions, customers must be capable and willing to pay the high costs to cover a properly designed and managed plant.

When the barriers to desalination are overcome, carefully regulated and monitored construction of desalination facilities may advance quickly. Regulators will have to work to develop comprehensive, consistent, and clear rules for desalination proposals, so that inappropriate proposals can be swiftly rejected and appropriate ones identified and facilitated. It is also a question of whether private companies, local communities, and public water districts that push for desalination facilities will do so in an open and transparent way. Public participation and input in decision making will be a crucial part of making such facilities succeed.

As part of that, we offer a set of recommendations at the end of the chapter to help water users and planners interested in making desalination a more significant part of international, national, and local water policy. This chapter provides information to help the public and policy makers understand and evaluate the arguments being put forward by both proponents and opponents of the current proposals.

Background to Desalination

The Earth's hydrologic cycle naturally desalinates water using the energy of the sun. Water evaporates from oceans, lakes, and land surfaces leaving salts behind. The resulting freshwater vapor forms clouds that produce precipitation, which falls to earth as rain and snow and moves through soils, dissolving minerals and becoming increasingly salty. The oceans are salty because the natural process of evaporation, precipitation, and runoff is constantly moving salt from the land to the sea, where it builds up over time.

Desalination refers to the wide range of technical processes designed to remove salts from waters of different qualities (Box 3.1 and Table 3.1). Desalination technology is in use throughout the world for a wide range of purposes, including providing potable fresh water for domestic and municipal purposes, treated water for industrial processes, and emergency water for refugees or military operations.

Desalination facilities in many arid and water-short areas of the world are vital to economic development. In particular, desalination is an important water source in

Box 3.1 What's in a Name? Desalination? Desalinisation? Desalinization? Desalting?

No single technical term (or spelling) is consistently used for the process of removing salt from water, though most water engineers and professional organizations use the term *desalination*. When one "googles" the term *desalinization,* Google asks "Did you mean *desalination*"? Conversely, The New Dictionary of Cultural Literacy, Third Edition (2002) has an entry for *desalinization,* but nothing for *desalination.* The Commonwealth countries spell it with an "s" in place of the "z." The diversity of professional associations and organizations (*organisations?*) in this field reflect the diversity of terms used, including the International Desalination Association, the Australian Desalination Association, the European Desalination Association, the Southeast Desalting Association, the American Desalting Association, and Middle East Desalinisation Research Center. In this report, we use *desalination* and *desalting* interchangeably: why use six syllables when three (or five) will do?

TABLE 3.1   Salt Concentrations of Different Waters

Water Source or Type	Approximate Salt Concentration (grams per liter)[a]
Brackish waters	0.5 to 3
North Sea (near estuaries)	21
Gulf of Mexico and coastal waters	23 to 33
Atlantic Ocean	35
Pacific Ocean	38
Persian Gulf	45
Dead Sea	~300

Notes: a. Slight spatial variations in salt content are found in all major bodies of water. The values in the table are considered typical. A gram per liter is approximately equal to 1000 parts per million.
Sources: OTV 1999; Gleick 1993.

parts of the arid Middle East, Persian Gulf,[1] North Africa, Caribbean islands, and other locations where the natural availability of fresh water is insufficient to meet demand and where traditional water-supply options or transfers from elsewhere are implausible or uneconomic. Increasingly, other regions are exploring the use of desalination as a potential mainstream source of reliable, high-quality water as the prices slowly drop toward the cost of more traditional alternatives.

---

1. As noted by the National Geographic Society: "Historically and most commonly known as the Persian Gulf, this body of water is referred to by some as the Arabian Gulf."

History of Desalination

The idea of separating salt from water is an ancient one, dating from the time when salt, not water, was a precious commodity. As populations and demands for fresh water expanded, however, entrepreneurs began to look for ways of producing fresh water in remote locations and, especially, on naval ships at sea. The history of desalination through the twentieth century has been reviewed previously in *The World's Water 2000–2001*. But efforts have been expanding in the past five years, both worldwide, and in the United States.

Despite a hot and cold approach to research and development, by the early twenty-first century, the U.S. government alone had spent nearly $2 billion on the basic research and development framework for many of the technologies now used for desalting seawater and brackish waters. Other investments are being made by private companies and many other governments. These investments helped stimulate the global desalination market, and many private commercial efforts are now advancing the technology and expanding operating experience.

Desalination Technologies

No single "best" method of desalination exists. A wide variety of desalination technologies effectively remove salts from salty water (or fresh water from salty water), producing a water stream with a low concentration of salt (the product stream) and another with a high concentration of remaining salts (the brine or concentrate). Most of these technologies rely on either distillation or membranes to separate salts from the product water (USAID 1980; Wangnick 1998, 2002; Wangnick/GWI 2005). Ultimately, the selection of a desalination process depends on site-specific conditions, the salt content of the water, economics, the quality of water needed by the end user, and local engineering experience and skills. Details of various desalination technologies are provided in Gleick (2000).

The earliest plants were based on mostly large-scale thermal evaporation or distillation of seawater, mimicking the natural hydrologic cycle by producing water vapor that is then condensed into fresh water. Even today, around 40 percent of the world's desalted water is produced with processes that use heat to distill fresh water from seawater or brackish water, particularly in the Persian Gulf region, where inexpensive or excess energy is available.

Beginning in the 1970s, increasing numbers of plants were installed, using membranes that mimic the natural biological process of osmosis. Membrane technologies, primarily based on RO, can desalinate both seawater and brackish water, but they are more commonly used to desalinate brackish water because costs increase along with the salt content of the water (Box 3.2). Another advantage of membrane technologies is their ability to remove microorganisms and many organic contaminants. Membrane technologies generally have lower capital costs and require less energy than distillation and thermal systems, but thermal desalination systems can produce water with much lower salt content than membrane systems (typically below 25 parts per million [ppm] total dissolved solids compared to under 500 ppm) (USBR 2003).

Box 3.2 Filtration/Membrane Systems

Reverse osmosis (RO) membranes are used for both brackish water and seawater desalination and are capable of removing some organic contaminants.

Nanofiltration (NF) membranes are used for water softening, removal of organics and sulfates, and elimination of some viruses. Removal is by combined sieving and solution diffusion.

Ultrafiltration (UF) membranes are used for removal of contaminants that affect color, high-weight dissolved organic compounds, bacteria, and some viruses. UF membranes also operate via a sieving mechanism.

Microfiltration (MF) membranes are used to reduce turbidity and remove suspended solids and bacteria. MF membranes operate via a sieving mechanism under a lower pressure than either UF or NF membranes.

Sources: Heberer et al. 2001; NAS 2004; Sedlak and Pinkston 2001.

Current Status of Desalination

Global Status

Some form of desalination is now used in around 130 countries. By January 2005, more than 10,000 desalting units larger than a nominal 100 cubic meters per day (m^3/d) had been installed or contracted worldwide. These plants have a total capacity to produce around 35.6 million m^3/d of fresh water from all sources, or around 13 cubic kilometers per year (km^3/yr).[2] In 2000, the cumulative installed desalination capacity was around 9.5 km^3/yr (Figure 3.1), implying a growth rate of around 7 percent per year. Actual production appears to be considerably less, because the database of desalination facilities adds plants when they are commissioned (or sometimes just planned) but does not have reliable information on plants that no longer operate. Although desalination provides a substantial part of the water supply in certain oil-rich Middle Eastern nations, globally, installed desalination plants have the capacity to provide just three one-thousandths of total world freshwater use.

---

2. Good data on installed or contracted desalination capacity exist (Wangnick/GWI 2005). Some of these plants, however, either do not operate at all or operate at less than full capacity. Figures on actual production of desalinated water are not collected, but it is certainly less than the total capacity.

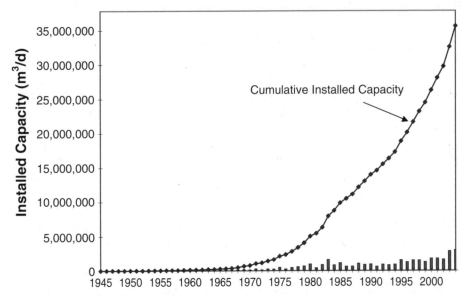

FIGURE 3.1   Time-series of global desalination capacity, January 2005.
Note: The bars show the installed capacity by year, and the line shows the cumulative installed capacity.
Source: Wangnick/GWI 2005.

Available technologies can desalinate water from various sources. Figure 3.2 shows the breakdown of water sources as of January 2005 (Wangnick/GWI 2005). Around 20 million m^3/d (or 56 percent) of desalination capacity is designed to process seawater. Another 8.5 million m^3/d (24 percent of total capacity) can process brackish water. The remaining capacity is used to desalinate waters of other kinds (Table 3.2).

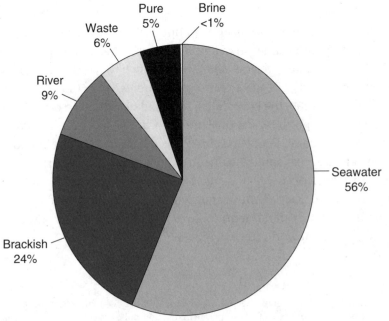

FIGURE 3.2   Global desalination capacity by source of water, January 2005.
Source: Wangnick/GWI 2005.

TABLE 3.2   Global Desalination Capacity by Source Water, January 2005

Water Source	Percent of Worldwide Installed Capacity
Seawater	56
Brackish	24
River	9
Waste water	6
Pure	5
Brine	<1

Source: Wangnick/GWI 2005.

Figure 3.3 shows total desalination capacity by process. The trend over the last decade shows a steady shift toward the construction of RO facilities, and RO plant capacity now exceeds multistage flash (MSF) capacity. Around 35 percent of total installed or contracted capacity is based on MSF and nearly half of all capacity is based on RO.

Half of all desalination capacity is in the Middle East/Persian Gulf/North Africa regions. Figure 3.4 shows those countries with more than 1 percent of global desalination capacity, as of January 2005. Eighteen percent of global capacity is in Saudi Arabia, followed by 17 percent in the United States, 13 percent in the United Arab Emirates, 6 percent in Spain, and 5 percent in Kuwait (Wangnick/GWI 2005). Most plants installed in Saudi Arabia, Kuwait, and the United Arab Emirates use distillation; most

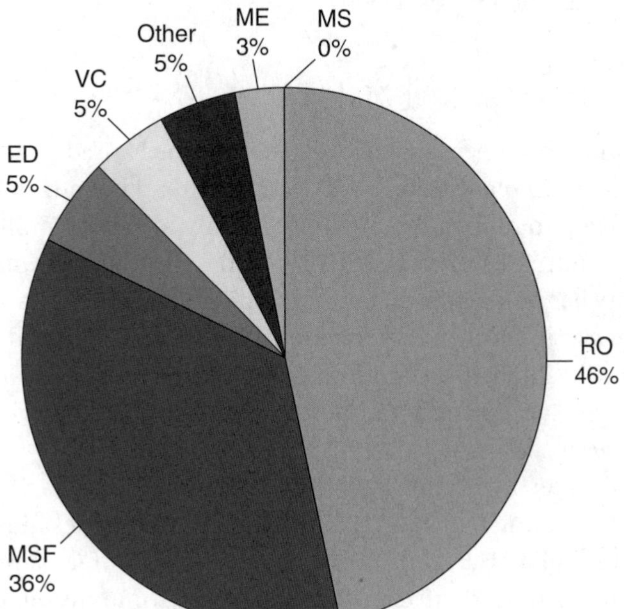

FIGURE 3.3   Global desalination capacity by process, January 2005.
Source: Wangnick/GWI 2005.

Note: Electrodialysis (ED)
    Multi-effect distillation (ME)
    Multistage flash distillation (MSF)
    Freeze, hybrid, nanofiltration, thermal and all other processes (Other)
    Reverse osmosis (RO)
    Vapor compression (VC)

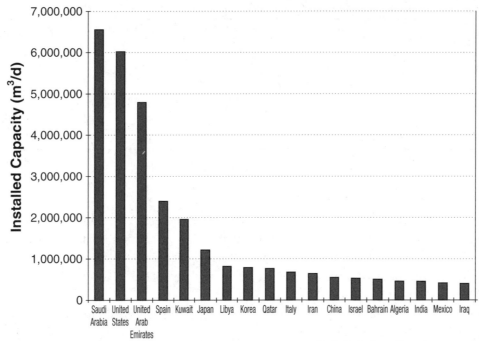

FIGURE 3.4   Countries with more than one percent of global desalination capacity, January 2005. Total installed capacity in cubic meters per day.
Source: Wangnick/GWI 2005.

plants in the United States rely upon RO plants and vapor compression. Importantly, many smaller island communities, not shown in this figure, rely on desalination for a large fraction of their total water need.

Desalination in the United States

Desalination plants have been built in every state in the United States. By January 2005, over 2,000 desalination plants larger than 100 m³/d had been installed or contracted. These plants have a total installed capacity of only around 6 million m³/d—around three one-thousandths of total U.S. use[3]—and as noted elsewhere, not all of the reported capacity has actually been built or is still in operation. Reported capacity has increased somewhat in recent years (Figure 3.5); between 2000 and 2005, the reported installed capacity has increased by around 30 percent, but the Wangnick/GWI (2005) database includes plants contracted but never built, built but never operated, and operated but now closed.

The source of water treated in the U.S. plants differs from that of the rest of the world (Figure 3.6). Around half of all U.S. capacity is used to desalinate brackish water. Twenty-five percent of all U.S. capacity desalinates river water—a relatively easy, cost-effective thing to do for industrial, power plant, or some municipal use. Although, globally, seawater is the largest source, less than 0.5 million m³/d of seawater is desalinated in the United States, or less than 10 percent of U.S. capacity.

---

3. The U.S. Geological Survey reports total U.S. water withdrawals in 2000 at around 565 cubic kilometers per year or around 1500 gallons per person per day for all uses.

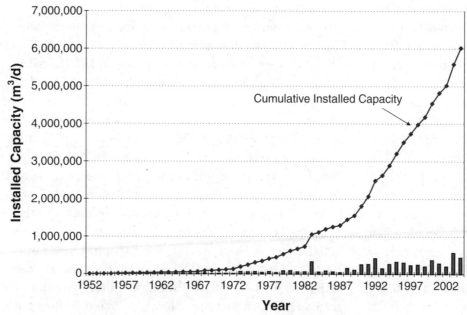

FIGURE 3.5   Time-series of U.S. desalination capacity, January 2005.

Note: The bars show the installed capacity by the year, and line shows the cumulative installed capacity.

Source: Wangnick/GWI 2005.

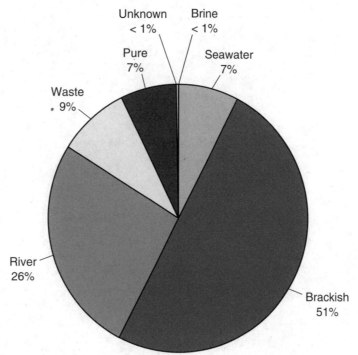

FIGURE 3.6   U.S. desalination capacity by source of water, January 2005.

Source: Wangnick/GWI 2005.

Like the rest of the world, RO is the most common desalination technology used in the United States, accounting for nearly 70 percent of the installed desalination capacity in the United States, or over 4 million m³/d (Figure 3.7). MSF, the second most common desalination technology globally, is uncommon in the United States. Only 1 percent of the total U.S. desalination capacity is based on MSF. Nanofiltration, however, is much more common in the United States. Of the 1.4 million m³/d of water that is desalinated worldwide, using nanofiltration, nearly 1 million m³/d occurs in the United States, accounting for around 15 percent of total U.S. capacity.

Four states—Florida, California, Arizona, and Texas—have the majority of the installed capacity. Figure 3.8 shows the states with more than 1 percent of the U.S. total installed capacity. Three of the four states with the greatest installed capacity are coastal, whereas the fourth, Arizona, is an arid state with limited water supply sources. A large plant built by the U.S. government in Yuma, Arizona, to desalinate Colorado River water, is included in this estimate, but this plant has never operated outside of short test periods. One of the largest desalination plants ever proposed for the United States is the Tampa Bay, Florida, plant. Touted as a breakthrough in low-cost desalination, this plant has been rife with problems, as noted in Box 3.3. Like the Yuma desalter and the Santa Barbara desalination plant (Box 3.4), the Tampa Bay plant is included in the national inventory but has never operated commercially or reliably.

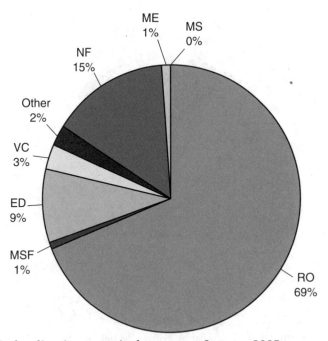

FIGURE 3.7   U.S. desalination capacity by process, January 2005.

Source: Wangnick/GWI 2005.

Note: Electrodialysis (ED)
      Multi-effect distillation (ME)
      Multistage flash distillation (MSF)
      Nanofiltration (NF)
      Freeze, hybrid, and all other processes (Other)
      Reverse osmosis (RO)
      Vapor compression (VC)

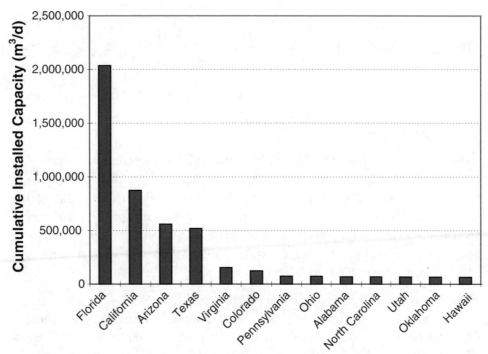

FIGURE 3.8 U. S. states with more than one percent of the total U.S. installed capacity, January 2005. Total installed capacity in cubic meters per day.
Sources: Wangnick/GWI 2005.

Box 3.3 The Experience of the Tampa Bay Desalination Plant[1]

In March 1999, regional water officials in Florida approved plans to build a reverse-osmosis plant with a capacity of 95,000 m^3/d. Claims were made that the cost of water would be very low and competitive with other local sources. Desalination advocates were extremely excited by the project and by the apparent breakthrough in price. The desalination facility was to be privately owned and operated and on completion would supplement drinking water supplies for 1.8 million retail water customers. The plant was considered necessary to help reduce groundwater overdraft and to help meet future demands.

The planning process for the plant began in October 1996. In early 1999, Tampa Bay Water selected S&W Water, LLC, a consortium between Poseidon Water Resources and Stone & Webster. Their proposal called for construction of the plant on the site of the Big Bend Power Plant on Tampa Bay to begin in January 2001 and for operation to begin in the second half of 2002 (Heller 1999; Hoffman 1999). A total of 167,000 m^3/d (44 million g/d) of feed

_____

1. A review of this project was provided in Gleick (2000). This updates the status of the Tampa Bay project as of early 2006.

continues

Box 3.3 *continued*

water would be used to produce around 95,000 m^3/d of potable water and 72,000 m^3/d of brine. The desalinated water would then be added to the municipal supply.

The agreement called for desalinated water to be delivered at an unprecedented wholesale cost of $0.45/m^3 for the first year, with a thirty-year average cost of $0.55/m^3 (Heller 1999). Southwest Florida Water Management District (SWFWMD) agreed to provide 90 percent of the projected $110 million in capital costs for construction of the plant and the cost of the pipeline needed to transport the water to the water-distribution system (U.S. Water News 2003; Heller 1999).

The project has been fraught with difficulties, and, as of early 2006, the plant is still not in operation. Serious management and technological failures have occurred. Several contractors declared bankruptcy, forcing Tampa Bay Water to purchase the plant and assume full risk. Excessive membrane fouling also became problematic. Fouling decreased the life of the membranes and increased costs. In addition, additional chemicals were required to clean the membranes, causing a violation of the sewer discharge permit issued by Hillsborough County.

Because of membrane fouling, Tampa Bay Water agreed to a $29-million, two-year contract with American Water/Pridesa (both owned by Thames Water Aqua Holdings, a wholly owned subsidiary of RWE) in November 2004 to get the plant running. Tests revealed that membrane fouling was still a problem, and many of the water pumps developed rust and corrosion problems because of cost-cutting (Pittman 2005). SWFWMD also threatened to withhold financing for the plant because of a disagreement with Tampa Bay Water about the capacity at which the plant would operate. In January 2006, the water authorities agreed that the plant could be operated at less than full capacity as long as groundwater pumping was reduced. Environmentalists and activists strongly opposed the deal (Skerritt 2006).

The plant is expected to reopen in late 2006 for another assessment period, after $29 million in repairs are finished, and is currently scheduled to be fully operational in January 2008, six years late. In a press release issued in early 2004, the new cost was described as ($0.67 per m^3), up from an initial expected cost of between $0.45 to 0.55/m^3 (Business Wire 2004). The recent decision to reduce the amount of water that the plant will produce as well as unforeseen problems will likely drive the price up further.

Careful examination of the project's cost claims should caution desalination advocates against excessive optimism on price, and, indeed, the project has

Box 3.3

been fraught with difficulties, in part, related to cost-cutting. Moreover, the project had several unique conditions that may be difficult to reproduce elsewhere. For example, energy costs in the region are very low—around $0.04 per kWhr—compared to other coastal urban areas. The physical design of the plant, sited at a local power plant, permitted the power plant to provide infrastructure, supporting operations, and maintenance functions. Cooling water discharge from the power plant was designed to provide intake water for the desalination plant and to dilute the brine before discharge back into the bay. Salinity of the source water from Tampa Bay is substantially lower than typical seawater: only about 26,000 ppm instead of 33,000 to 40,000 ppm typical for most seawater. Financing was to be spread out over thirty years, and the cost of money was only 5.2 percent (Wright 1999).

Box 3.4   The Experience of the Santa Barbara Desalination Plant

The city of Santa Barbara's experience with desalination should caution local communities that are planning desalination facilities. Between 1987 and 1992, California experienced an extended drought, which strongly impacted particularly the coastal Santa Barbara region, where water resource options are particularly limited. Santa Barbara relies extensively on rainfall and local groundwater to meet its water needs. By 1991, Santa Barbara residents were faced with a severe shortage. The city's few reservoirs were rapidly drying up, despite successful conservation efforts that reduced water use by nearly 40 percent. City officials requested proposals to identify a new water source. As fears mounted, Santa Barbara residents overwhelmingly approved construction of an emergency desalination plant as well as a piped connection to the proposed Central Coast Branch of the State Water Project.

In 1991, Santa Barbara partnered with the Montecito and Goleta Water Districts to construct a 25,000 m^3/d reverse-osmosis desalination facility at a cost of $34 million. The cost of the water was estimated to be roughly $1.22/$m^3$, which was substantially more expensive than local supplies.

continues

Box 3.4 *continued*

Construction of the desalination plant began in May 1991. The plant was completed in March 1992 and successfully produced water during start-up and testing. Shortly after construction was completed, however, the drought ended. The desalination facility was placed in an active standby mode, because the cost to produce the water was too high to warrant use during nondrought periods. In addition, the high cost of building the plant and connecting to the State Water Project raised water prices high enough to encourage substantial additional conservation, further decreasing need for the plant.

At the end of the initial contract, Santa Barbara became the sole owner of the facility. In January 2000, Santa Barbara sold half of the plant's capacity to a company in Saudi Arabia. The new capacity of the desalination plant is 11,000 m^3/d but it has never operated and has been decommissioned. The facility now "serves as a sort of insurance policy, allowing the City to use its other supplies more fully" (City of Santa Barbara 2005). Restart costs and timing remain uncertain.

Desalination in California

Traditionally, desalination has been a minor component of California's water-supply portfolio. The IDA database (Wangnick/GWI 2005) lists around twenty seawater desalination plants (mostly small in size), installed or contracted in California since 1955, with a cumulative installed capacity of around 150,000 m^3/d, but these numbers appear to include old plants built and now shut down, built and never operated, or simply contracted and never built. In a more recent and detailed report, the California Coastal Commission (CCC) compiled a list of the desalination facilities that are currently operational along the California coast. The CCC lists about a dozen, mostly small, desalination facilities located along California's coast with a total capacity of 23,000 m^3/d (Table 3.3). California's desalination plants rely largely upon RO and other membrane systems to treat water. Nearly 85 percent of the reported capacity is desalinated with RO.

One of the largest ocean desalination plants in California was built in the 1990s in Santa Barbara, in response to serious water-supply constraints and a persistent drought. This plant also never operated commercially and proved to be a "white elephant" (Box 3.4).

California's Proposed Expansion

California is currently in the midst of a great surge in interest in desalination, far exceeding the efforts of any time during the last few decades. This new interest is the

TABLE 3.3 Desalination Facilities Located Along the California Coast

Operator/ Location	Purpose	Ownership	Maximum Capacity (m³/d)	Status
Chevron/ Gaviota	Industrial processing	Private	1,550	Active
City of Morro Bay	Municipal/ domestic	Public	2,270	Intermittent use
City of Santa Barbara	Municipal/ domestic	Public	10,600	Decommissioned
Duke Energy/ Morro Bay	Industrial processing	Private	1,630	Not known
Duke Energy/ Moss Landing	Industrial processing	Private	1,820	Active
Marina Coast Water District	Municipal/ domestic	Public	1,140	Temporarily idle
Monterey Bay Aquarium	Aquarium visitor use	Nonprofit	150	Active
PG&E/Diablo Canyon	Industrial processing	Private	2,180	Not known
Santa Catalina Island	Municipal/ domestic	Public	500	Inactive
U.S. Navy/ Nicholas Island	Municipal/ domestic	Government	90	Not known
Oil and gas companies	Platform uses	Private	8-110	Active

Source: California Coastal Commission 2004, Baucher, 2006

result of several factors, including improvements in desalination technology, dropping costs, ongoing water management and scarcity concerns, and an increased commercialization effort on the part of private desalination companies and promoters. More than twenty proposed desalination plants are currently along California's coast (Figure 3.9), most of which are considerably larger than any previously built in the state. Table 3.4 lists the major proposed projects as of early 2006. All of the proposed plants use RO to treat ocean or estuarine water. The total capacity of the proposed plants is between 1.9 and 2.3 million m³d, which would represent a massive 100-fold increase over current seawater desalination capacity. If all of these plants are built, seawater desalination would supply 7 percent of California's year 2000 urban water use. It remains to be seen whether or not the proposed expansion of desalination in California occurs, whether it is premature, or whether other solutions to California's long-term water challenges will be found.

In the following sections, we review the arguments made for, and against, desalination. These arguments are being made throughout the world where water supply, quality, or reliability problems exist. The issues discussed in the next few sections highlight several critical conditions that will have to be met before large-scale desalination can become a reality (for a more comprehensive discussion, see Cooley et al. 2006).

FIGURE 3.9 Map of proposed desalination plants in California, Spring 2006.
Note: MGD/million gallons per day

Advantages and Disadvantages of Desalination

Economics and Costs

Economics is one of the most important factors determining the ultimate success and extent of desalination. Yet the discussion to date of actual costs has been muddled and muddied because estimates have been provided in various units, years, and ways that are not readily comparable. For example, some authors report the cost of desalinated water delivered to customers (NAS 2004), whereas others present the cost of produced water before distribution. Some report the costs for desalinating brackish water, some ocean water (Wilf and Bartels 2005; Figure 1.6 in NAS 2004; Karnal and Tusel 2004; Segal 2004; Chaudhry 2004; Semiat 2000). Often it is not clear what values are being reported, as in a story in the *Sydney Morning Herald* (2006).[4] The year and type of estimate (actual operating experience, bid, or engineer's estimate), interest rate, amortization period, energy cost, environmental conditions, and presence or absence of subsidies, are often obscure. The effect of some of these variables on the cost of desalination will be discussed in greater detail.

Table 3.5 is an effort to standardize the reported costs of *produced* water from seawater desalination plants around the world (excluding distribution costs). When necessary, we have converted the estimates to U.S. dollars per cubic meter (US\$/m³) but have not adjusted the apparent year of each reported cost for inflation because

---

4. A cost of AUD\$1.44 per cubic meter is presented for seawater desalination, and compared with a cost of AUD\$1.35 per cubic meter for recycled wastewater. The description of the latter project includes separate distribution to customers, but it is not clear if the former number includes distribution.

TABLE 3.4 Proposed Plants in California as of Spring 2006

Operator	Location	Co-located?	Max Capacity (m³/d)	Intake	Discharge
Marin Municipal Water District	San Rafael	No	38,000–57,000	Surface	Mixed with WW
East Bay Municipal Utility District/	Pittsburg/Oakland/	Likely	76,000–300,000	Surface	Not known
San Francisco Public Utility Commission/ Contra Costa Water District/Santa Clara Valley Water District	Oceanside				
East Bay Municipal Utility District	Crockett	No	5,700	Surface	N/A
Montara Water and Sanitary District	Montara	No	N/A	N/A	N/A
City of Santa Cruz	Santa Cruz	No	9,500, possible expansion to 17,000	Surface	Mixed with WW
Cal Am Water Company	Moss Landing	Yes	42,000–45,000	Surface	Surface
Pajaro-Sunny Mesa/Poseidon Resources	Moss Landing	Yes	76,000–95,000	Surface	Surface
City of Sand City	Sand City	No	1,100	Subsurface	Subsurface
Monterey Peninsula Water Management District	Sand City	No	28,000	Subsurface	Subsurface
Marina Coast Water District	Marina	No	4,900	Subsurface	Subsurface
Ocean View Plaza	Cannery Row	No	190	Surface	Surface
Cambria Community Services District/ Department of the Army	Cambria	No	1,500	Subsurface	Subsurface
Arroyo Grande/Grover Beach/ Oceano Community Services District	Oceano	No	7,100	Subsurface	Mixed with WW
Los Angeles Department of Water and Power	Playa Del Rey	Yes	45,000–95,000	Surface	Mixed w/cooling water or WW
West Basin Municipal Water District	El Segundo	Yes	76,000	Surface	Surface
Long Beach Water Department	Long Beach	No	34,000	Subsurface	Subsurface
Poseidon Resources	Huntington Beach	Yes	190,000	Surface	Surface
Municipal Water District of Orange County	Dana Point	No	95,000	Subsurface	Mixed with WW
San Diego County Water Authority/ Municipal Water District of Orange County	Camp Pendleton	Yes	190,000, expanding to 380,000	Surface	Surface
Poseidon Resources	Carlsbad	Yes	190,000, possible expansion to 300,000	Surface	Surface
San Diego County Water Authority	Carlsbad	Yes	190,000, possible expansion to 300,000	Surface	Surface

Note: ww, waste water; N/A, not available.

Source: Cooley etal. 2006.

TABLE 3.5   Summary of Reported First Year Cost of Produced Water for RO Plants

Facility or Location	US$/m³ (first year)	Operational?	Year	Source
Ashkelon, Israel	0.54	Yes	2002	EDS 2004; Segal 2004; Zhou & Tol 2005
Ashkelon, Israel	0.53	Yes	2003	NAS 2004
Ashkelon, Israel	0.55	Yes	2004	Wilf & Bartels 2005
Ashkelon, Israel	0.62	Yes	2005	Semiat 2006
Bahamas	1.48	Yes	2003	NAS 2004
Carlsbad, CA (Poseidon)	0.77	No	2005	*San Diego Daily Transcript* 2005
Dhekelia, Cyprus	1.09	Yes	1996	Segal 2004
Dhekelia, Cyprus	1.43	Yes	2003	NAS 2004
Eilat, Israel	0.74	Yes	1997	Wilf & Bartels 2005
Hamma, Algiers	0.84	No	2003	EDS (2004); Segal (2004)
Larnaca, Cyprus	0.75	Yes	2000	Segal 2004
Larnaca, Cyprus	0.85	Yes	2003	NAS 2004
Larnaca, Cyprus	0.85	Yes	2001	Wilf & Bartels 2005
Moss Landing, CA (Cal American)	1.28[1]	No	2005	MPWMD 2005
Moss Landing, CA (Poseidon)	0.96	No	2005	MPWMD 2005
Perth, Australia	0.92	No	2005	WTNet 2006 2006
Singapore	0.46	Yes	2002	Segal 2004
Singapore	0.45	Yes	2003	NAS 2004
Sydney, Australia	1.11[2]			
Tampa Bay, Florida	Four bids from 0.46 to 0.58	No	1999	Semiat 2000
Tampa Bay, Florida	0.55	No	2003	Segal 2004
Tampa Bay, Florida	0.58	No	2004	Wilf & Bartels 2005
Tampa Bay, Florida	0.66	No	2004	Arroyo 2004
Trinidad	0.73	Yes	2003	Segal 2004
Trinidad	0.74	Yes	2003	NAS 2004

Notes: 1. May include conveyance costs from the desalination facility to the existing distribution mains.
2. May include some or all distribution costs.

inflation varies from country to country. Even without adjustment to current-year dollars, costs vary far more widely than can be explained by time value of money considerations.

Desalination Cost Trends

Many authors report a declining trend in the cost of seawater desalination. Chaudhry (2004) shows a decline from around $1.60/m³ in 1990 to about $0.63/m³ in 2002, in California, and replicated a graph from the Southern Regional Water Authority in Texas that shows a decline from $1.60/m³ in 1980 to a projected cost of about $0.80/m³ in 2010. Awerbuch (2004) reports that Abu Dhabi recently completed a 190,000 m³/d MSF plant and claims the plant produces water at $0.70/m^3$. In contrast, the cost of desalination in Kuwait is reportedly between $1.33 and $1.83 per m^3 (Al Fraij et al. 2004). Zhou and Tol (2005) show that capital costs have been decreasing over time by performing regressions on a worldwide dataset compiled by Wangnick (2002). Assembly of individual membrane components into large membrane modules, or packaging along with valves, pumps, and so forth, in so-called "package plants," may allow costs to fall somewhat further, though past trends are no indication of future ones.

A notable counter-trend is emerging, and some experts think that membrane costs are unlikely to fall much further in the near term (AWWA 2006). All of the newer cost estimates are notably higher than similar plant bids just a few years ago. The director general of the majority owner of the consortium operating the Ashkelon plant stated last year that more recent tenders for plants in Israel and elsewhere were in the range of $0.82 to $1.02/m^3$ due to increases in the cost of raw materials (e.g., steel) and energy and rising interest rates (*Jerusalem Post* 2005). Plants under construction in Hamma (Algiers) and Perth (Australia) were bid in the same range (Cooley et al. 2006). Notably, the Hamma plant is similar in size and other features (e.g., water temperature and salinity) to the Ashkelon plant, but water from the Hamma plant is priced about 35 percent higher. Costs currently in discussion at Moss Landing, California, and in Sydney, Australia, are even higher, exceeding $1/m^3$ in two of three reported estimates. Higher capital and energy costs appear to have created an upward trend. Future trends are difficult to forecast and policy makers should treat any predictions with skepticism.

Visible and Hidden Subsidies Hide True Costs

Hidden and visible subsidies affect the reported and actual costs. For example, all four bids for the Tampa Bay project were in the range of $0.46 to $0.58/m^3$ in 1999 (Semiat 2000). They were among the lowest costs ever proposed for a significant desalination project, in part, because a Florida regulatory entity provided low-cost capital and because of other subsidies (Box 3.3).

Similarly, five projects in Southern California have qualified for a $0.20/m^3$ subsidy from the Metropolitan Water District of Southern California (MWD). The proposed Poseidon project in Carlsbad is reported to cost about $0.77/m^3$ (*San Diego Daily Transcript* 2005) before this subsidy. Some project advocates, however, have been touting the cost at the post-subsidy level of $0.57/m^3$ as the actual cost.

Sometimes the subsidies are less obvious and more difficult to quantify. The Ashkelon, Israel, desalination plant that opened in August 2005 involved initial payments of about $0.53/m^3$. But the land on which the plant is constructed was provided at no cost by the Israeli government because the market value of the land is very low due to its proximity to the Gaza strip (Semiat 2006). As a result, comparing the Ashkelon cost with a new facility on the California coast, where land is expensive, is misleading.

Energy Costs

Energy is the largest single variable cost for a desalination plant, varying from one-third to more than one-half of the cost of produced water (Chaudhry 2004). If energy prices rise, the share of desalination costs attributable to energy will also continue to rise, unless there is a way to greatly reduce the actual amount of energy used in desalination processes. Semiat (2000) reports the following cost breakdown for an RO plant: 44 percent of cost in electrical power, 19 percent in other operation and maintenance expenses, and 37 percent in fixed charges (amortization of capital) (Figure 3.10). Thermal plants use even more energy. Wangnick (2002) reports that energy use represents nearly 60 percent of the typical water costs from a very large thermal seawater desalination plant (Figure 3.11), with another third coming from fixed capital costs, but

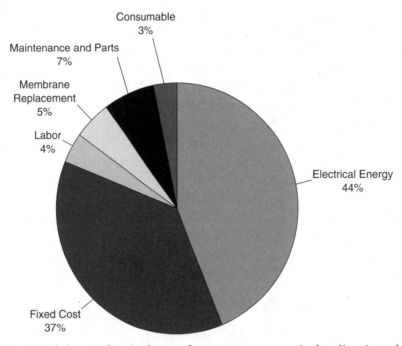

FIGURE 3.10   Breakdown of typical costs for a reverse osmosis desalination plant.
Sources: USBR and SNL 2003.

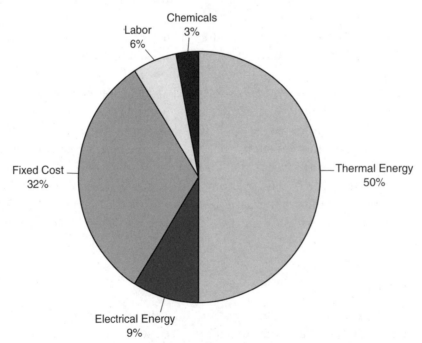

FIGURE 3.11 Breakdown of typical costs for a very large seawater thermal desalination plant.
Source: Wagnick 2002.

even this varies greatly from plant to plant. At these percentages, a 25 percent increase in energy cost would increase the cost of produced water by 11 percent and 15 percent for RO and thermal plants, respectively.

Although opportunities for reductions exist, there are ultimate limits beyond which energy efficiency improvements cannot be made (NAS 2004). The theoretical minimum amount of energy required to remove salt from a liter of seawater, using RO, is approximately 2.8 kilojoules (kJ) (or approximately 1 kWh/m^3).[5] Even the best plants now operating use as little as four and as much as twenty-five times the theoretical minimum (Chaudhry 2004, WTNet 2006, Wilf and Bartels 2005, EDWR 2006),[6] so efforts to reuse energy or minimize energy demands will help reduce overall costs. On energy costs alone, if current best practice uses approximately 3 kWh/m^3, the minimum energy cost alone will be \$0.30/m^3, if electricity is \$0.10/kWh. Energy prices already significantly exceed this in California, unless special energy contracts are signed.

Although Ashkelon was constructed on budget and within the original time frame, the cost of produced water there has already increased about 17 percent, at least in part, due to energy price increases (*Jerusalem Post* 2005). The Singapore contract contains an energy price escalator that does not take effect until the fourth year of operation. This means the cost shown in Table 3.5 reflects energy prices before the large energy price increase of recent years. The actual cost of produced water at Tampa Bay is still uncertain, as noted earlier, and recent energy price increases may drive it even higher.

Plant size also has an effect on costs. Economies of scale exist for all desalination technologies that result from fixed costs that can be spread over larger production in larger plants, though these economies of scale are limited. A doubling of plant size from 9,500 to 19,000 m^3/d, for example, might reduce cost by 30 percent, whereas a doubling from 95,000 to 190,000 m^3/d might reduce cost by only 10 percent. This implies that water produced by smaller plants is much more expensive than water produced by medium-to-large plants of the same design. For example, a small proposed plant in southern San Luis Obispo County (Hill 2006), has an estimated cost of nearly \$2.00/m^3, and the small proposed plant in Sand City, Monterey County (MPWMD 2005), has an even higher estimated cost.

Other Cost Factors

Several other cost factors, including environmental damages or the costs of environmental protection, are not yet well understood, especially in sensitive coastal settings like California and the Persian Gulf. The experience of developers, the amortization period, the cost of money, and regulatory issues also affect final costs.

Many newly proposed facilities are co-located with power plants or wastewater treatment plants to minimize construction and environmental mitigation costs associated with intake and brine discharge pipelines. Because these pipelines can be 5 percent to 20 percent of the capital cost of a new facility (Voutchkov 2005), co-location can potentially reduce costs by up to 10 percent (e.g., 20 percent of capital costs that are 50 percent of total costs). Advantages and disadvantages of co-location will be discussed in more detail.

---

5. Not accounting for the inefficiency of conversion from thermal to electrical energy, as required by some desalination systems.

6. Encyclopedia of Desalination and Water Resources, http://www.desware.net/desa4.aspx.

An often overlooked cost factor is the period over which the facility investment is amortized. A twenty-year rather than thirty-year amortization period at 6 percent interest, for example, would increase the cost per unit water produced by about 20 percent. Development and operating experience affect costs, although there is no clear trend, and we were unable to quantify the impact of experience. In some cases, experience may lower cost or may increase the likelihood of winning a contract. A team that had previous experience in Eilat (Israel) and Larnaca (Cyprus) developed the Ashkelon facility in Israel (Semiat 2006), which is among the plants with the lowest produced water cost. The Algiers facility is only somewhat more expensive than the plant in Trinidad, despite the upward trend in energy and capital costs described previously. Ionics/GE is the developer of these facilities, and successful experience in Trinidad may have helped win the contract in Algiers and temper the price increase.

In contrast, a lack of experience may also result in unrealistic, and unobtainable, cost estimates. The development team in Tampa Bay, Florida, for example, did not have much previous experience. Problems with design, construction, and management of that plant have led to delays of nearly six years and much higher estimated costs (Arroyo 2004).

Future Costs of Desalination

Extreme caution, even skepticism, should be used in evaluating different estimates and claims of future desalination costs. Predictions of facilities costs tend to conflict with actual costs once plants are built, and many cost estimates are based on so many fundamental differences that direct comparisons are invalid or meaningless. The cost of water depends on assumptions about energy costs, technological improvements, capital and labor costs, debt rates, and financing periods. Comparison years are rarely normalized. Sometimes costs to users are compared to costs of production. Additional costs arise if the water must be moved a great distance to the point of use.

Experience to date suggests that desalinated water can be delivered to users (as opposed to "produced") at prices between $1 and $3/m^3, with the great difference due to the factors listed previously. Even the low end of this range remains above the price of water typically paid by urban water users and far above the price paid by farmers. For example, growers in the Western United States usually pay $0.01 and $0.05/m^3 for water. Even urban users rarely pay more than $0.30 to $0.80/m^3, and that price usually includes a wastewater treatment fee.

The capital and operating costs for desalination have decreased recently, in part, because of declining real energy prices, but also because of technological improvements, economies of scale associated with larger plants, and improved project management and experience. The greatest progress in cost reduction has been associated with improvements in RO technology. Salt rejection, a measure of the ability to remove salt from feed water, can be as high as 99.7 percent today, up from 98.5 percent a decade ago, whereas the output of product from a unit of membranes has risen from 60 m^3/d to 84 m^3/d (Glueckstern 1999). Membrane manufacturers are now offering longer guarantees on membrane life, reflecting greater confidence in design and performance of the most sensitive technical component of the process. Other advances that could lead to cost savings are the development of inexpensive corrosion-resistant heat-transfer surfaces, using off-peak energy produced by base-load plants, cogeneration, and co-locating desalination and energy plants to take advantage of economies of production.

Despite optimistic projections from desalination proponents, the long-term objectives of reducing costs 50 percent by 2020 (USBR and SNL 2003) seem unlikely to be achieved with incremental improvements. Radically new technologies or breakthroughs in both materials and energy costs seem necessary, and, although these are possible, they do not appear easy to achieve or likely to occur in the short term.

Ultimately, no one can predict the actual cost of seawater desalination in coming years. Unless energy prices decline substantially, it seems unlikely that production costs in the next few years in the Western United States and California specifically will be less than $0.80 to $0.90/m^3. They will be considerably higher for small plants. Environmental restrictions, land costs, and other factors unique to different regions (e.g., cold ocean waters) may further increase costs.

Reliability of Water Supply

Proponents of desalination argue that one of the important benefits of desalination is the supply reliability provided by diversifying sources, especially in arid and semiarid climates where climatic variability is high. The production of desalinated water is largely independent of climate and, instead, depends on ensuring the continued operation of desalination infrastructure. There is also a value to new "supply" under local control and to increased diversity of supply as a way to increase resilience to natural disasters, terrorism, or other threats to water systems. The issues of climate change and local control are discussed elsewhere in this report.

In a region like California, the reliability value of desalination appears to be especially high. Water allocations, rights, and use are often in flux, or even dispute. Renewable natural water supplies are highly variable and increasingly overallocated or overused. Population is growing rapidly, and ecosystems are increasingly being seen as deserving of water once taken for human uses. Increased demands on such limited supplies affect reliability, especially during dry periods, and regional controversies are threatening continued large-scale diversion of water from the North to the South, from the Colorado River basin, and from the mountains to the coastal regions (Cooley et al. 2006).

Various definitions of water-supply reliability exist, but the most general characteristic is consistent availability on demand. Infrastructure is often built with local reliability concerns in mind, and water utilities often invest in multiple sources of supply with different levels and kinds of risks. Water utility companies invest substantial amounts of money to reduce the risk of supply interruptions because they understand that the cost to their customers of supply disruptions is often far greater than the cost of improved system reliability. For example, the East Bay Municipal Utilities District (EBMUD), which serves 1.2 million people on the east side of San Francisco Bay, has recently invested over $200 million in a Seismic Improvement Program to strengthen the ability of their reservoirs, treatment plants, and distribution systems to continue to function and provide post-earthquake firefighting capability after a large earthquake on the local Hayward Fault. EBMUD estimated that an earthquake that damaged the water system would cause nearly $2 billion in water-related losses (EBMUD 2005). This investment had no effect on the quality of current supply but had the sole effect of improving reliability in the event of an earthquake. The cost of this program is explicitly, and separately, reflected in customer rates.

Similarly, dams and reservoirs are used to reduce the risk of supply interruption due to drought. Other threats to water-supply reliability include climate change, changes in runoff patterns and groundwater recharge, as more impermeable surfaces are created by land development, changes in water quality or environmental regulations, variation in important cost factors (e.g., interest rates, labor, or energy), legal issues related to water rights or contracts for water deliveries, and cultural and political factors.

Also, an implicit opportunity exists to provide more water for the environment buried in most reliability improvements. The opportunity is not easy to capture and not uniform for all regions. But this opportunity should be analyzed for each proposed desalination project. For example, a desalination or water recycling facility may make it possible to release water from storage facilities for critical environmental purposes during average or even drier than average years; or it may make it possible to pump less groundwater or take less surface water from natural ecosystems at all times.

For example, proponents argue that a proposed desalination plant at Moss Landing has the potential to reduce withdrawals from the Carmel River. California-American Water Company (Cal Am) supplies water from the river to users in the Monterey Peninsula. In 1995, a state agency found that Cal Am's water diversions from the Carmel River exceeded their water rights and that these excess diversions were causing damage to the environment. They then ordered Cal Am to reduce water diversions from the Carmel River by nearly 36,000 m^3/d, a 70 percent reduction of withdrawals at that time, or obtain new permits. Several new supply sources have been examined, including a dam, desalination, groundwater recharge, and reclamation. If Cal Am builds a desalination plant for the purpose of satisfying the state requirement, the plant will have explicit environmental benefits.

In many regions of the world, water resources are increasingly transferred from one place to another, especially from rural to urban communities, from water-rich to water-poor regions, and toward economic interests willing and able to pay for water. This has raised two separate issues of local control of resources. The first is the worry of rural—often agricultural—areas that distant urban or economic powers will steal local resources. The classic example is the efforts of the city of Los Angeles in the early part of the twentieth century to obtain water from farming communities hundreds of miles away, which has colored California water politics ever since (Reisner 1986). Las Vegas is currently contemplating major investments in water systems capable of taking groundwater from distant towns and farms to diversify their water-supply options and reduce dependence on limited supplies from the Colorado River—a move strongly opposed by some of those rural communities. Other examples can be found around the world, including in India and China. The second concern is that urban centers will become dependent on distant resources and increase their vulnerability to supply disruptions over which they have limited control.

Desalination may offer a solution to both of these problems by providing a reliable, high-quality source of water under direct local control, reducing the need for imported water at the same time that it reduces the vulnerability to outside disruptions.

Water Quality and Health

Although the quality of desalinated water is typically very high, several potential health concerns have been identified. End-use water quality of desalinated water is a function of source water quality, treatment processes, and distribution of the product water. Harmful contaminants can be introduced at each of these three stages.

The water fed into a desalination system may introduce biological and chemical contaminants that are hazardous to human health. Biological contaminants include viruses, protozoa, and bacteria. Chemical contaminants include regulated and unregulated chemicals, xenobiotics (including endocrine disruptors, pharmaceuticals, and personal care products), and algal toxins (MCHD 2003). These contaminants are of particular concern if they are not removed during subsequent treatment processes.

Boron, for example, is normally found in seawater at concentrations of between 4 and 7 milligrams per liter (mg/L) and is known to cause reproductive and developmental toxicity in animals and irritation of the digestive tract. RO membranes can remove between 50 and 70 percent of this element but pass the rest, where it is then concentrated in the product water. Concern has been expressed that boron may be found in desalinated water at levels greater than the World Health Organization's provisional guideline of 0.5 mg/L and the California Action Level of 1 mg/L (WHO 2003; CDHS 2005). Some membranes are being developed to improve boron removal (Toray 2005). Other methods for addressing boron include blending the desalinated product water with water containing low boron levels. Arsenic, small petroleum molecules, and some microorganisms unique to seawater are also capable of passing though RO membranes and reaching the product water (Cotruvo 2005).

Treatment may also introduce new contaminants or remove essential minerals. For example, brominated organic by-products may be found in the product water as a result of disinfection. These by-products "have greater carcinogenic or toxic potential than many chlorinated by-products" (Cotruvo 2005). In addition, essential minerals, such as magnesium and calcium, are often stripped from the product water. When ingested, water with a low mineral content can actually leach essential nutrients from the human body. Similar leaching can occur within a distribution system, introducing contaminants into the product water (as noted later). Post-treatment can replace some of these minerals, and the World Health Organization suggests that remineralization with calcium and magnesium can have positive health effects, such as is the case with fluoridation (WHO 2004).

Monitoring and appropriate regulation are required of all desalination facilities to assure public health and environmental protection, as Cotruvo (2005) noted, ". . . monitoring of source water, process performance, finished water and distributed water must be rigorous to assure consistent quality at the customer's tap. Moreover, additional water quality or process guidelines specific to desalination are needed to assure water quality, safety, and environmental protection" Such protections can be provided by regulatory oversight by public utility commissions or health departments, or new legislation.

Co-Locating Desalination and Energy Facilities

Integrating desalination systems with existing power plants (or building joint facilities), offers several possible advantages, including making use of discarded thermal energy from the power plant, using the existing intake and outfall structures to obtain seawater and discharge brine, and reducing construction costs. In addition, building on existing sites may prevent impacts at more pristine or controversial locations. Co-location is common for distillation plants built in the Persian Gulf, was proposed by Poseidon Resources for the Tampa Bay desalination plant, and is being considered for nearly half of the proposed plants in California (Filtration and Separation 2005).

The energy and economic advantages of co-location can be substantial. Co-location may allow a desalination plant to take advantage of lower-cost electricity during off-peak periods and to avoid power grid transmission costs. Thermal desalination plants, for example, use thermal energy discarded from the power plant to produce desalinated water, a process referred to as cogeneration. Many of the distillation plants installed in the Middle East and North Africa operate under this principle. In contrast, most of the proposed co-located plants along the California coast share physical infrastructure like the intake and outfall pipes and are only loosely thermally coupled to the power plant; a portion of the power-plant cooling water may be pumped to the adjacent desalination plant, where it undergoes treatment; because warm water is easier to desalinate, treatment costs are lower. The brine is then returned to the outfall and diluted with cooling water from the power plant.

Co-location, however, may also have drawbacks that require careful review and consideration. Opponents argue that co-location will prolong the life of antiquated power plants that use once-through cooling systems. Once-through cooling (OTC) is an inexpensive, simple technology in which seawater is pumped through the heat exchange equipment once and then discharged. These cooling systems can impinge and entrain marine organisms and discharge warm water laced with antifouling chemicals, resulting in significant environmental damage. Many of the power plants that use OTC systems were constructed before 1980, when the marine impacts of this technology were not well understood or regulated, and efforts are under way in the United States to phase out, or more strictly regulate, OTC.

Environmental Effects of Desalination

Desalination, like any other major industrial process, has environmental impacts that must be understood and mitigated. These include effects associated with the construction of the plant, and especially, its long-term operation, including the effects of withdrawing large volumes of brackish water from an aquifer or seawater from the ocean, and discharging large volumes of highly concentrated brine. Indirect impacts associated with the substantial use of energy must also be considered. Later we discuss several of the most important environmental impacts of desalination, but this discussion is not meant to be comprehensive. Each desalination facility must be individually evaluated in the context of location, plant design, and local environmental conditions.

Water Intakes: Impingement and Entrainment

Desalination plants require substantial amounts of source water to produce high-quality product and for other plant operations. Most thermal seawater desalination processes require substantial amounts of cooling water and have significantly greater seawater intake flow rates than comparably sized membrane systems.

Intake water design and operation have environmental and ecological implications. As noted in a recent report by the California Energy Commission on power-plant cooling-water intake structures, "Seawater . . . is not just water. It is habitat and contains an entire ecosystem of phytoplankton, fishes, and invertebrates" (York and Foster 2005). Large volumes of seawater are sucked into the coastal plants during operation. Intake pipes generate a current, which some marine organisms are unable to escape. Large marine organisms, such as adult fish, invertebrates, birds, and even mammals, can be killed on the intake screen (impingement); organisms small enough to pass through the intake screens, such as plankton, eggs, larvae, and some fish, are killed during the processing of the salt water (entrainment). The impinged and entrained organisms are then disposed of in the marine environment. Decomposition of these organisms can reduce the oxygen content of the water near the discharge point, creating additional stress on the marine environment.

The effects of impingement and entrainment are species- and site-specific, and only limited research on the implications of desalination facilities has been done. The magnitude and intensity of these effects depend upon several factors, including the percent mortality of the vulnerable species, the mortality rate of the organism relative to the natural mortality rate, and the standing stock in the area of interest (Edinger and Kolluru 2000).

Recent studies suggest that impingement and entrainment may have greater economic and ecological effects on the marine environment than previously understood. A recent overview of desalination seawater intakes asserts that "[e]nvironmental impacts associated with concentrate discharge have historically been considered the greatest single ecological impediment when siting a seawater desalination facility. However, recent analyses have noted that marine life impingement and entrainment associated with intake designs were greater, harder-to-quantify concerns and may represent the most significant direct adverse environmental impact of seawater desalination" (Pankratz 2004).

Several technological and operational measures can reduce impingement and entrainment, including use of subsurface intake wells and careful design and location of intake structures. Subsurface intake wells, for example, use sand as a natural filter and can reduce or eliminate impingement and entrainment of marine organisms. Subsurface intake wells, however, have some limitations; they require a gravelly or sandy substrate and appear to be limited to intake volumes of 380 to 5,700 m^3/d of water per well (Pankratz 2004). They can also damage freshwater aquifers and the beach environment. The CCC recommends that "[b]each wells should only be used in areas where the impact on aquifers has been studied and saltwater intrusion of freshwater aquifers will not occur. Infiltration galleries are constructed by digging into sand on the beach, which could result in the disturbance of sand dunes" (1993).

The design and location of intake pipes also play a role in the severity of impingement and/or entrainment. Passive screens can reduce intake velocities, which also reduce impingement and entrainment. Velocity caps convert vertical flow into horizontal flow, creating a rapid change in flow direction that some fish avoid. Pipes can also be placed outside of areas with high biological productivity.

Ultimately, the individual volumes, designs, locations, and local ocean conditions will determine the impacts of desalination impingement and entrainment. As a result, careful siting, design, and monitoring are required.

Brine Composition and Discharge

The other most significant environmental problem associated with desalination is the adequate and safe disposal of the concentrated brine produced by the plant. The most obvious characteristic of all brine is an elevated salt concentration, but desalination brines also contain other contaminants. Brine salinity depends on the salinity of the feedwater, the desalination method used, and the recovery rate of the plant. Typical brines contain twice the salt as the feedwater and have a higher density. In addition to high salt levels, brine from seawater desalination facilities can contain concentrations of constituents typically found in seawater, such as manganese, lead, and iodine as well as chemicals of anthropogenic origin, such as nitrates, introduced via urban and agricultural runoff (Talavera and Quesada Ruiz 2001), and impinged and entrained marine organisms killed during the desalination process, as noted previously.

Chemicals used throughout the desalination process may also be discharged with the brine. The majority of these chemicals are applied during pretreatment to prevent membrane fouling (Amalfitano and Lam 2005).[7] For example, chlorine and other biocides are applied continuously to prevent organisms from growing on the plant's interior, and sodium bisulfite is then often added to eliminate chlorine, which can damage membranes. Antiscalants, such as polyacrylic or sulfuric acid, are also added to prevent salt deposits from forming on piping. In addition, coagulants, such as ferric chloride and polymers, are added to the feedwater to bind particles together. The feedwater then passes through a filter, which collects the particulate matter. The RO membranes reject the chemicals used during the desalination process into the brine. The particulate matter on the filter is also discharged with the brine or collected and sent to a landfill.

In addition to using chemicals for pretreatment, chemicals are required to clean and store the RO membranes. Industrial soaps and dilute alkaline and acid aqueous solutions are commonly used to clean the membranes every three to six months. The membranes are then rinsed with product water. The first rinse, which contains a majority of the cleaning solution, is typically neutralized and disposed of in local treatment systems. Subsequent rinses, however, are often discharged into the brine. Frequent cleaning and replacement of the membranes due to excess membrane fouling may lead to discharges in violation of sanitary system discharge permits. This problem has occurred in Tampa Bay.

Brine also contains heavy metals introduced during the desalination process. Corrosion of the desalination equipment leaches several heavy metals, including copper, lead, and iron, into the waste stream. In an early study of a desalination plant in Florida, Chesher (1975) found elevated copper and nickel levels in the water column and in sediments near the brine discharge point. Copper levels were particularly high during unstable operating periods and immediately following maintenance. Engineering changes made at the plant, however, permanently reduced copper levels.

Perhaps the best way to reduce the effects of brine disposal is to reduce the volume of brine that must be discharged and minimize the adverse chemicals found in the brines. Both man-made filters and natural filtration processes can reduce the amount of chemicals applied during the pretreatment process. Ultrafiltration, for example, can replace coagulants, effectively removing silt and organic matter from feedwater (Dudek & Associates 2005). Ultrafiltration also removes some of the guesswork involved in balancing the pretreatment chemicals, because pretreatment "must be continuously optimized to deal with influent characteristics" (Amalfitano and Lam 2005). These filters, however, are backwashed periodically to remove sludge buildup and cleaned with the same solution used on RO membranes. Backwash can be disposed of with the waste brine or dewatered and disposed of on land. Additionally,

---

7. Fouling is caused by the deposition of organic and inorganic matter on the sensitive RO membranes. Fouling reduces the efficiency of the membranes, resulting in more frequent cleaning and replacement of the membranes and higher operating costs.

subsurface intake wells reduce chemical usage during pretreatment by reducing the biological organisms that cause biofouling. Subsurface intake wells, including infiltration galleries and horizontal and vertical beach wells, use sand as a natural filter.

Several brine disposal options exist. For desalination plants located on the coast, disposal methods include discharge to evaporation ponds, the ocean, confined aquifers, or saline rivers that flow into an estuary. Options for inland disposal of brines and concentrates include deep-well injection, pond evaporation, solar energy ponds, shallow aquifer storage for future use, and disposal to a saline sink via pipeline or injection to a saline aquifer (NAS 2004).

Each disposal method, however, has a unique set of advantages and disadvantages. Large land requirements make evaporation ponds uneconomical for many developed or urban areas. Sites along coastlines, for example, tend to have high land values, and coastal development for industrial processes is discouraged. Injection of brine into confined groundwater aquifers is technically feasible, but it is both expensive and hard to ensure that other local groundwater resources remain uncontaminated. Unless comprehensive and competent groundwater surveys are done, there is a risk of unconfined brine plumes appearing in freshwater wells. Direct discharge into estuaries disrupts natural salinity balances and causes environmental damage of sensitive marshes and fisheries. All of these methods add to the cost of the process, and some of them are not yet technically or commercially available. As noted by the 2003 U.S. Desalination Roadmap, "Finding environmentally-sensitive disposal options for this concentrate that do not jeopardize the sustainability of water sources is difficult, and, thus, next-generation desalination plants will have to be designed to minimize the production of these concentrates, or find useful applications for them" (USBR and SNL 2003).

Ocean discharge is the most common and least expensive disposal method for coastal desalination plants (Del Bene et al. 1994), although this approach can have significant impacts on the marine environment. Brine discharged into the ocean can be pure, mixed with wastewater effluent, or combined with cooling water from a co-located power plant. Ocean discharge assumes that dilution of brine with much larger volumes of ocean water will reduce toxicity and ecological impacts. The notion that diluting brine with cooling water reduces the toxicity of the brine is based on the old adage "dilution is the solution to pollution." Although this may be true for some brine components, such as salt, it does not apply to others. The toxicity of persistent toxic elements, including some subject to bioaccumulation, such as heavy metals, is not effectively minimized by dilution.

Certain habitat types and organisms are at greater risk than others. Along California's coast, rocky habitat and kelp beds are particularly rich and sensitive ecosystems, and effort should be made to avoid these areas. Benthic organisms in the immediate vicinity of the discharge pipe are at the greatest risk from the effects of brine discharge. These can include crabs, clams, shrimp, halibut, and ling cod. Some have limited mobility and are unable to move in response to altered conditions. Many benthic organisms are important ecologically because they link primary producers, such as phytoplankton, with larger consumers (Chesapeake Bay Program 2006).

In 1979, Winters et al. noted the risks that the chemical constituents and physical behavior of brine may pose to the marine environment:

> At this stage of development the desalination industry is characterized essentially by an absence of truly significant data on the possible impact of

the effluents on marine ecosystems. . . . It is impossible to determine the extent of ecological changes brought about by some human activity (e.g., desalination) without totally studying the system involved. Ideally such studies should involve a thorough investigation of both the physical and biological components of the environment. These studies should be done over a long period of time. Baseline data should actually be gathered at the site prior to construction for subsequent comparative uses. This will allow for a thorough understanding of the area in its 'natural' state. Once the plant is in operation monitoring should be continued on a regular basis for a period of at least one year but preferably for two or three years.

Yet twenty-five years later, only a few studies have performed a comprehensive analysis of the effects of brine discharge on the marine environment; the majority of studies conducted thus far focus on a limited number of species over a short time period with no baseline data. More comprehensive studies are needed to adequately identify and mitigate the impacts of brine discharge.

Coastal Development and Land Use

In addition to affecting the coastal environment through water intake and discharge, desalination can also affect the coast through impacts on developments, land use, and local growth, which are often controversial and contentious topics. Rapid, unplanned growth can damage local environmental resources as well as the social fabric of a community anywhere, but coastal developments are often particularly divisive. For example, building new homes and businesses without investing in infrastructure can cause overcrowded schools, traffic, and water shortages. Urban and agricultural runoff and increases in wastewater flows create water-quality problems in local rivers, streams, and/or the ocean. Some developments can change the nature of views, beach access, and other environmental amenities.

In coastal areas throughout California, clean, potable water is sometimes considered a limiting resource and used to constrain development. Some coastal communities have reached "build out"—the level of development considered to be the maximum that a region can sustain; others are well below this level. As a result, some desalination opponents worry that water provided by desalination may facilitate growth in these regions. In 2004, the CCC, which administers the Coastal Act, highlighted the importance of the growth-inducing impacts of desalination. This report concludes: "[a] desalination facility's most significant effect could be its potential for inducing growth."

Desalination and Climate Change

Climate change will result in significant changes to water resources and coastal ocean conditions. These changes have important implications—both good and bad—for desalination. Literature reviews of the effect of climate change on water resources (Gleick et al. 2000; Kiparsky and Gleick 2003) indicate that climate change will likely increase temperatures, increase climate variability, including storm intensity and drought frequency, raise sea level, and alter the effects of extreme events, such as the El

Niño/Southern Oscillation. Although some uncertainty remains about how precipitation patterns, timing, and intensity will be altered, there is general consensus for mountain regions that depend on snowfall and snowmelt that climate change will "increase the ratio of rain to snow, delay the onset of the snow season, accelerate the rate of spring snowmelt, and shorten the overall snowfall season, leading to more rapid and earlier seasonal runoff" (Gleick 2000; Kiparsky and Gleick 2003).

These climatic changes will affect the supply of, and demand for, water resources. According to the Intergovernmental Panel on Climate Change (IPCC) (2001):

> Increases in average atmospheric temperature accelerate the rate of evaporation and demand for cooling water in human settlements, thereby increasing overall water demand, while simultaneously either increasing or decreasing water supplies (depending on whether precipitation increases or decreases and whether additional supply, if any, can be captured or simply runs off and is lost).

In addition, rising sea levels may exacerbate seawater intrusion problems in coastal aquifers or rivers that communities depend on for water.

Some view desalination as a means of adapting to climate change and argue that desalination facilities can reduce the dependence of local water agencies on climate-sensitive sources of supply. As climate change begins to alter local hydrology, the resilience of water-supply systems may be affected. When variability of supply goes up, the risk of extreme events increases. A reliable supply of high-quality water from desalination systems that are independent of hydrologic conditions can provide a buffer against this variability.

The IPCC lists desalination plants as a supply-side adaptive measure available to meet potential increases in urban water demand associated with climate change (2001). In a recently released water plan, the Australian government contends that desalination plants "provide a reliable supply and a good quality water and are immune from drought and climate change impacts" (NSW 2004). Climate change has been used as an argument to justify a desalination plant in London, where a Thames Water representative recently said, "We've two challenges. One is population. . . . At the same time we've got climate change" (Barkham 2004). This advantage must be considered in any long-term water plan and evaluated in the context of system reliability, as discussed previously.

At the same time, however, desalination facilities are likely to have some special vulnerability to climate impacts. Ocean desalination plants are constructed on the coast and are particularly vulnerable to changes associated with rising sea levels, storm surges, and increased frequency and intensity of extreme weather events. Intake and outfall structures are affected by sea level. Over the expected lifetime of a desalination facility, sea levels could plausibly rise by as much as a foot or more, and storm patterns are also likely to change on a comparable time scale. All of these impacts have the potential to affect desalination plant design and operation and should be evaluated before plant construction and operation are permitted.

Public Transparency

Desalination plants are often subject to extensive review, but the regulatory and oversight process for desalination can be unclear and contradictory. Adequate review is essential to ensure environmental protection, public health, and appropriate use of our resources. But uncertainty about the project review process can act as a barrier to project development. Standard, clear regulations at the local and national level could help ensure that desalination plants are built in appropriate places to appropriate standards. These policies must not hinder desalination plants by inappropriate regulation or accelerate them via regulatory exemptions.

In addition, it is critical that any public subsidies be tied to direct public benefit, such as environmental restoration. Most of the proposed desalination plants are likely to be located in existing industrial areas to take advantage of infrastructure and local resources. Because low-income populations tend to live in these areas, desalination plants may have a disproportionate impact on those communities. These communities have traditionally borne significant air-quality impacts from local facilities, higher exposure to noise and industrial chemicals, and truck traffic. When desalination facilities are built as co-located plants, the on-site energy plant may be forced to operate at a higher capacity or continuously. Local communities may also suffer as a result of the desalination plant's water-quality impacts. Brine discharges can affect those who swim and fish in the area. Fish may have elevated levels of metals or other toxins with human health implications. Low-income and minority populations may also bear disproportionate effects of increases in water rates (EJCW 2005).

Taking account of local conditions, opinions, and sentiment during decisions about siting, building, and operating desalination facilities is vital. Open and early access to draft contracts, engineering designs, and management agreements is necessary for public review. Further, contracts with private companies must include provisions about who assumes the risk associated with the project if one or more of the contractors declare bankruptcy, as occurred in Tampa Bay. Adequate comment periods and appropriate public hearing schedules are also necessary to ensure that decisions about desalination plants are fair and equitable. Among the principles recommended in a recent comprehensive report on environmental justice and water are

- State legislatures should establish independent reviews of social, economic, and environmental inequities associated with current water rights and management systems.

- There should be independent review of the social and economic impacts of water development on local communities.

- Local public review and approval should be required for any proposals to introduce private control, management, or operation of public water systems.

- All water and land-use projects should be planned, implemented, and managed with participation from impacted community members.

- Actions are required to clean up pollution of water bodies upon which low-income populations rely for subsistence fishing (EJCW 2005).

Summary

In energy-rich arid and water-scarce regions of the world, desalination is already a vitally important option. Desalinated water is being used as a source of municipal supply in many areas of the Caribbean, North Africa, Pacific Island nations, and the Persian Gulf. In some regions of the world, nearly 100 percent of all drinking water now comes from desalination—providing a vital and irreplaceable source of water. But the goal of unlimited, cheap, fresh water from the oceans continues to be an elusive dream for most. Despite much progress over the past several decades, and despite recent improvement in economics and technology, desalination still makes only modest contributions to overall water supply. By 2005, the total amount of desalinated water produced in one year was about as much as the world used in a few hours.

Growing numbers of regions are seriously considering desalination as a part of their water future, and there is no doubt that plants will be built in the future. Far more doubt exists, however, about its current technological and economic competitiveness and the ultimate impacts that desalination plants will have. This chapter has identified both advantages and disadvantages of desalination that must be carefully evaluated before any plant should be built. Following on from Cooley et al. (2006), we also offer a set of recommendations and standards that should be met before desalination facilities are permitted and built.

Perhaps the greatest barrier to desalination remains its high economic cost compared to alternatives, including other sources of supply, improved wastewater reuse, and especially more efficient use and demand management. We do not believe that the economic evaluations of desalination that are commonly presented to regulators and the public adequately account for the complicated benefits and costs associated with issues of reliability, quality, local control, environmental effects, and impacts on development. In general, significant benefits and costs are often both excluded from the costs presented publicly.

Is desalination the ultimate solution to our water problems? No. Is it likely to be a vital piece of our water management puzzle? Yes. In the end, decisions about desalination developments will revolve around complex evaluations of local circumstances and needs, economics, financing, environmental and social impacts, and available alternatives. Such decisions should be transparent, open, public, and systematic.

Desalination Conclusions and Recommendations

The process of permitting desalination facilities rarely includes adequate comment periods and appropriate public hearing schedules.

- Decisions about developing desalination plants must include fair and equitable processes.
- Public comment on desalination proposals should be both encouraged and solicited.
- Comment periods and hearing schedules must be widely advertised.

The cost of desalination has fallen in recent years, but it remains a high-cost water option.

- Cost estimates should clearly indicate whether they refer to produced or delivered water, and if delivered, to what location(s).
- Desalination cost estimates and assumptions must be clearly described in all proposals.

The assumption that desalination costs will continue to fall may be wrong. Recent increases in energy and construction costs, and diminishing potential for gains in membrane performance, suggest that further cost declines may be limited.

- The historical downward trend in the cost of produced water should not be projected forward without specific rationale for each specific project proposal.

- All cost estimates should both explicitly state the assumptions that underlie them and justify those assumptions over the lifetime of the facility.

Desalination is the most energy-intensive water supply or demand management option in California. The future cost of desalinated water will be more sensitive to changes in energy prices than will other sources of water.

- Energy price risk should be explicitly evaluated by project proponents, including year-to-year variation and trends over time, in the revenue requirements of water utilities.

Public subsidies for desalination plants are inappropriate unless explicit public benefits are guaranteed.

- No public subsidies should be offered to desalination facilities unless they come with a guarantee of public benefits, such as restoration of ecosystem flows.

Desalination plants offer both system-reliability and water-quality advantages, but these advantages are neither guaranteed nor automatic.

- The value of reliability and water quality advantages of desalination plants should be estimated.

- The value of reliability to customers in general, and the reliability value of a particular option like desalination, should be evaluated separately.

Impingement and entrainment of marine organisms are among the most significant environmental threats associated with seawater desalination.

- The effects of impingement and entrainment are species- and site-specific and require detailed site-specific baseline ecological assessments and impact studies.

- Intake pipes should be located outside of areas with high biological productivity.

Subsurface and beach intake wells may mitigate some of the environmental advantages of open ocean intakes.

- Site-specific analysis of the advantages and disadvantages of subsurface and beach intake wells is needed.

- Such wells should be used only in areas where the impact on freshwater aquifers has been studied and saltwater intrusion will not occur.

Highly concentrated salt brines are a serious pollutant, and desalination brines often contain other chemical pollutants that require regulation, monitoring, and transparent reporting. Safe disposal of this effluent is a challenge.

- Concentrations of chemicals in brine discharges should be carefully regulated, monitored, and minimized.

- Disposal of brine in sensitive wetlands should be prohibited under all circumstances.

- Disposal of brine in underground aquifers should be prohibited unless comprehensive and competent groundwater surveys are done, and there is no reasonable risk of brine plumes appearing in freshwater wells.

Desalination also offers the opportunity to return water to the environment.

- Substituting desalinated water for water from rivers, lakes, or groundwater aquifers can reduce pressure on these natural sources.

- Environmental benefits claimed by desalination proposals must come with binding mechanisms to ensure that the benefits promised are delivered in the form, degree, and consistency required.

Co-location of desalination facilities at existing power plants has potential economic and environmental advantages and disadvantages.

- Co-location can reduce infrastructure costs through the use of shared facilities.

- Co-location should not be encouraged at the expense of maintaining old, inefficient, and environmentally damaging facilities.

- Desalination should not be used as an excuse to keep once-through cooling systems in operation longer than currently permitted.

- Because of the uncertainty associated with once-through cooling systems, the effects of the desalination must be assessed independently of the power plant.

Desalination facilities on the coast will affect coastal development and land use, often in unexpected ways.

- The effects of desalination on coastal development activity and planning must be considered a fundamental part of impact assessments and not assumed to be incidental, minimal, or secondary.

- Growth-inducing impacts of desalination facilities must be evaluated on a case-by-case basis through appropriate laws and regulations.

Desalination plants will be vulnerable to climate change. Coastal desalination facilities will be subject to rising sea levels, storm surges, and greater frequency and intensity of extreme weather events.

- All desalination facilities should be designed and constructed, using estimates of future, not present, climate and ocean conditions.

- No desalination facilities should be permitted unless consideration of climate change factors has been integrated into plant design.

Desalination facilities will affect local and regional energy use, air quality, and greenhouse gas emissions.

- Plans for desalination must explicitly describe the energy implications of the facility and how it fits into regional efforts or requirements to reduce greenhouse gas emissions or meet regional, state, or national clean air requirements.

- Regulatory agencies should consider requiring desalination plants to offset their greenhouse gas emissions.

The regulatory and oversight process for desalination is sometimes unclear and contradictory.

- National, state, and local policies should standardize and clarify the regulation of desalination.

- Desalination should not be hindered by inappropriate regulation or accelerated by regulatory exemptions.

Decisions about siting, building, and operating desalination facilities will be controversial if local conditions, opinions, and sentiment are not taken into account.

- The process of designing, permitting, and developing desalination facilities must be transparent and open.

- Access to draft contracts, engineering designs, and management agreements must be widely available for public review beginning in the early stages of project development.

- Contracts with private desalination developers must include explicit provisions about financial risk in the event of project failure.

- Independent review of the social and economic impacts of desalination facilities on local communities should be commissioned and made publicly available.

- Local public review and approval should be required for any proposals to introduce private desalination facilities into public water systems.

- All desalination projects should be planned, implemented, and managed with early participation from affected community members.

More research is needed to fill gaps in our understanding, but the technological state of desalination is sufficiently mature and commercial to require the private sector to bear most additional research costs.

- Public research funds should be restricted to analyzing the public aspects of environmental impacts, mitigation, and protection.

REFERENCES

Al Fraij, K. M., Al Adwani, A. A., Al Romh, M. K. 2004. The future of seawater desalination in Kuwait. In *Desalination and water re-use.* 83–84. Nicklin, S. (editor). United Kingdom: Tudor Rose.

Amalfitano, A., Lam, K. 2005. Power to seawater RO in China. *World Water and Environmental Engineering,* 28(4):16–17.

American Water Works Association (AWWA). 2006. Costs fall as membrane system numbers rise. *Main Stream,* 50:1. Denver, CO: American Water Works Association.

Arroyo, J. July 2005. Briefing on TWDB desalination activities for the South Central Desalting Association. http://www.twdb.state.tx.us/Desalination/Desal/PPT%20index.asp.

Awerbuch, L. 2004. The status of desalination in today's world. In *Desalination and water re-use.* 9–12. Nicklin, S. (editor). United Kingdom: Tudor Rose.

Barkham, P. 2004. Thames tides set to top up London tap water. *Guardian Unlimited,* June 14.

Baucher, B. 2006. Personal communication. City of Morro Bay.

Business Wire. 2004. American water-pridesa secures contract to remedy and operate the Tampa Bay seawater desalination plant. Business Wire, November 16, 2004.

California Coastal Commission (CCC). 1993. *Seawater desalination in California.* http://www.coastal.ca.gov/desalrpt/dtitle.html.

California Coastal Commission (CCC). 2004. *Seawater desalination and the California Coastal Act.* http://www.coastal.ca.gov/energy/14a-3-2004-desalination.pdf.

California Department of Health Services (CDHS). 2005. *Drinking water notification levels.* http://www.dhs.ca.gov/ps/ddwem/chemicals/AL/notificationlevels.htm.

Chaudhry, S. 2004. *Unit cost of desalination.* Sacramento, CA: California Desalination Task Force, California Energy Commission, California Department of Water Resources. http://www.owue.water.ca.gov/recycle/desal/Docs/UnitCostofDesalination.doc.

Chesapeake Bay Program. 2006. http://www.chesapeakebay.net/info/benthos.cfm.

Chesher, R. H. 1975. Biological impact of a large-scale desalination plant at Key West, Florida. In *Tropical marine pollution.* 99–181. Ferguson Wood, E. J., Johannes, R. E. (editors). New York: Elsevier Scientific Publishing Company.

City of Santa Barbara. 2005. *City of Santa Barbara water: Water supply sources.* http://www.santabarbaraca.gov/Government/Departments/PW/SupplySources.htm?js=false.

Cooley, H., Gleick, P. H., Wolff, G. 2006. *With a grain of salt: A review of seawater desalination.* A report of the Pacific Institute, Oakland, CA. May.

Cotruvo, J. A. 2005. Water desalination processes and associated health and environmental issues. *Water Conditioning and Purification,* January, 13–17.

Del Bene, J. V., Jirka, G., Largier, J. 1994. Ocean brine disposal. *Desalination,* 97:365–372.

Dudek & Associates, Inc. 2005. *Environmental impact report for precise development plan and desalination plant.* EIR 03-05-SCH#2004041081. Prepared for the city of Carlsbad, CA.

East Bay Municipal Utilities District (EBMUD). 2005. *Urban water management plan.* Chapter 2. Oakland, CA. http://www.ebmud.com/water_&_environment/water_supply/urban_water_management_plan/2005_uwmp/default.htm.

Edinger, J. A., Kolluru, V. S. 2000. Power plant intake entrainment analysis. *Journal of Energy Engineering,* April, 1–14.

Encyclopedia of Desalination and Water Resources (EDWR). 2006. http://www.desware.net/desa4.aspx

Environmental Justice Coalition for Water. 2005. *Thirsty for justice: A people's blueprint for California water.* The Environmental Justice Coalition for Water, Oakland, CA.

European Desalination Society (EDS). 2004. "The cost of water." *European Desalination Society newsletter,* Vol. 20, (No. 2), May.

Filtration and Separation. 2005. *Cutting the costs and environmental impact of seawater desalination through power plant co-location.* http://www.filtsep.com/latest_features/webzine_feature/September_04_Poseidon.html.

Gleick, P. H. 1993. *Water in crisis: A guide to the world's fresh* water resources. Oxford University Press, Oxford, New York.

Gleick, P. H. 2000. *The world's water 2000–2001.* Washington, DC: Island Press.

Gleick, P. H. and others. 2000. *Water: The potential consequences of climate variability and change.* A report of the National Water Assessment Group, U.S. Global Change Research Program, U.S. Geological Survey, U.S. Department of the Interior and the Pacific Institute for Studies in Development, Environment, and Security. Oakland, CA.

Glueckstern, P. 1999. Desalination today and tomorrow. *International Water and Irrigation,* 19(2):6–12.

Heberer, T., Feldman, D., Redderson, K., Altmann, H., Zimmermann, T. 2001. *Removal of pharmaceutical residues and other persistent organics from municipal sewage and surface waters applying membrane filtration.* Proceedings of the National Groundwater Association, 2nd International Conference on Pharmaceuticals and Endocrine Disrupting Chemicals in Water, October 9–11, 2001, National Groundwater Association. Minneapolis, MN, and Westerville, OH.

Heller, J. 1999. Water board green-lights desalination plant on bay. *Tampa Bay Business News,* March 16. http://www.tampabay.org/press65.asp.

Hill, K. 2006. Reverse osmosis endorsed. *Santa Maria Times,* January 17.

Hoffman, P. 1999. (Personal communication, Stone and Webster Company).

Intergovernmental Panel on Climate Change (IPCC). 2001. *Climate change 2001: Impacts, adaptation, and vulnerability.* Cambridge, England: Cambridge University Press.

Jerusalem Post. 2005. Ashkelon desalination plant begins pumping potable water. August 5.

Karnal, I., Tusel, G. F. 2004. A comparison of options for the refurbishing of existing power/desalination plants. In *Desalination and water re-use.* Leicester, England: Tudor Rose Holdings Ltd.

Kiparsky, M., Gleick, P. H. 2003. *Climate change and California water resources: A survey and summary of the literature.* Department of Water Resources. California water plan update, 2005, vol. 4. Sacramento, CA.

Monterey County Health Department (MCHD). 2003. *Desalination and public health.* http://www.owue.water.ca.gov/recycle/desal/Docs/DesalHealthEffects.doc.

Monterey Peninsula Water Management District (MPWMD). 2005. MPWMD comparative matrix—Part 1, desalination. From a staff presentation at the September 8, 2005, meeting of the Monterey Peninsula Water Management District, Monterey, CA.

National Academy of Sciences (NAS). 2004. Review of the desalination and water purification technology roadmap. Water Science and Technology Board. Washington, DC: National Academies Press.

NSW Department of Infrastructure, Planning and Natural Resources. 2004. Meeting the challenges—Securing Sydney's water future. http://www.dipnr.nsw.gov.au/waterplan/.

OTV. 1999. "Desalinating seawater." *Memotechnique,* Planete Technical Section, No. 31:, p. 1, February.

Pankratz, T. 2004. An overview of seawater intake facilities for seawater desalination. *The future of desalination in Texas.* Vol. 2: Biennial report on seawater desalination. Texas Water Development Board. http://rio.twdb.state.tx.us/Desalination/The%20Future%20of%20Desalination%20in%20Texas%20-%20Volume%202/documents/C3.pdf.

Pittman, C. 2005. Desal delays boil into dispute. *St. Petersburg Times,* August 16. http://pqasb.pqarchiver.com/sptimes/882831921.html?MAC=6f1ad2567d7134e862bfbd6c2df13162&did=882831921&FMT=FT&FMTS=FT&date=Aug+16%2C+2005&author=CRAIG+PITTMAN&pub=St.+Petersburg+Times&printformat=&desc=Desal+delays+boil+into+dispute.

Reisner, M. 1986. *Cadillac desert: The American West and its disappearing water.* New York: Viking Penguin, Inc.

San Diego Daily Transcript. 2005. Valley Center, Poseidon Resources ink water purchase deal. December 20, 2005.

Sedlak, D. L., Pinkston, K. E. 2001. *Factors affecting the concentrations of pharmaceuticals released to the aquatic environment.* Proceedings of the National Groundwater Association, 2nd International Conference on Pharmaceuticals and Endocrine Disrupting Chemicals in Water, October 9–11, 2001, National Groundwater Association. Minneapolis, MN, and Westerville, OH.

Segal, D. 2004. Singapore's water trade with Malaysia and alternatives. http://www.transboundarywaters.orst.edu/publications/related_research/Segal-Singapore-Malaysia%2004_abstract.htm. Cited costs are originally from Morris, R. 2004. Technological trends in desalination and their impact on costs and the environment, presented at the World Bank Waterweek.

Semiat, R. 2000. Desalination: present and future. *Water International,* 25(1):54–65.

Semiat, R. 2006. (Personal communications, Deputy Director of the Technion Water Research Institute, Haifa, Israel).

Skerritt, A. 2006. Water deal has loophole the size of an aqueduct. *St. Petersburg Times,* January 27, 2006.

Sydney Morning Herald. 10 January 2006. Cheaper water—the simple solution at hand. http://www.smh.com.au.

Talavera, J. L., Quesada Ruiz, J. J. 2001. Identification of the mixing processes in brine discharges carried out in Barranco del Toro Beach, south of Gran Canaria (Canary Islands). *Desalination,* 139:277–286.

Toray. 2005. Press release. Toray awarded first order for its "High boron removal reverse osmosis membrane element for seawater desalination." http://www.toray.com/news/water/nr050714.html.

U.S. Agency for International Development (USAID). 1980. *The USAID Desalination Manual.* Washington, DC: CH2M HILL International for the U.S. Agency for International Development.

U.S. Bureau of Reclamation (USBR). 2003. *Desalting handbook for planners,* 3rd ed. Desalination and water purification research and development report #72. United States Department of the Interior, Bureau of Reclamation, Water Treatment Engineering and Research Group. Denver, CO.

U.S. Bureau of Reclamation and Sandia National Laboratories (USBR and SNL). 2003. Desalination and water purification technology roadmap: A report of the Executive Committee. Desalination & Water Purification Research & Development Report #95. United States Department of the Interior, Bureau of Reclamation, Water Treatment and Engineering Group. Denver, CO.

U.S. Water News. 2003. Tampa Bay tapping bay as new source of drinking water despite some concerns. http://www.uswaternews.com/archives/arcquality/3tambay4.html.

Voutchkov, N. 2005. Shared infrastructure benefits desalination economics. *Source Magazine,* 19:2. California/Nevada section of the American Water Works Association, Rancho Cucamonga, CA.

Wangnick, K. 1998. *1998 IDA worldwide desalting plants inventory, No. 15.* Produced by Wangnick Consulting for the International Desalination Association. Gnarrenburg, Germany.

Wangnick, K. 2002. *2002 IDA worldwide desalting plants inventory.* Produced by Wangnick Consulting for the International Desalination Association. Gnarrenburg, Germany.

Wangnick/GWI. 2005. *2004 worldwide desalting plants inventory.* Oxford, England: Global Water Intelligence. Data provided to the Pacific Institute.

WaterTechnologyNet (WTNet). 2006. Perth seawater desalination plant, Kwinana, Australia. http://www.water-technology.net/project_printable.asp?ProjectID=3415.

Wilf, M., Bartels, C. 2005. Optimization of seawater RO systems design. *Desalination,* 173:1–12.

Winters, H., Isquith, I.R., and R. Bakish. 1979. Influence of desalination effluents on marine ecosystems. *Desalination,* 30:403-410.

World Health Organization (WHO). 2003. *Boron in drinking water: Background document for development of WHO guidelines for drinking-water quality.* Geneva, Switzerland: World Health Organization.

World Health Organization (WHO). 2004. Rolling revision of the WHO guidelines for drinking-water quality. Consensus of the meeting: Nutrient minerals in drinking-water and the potential health consequences of long-term consumption of demineralized and remineralized and altered mineral content drinking-water. Geneva, Switzerland.

Wright, A.G. 1999. Tampa to tap team to build and run record-size U.S. plant. Engineering-News Record, March 8, 1999.

York, R., Foster, M. 2005. *Issues and environmental impacts associated with once-through cooling at California's coastal power plants.* Sacramento, CA: California Energy Commission. http://www.energy.ca.gov/2005publications/CEC-700-2005-013/CEC-700-2005-013.PDF.

Zhou, Y., Tol, R. S. 2005. Evaluating the costs of desalination and water transport. *Water Resources Research,* 41:W03003, doi:10.1029/2004WR003749.

Floods and Droughts

Heather Cooley

Introduction

Floods and droughts are extreme hydrologic events: they occur infrequently and are periods that are either wetter or drier than "normal" for a given region. Although the definitions of both depend upon natural variability in hydrologic conditions, floods and droughts are perceived as disasters or hazards only when they affect humans and their institutions. Ironically, humans have been attracted to both floodplains and drought-prone areas throughout their history because many of their characteristics make them attractive for development. Floodplains tend to be flat and have access to fertile soil, water supply, and a transportation corridor. Drought-prone areas, on the other hand, have less vegetation and more open space, making it easier to clear the land for development. In addition, fewer disease vectors are associated with drought-prone areas, and warm, dry climates are often considered attractive.

Although floods and droughts are natural events to which plants and animal species have adapted over time, their intensity, severity, and consequences can be worsened or lessened by human action (or inaction). Within an ecological context, droughts and floods are important ecosystem processes, referred to as *disturbances* that can promote long-term ecosystem diversity and stability. Some species have adapted to hydrologic extremes and, in some cases, require these extremes for survival. Human modification of the environment, however, has altered many natural systems to such a degree that full recovery after a large disturbance may not always be possible: "Just as not every natural disturbance is a disaster, not every disaster is completely natural. We have altered so many natural systems so dramatically that their ability to bounce back from disturbance has been greatly diminished" (Abramovitz 2001).

Floods and droughts are among the most common and damaging of all natural hazards, and their impacts continue to grow. Between 1900 and 2005, floods and droughts killed an estimated 17 million people and affected over five billion people (based on data from EM-DAT 2006).[1] And these numbers appear to be growing.

Economic losses from natural disasters, including floods and droughts, are greatest for the richest nations. However, losses as a percent of gross domestic product (GDP)

---

1. The actual number killed or affected by floods and droughts is likely substantially higher, as estimates are hampered by insufficient data, particularly during the early part of the 20th century.

are greatest for poor nations, and disasters in these countries threaten to hamper development.[2] Between 1985 and 1999, economic losses in the richest nations accounted for 57 percent of total measured global economic losses from natural disasters and represented 2.5 percent of their GDP; economic losses in the poorest nations accounted for 24 percent of measured global economic losses from natural disasters but over 13 percent of their GDP (Abramovitz 2001). Thus, losses in the poorest nations reinforce the cycle of poverty and likely increase vulnerability to future hazards.

Historically, flood and drought response has largely been based on crisis management. This approach has proven to have serious limitations, and, in recent years, managers have sought to reduce vulnerability by applying risk management principles. In addition, this approach can be used to harness the environmental and social benefits of floods and droughts to improve ecosystem functions. In this chapter, we first discuss the complicated definitions of droughts and floods and describe some of the effects of these events on humans and the environment. We then discuss disaster response strategies that have evolved over time, emphasizing a more recent approach that seeks to reduce vulnerability to these events.

Droughts

No simple, concise definition of drought exists. In general, a drought is a hydrological extreme caused by a persistent and abnormal moisture deficiency that has adverse impacts on vegetation, animals, and people over a relatively large area. A dry period is typically recognized as a drought when demand exceeds supply. It can be related to changes in the intensity, frequency, and/or timing of precipitation events, but, because of the potential for human activities to affect both water supply and demand, drought definitions increasingly incorporate human influences and impacts (National Drought Mitigation Center 2005a). For example, "policy droughts" can be caused by poor management decisions, as described below.

Drought characteristics vary by region and by water-use activity. For example, a three-month dry period during the growing season in the Midwest of the United States would adversely affect agriculture but would have little impact on most fish and wildlife populations. On the other hand, a dry period of that duration during the growing season in California is normal and would not constitute a drought. A complex irrigation system in California allows farming during hot, dry summers, whereas farmers in the Midwest largely rely on rain to irrigate their crops.

A drought is a temporary phenomenon and, as such, it is distinct from aridity, which is a climatic feature of a particular region. Although perceived as rare, droughts occur periodically in every climatic zone. Some areas, however, are clearly more drought-prone than others. Kenya, for example, experiences a major drought every decade and minor droughts every three to four years (UNEP and the Government of Kenya 2000). London, on the other hand, experiences drought much less frequently (although, ironically, as this volume was being prepared, London was experiencing one of its worst droughts in decades).

---

2. GDP is an imperfect, albeit common, measure of economic wealth or activity.

Determining the beginning of a drought is difficult, and it is often better defined in hindsight; what begins as an extended spell of good weather can become devastating, destroying crops and spawning vast wildfires. Scientists typically evaluate droughts based on precipitation, temperature, and soil-moisture data. The data are often encapsulated by a drought index. However, a dizzying array of indices is available, each possessing particular strengths or weaknesses. The Palmer Drought Severity Index, for example, is based on a soil-moisture algorithm and is best suited for homogenous landscapes. This index is not appropriate for regions with lags between precipitation and runoff, such as California and Colorado (Hayes n.d.). The Standard Precipitation Index is based on the long-term precipitation record for a specific location. This simple index provides early warning and monitoring of drought conditions but is sensitive to data availability and the distribution of these data over time (Meteorological Hazards and Seasonal Forecasting Group 2005).

Droughts can be short-lived or persistent. They can also be localized in a particular region or extend over an extremely large area. These characteristics are not necessarily correlated. For example, the Dust Bowl of the 1930s lasted nearly ten years, affected over 200,000 km^2 of land in the Central United States, and spawned legends of misery, human migration, and poverty (NWS 2006). A relatively short drought in Africa, in 1991 to 1992, affected over 6.7 million km^2 (NASA n.d.). Figure 4.1 shows the percent of the United States land area that has been affected by severe or extreme drought between 1895 and 2005 and highlights the temporal and spatial variability associated with drought within a single region. Since 1895, one or more areas of the United States has experienced severe or extreme drought at any given moment, suggesting that droughts are not as rare as commonly believed.

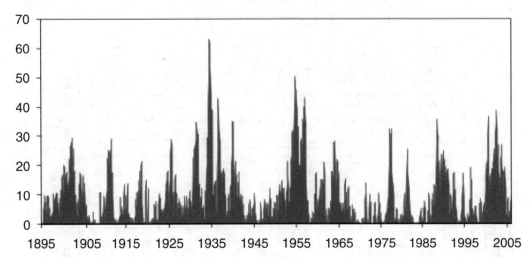

FIGURE 4.1 Percent area of the United States in extreme or severe drought between 1895 and 2006.

Source: National Drought Mitigation Center n.d.

Several disciplinary perspectives exist regarding droughts. The National Drought Mitigation Center in the United States identifies four drought types (2005a):

- Meteorological droughts occur when precipitation is below normal.

- Agricultural droughts occur when soil-moisture levels caused by below-normal precipitation are insufficient to meet a particular crop's needs at a particular time, causing reduced biomass and yield.

- Hydrological droughts refer to shortfalls in average surface and subsurface water flows.

- Socioeconomic droughts are those that occur when below-normal precipitation levels affect people.

As Figure 4.2 indicates, these drought types are temporally correlated. A meteorological drought, for example, occurs on a relatively short time-scale as precipitation levels fall below normal levels. Over time, soil moisture declines, reducing crop yield and resulting in an agricultural drought. As the meteorological drought persists, surface and groundwater levels decline, resulting in a hydrological drought. When demand exceeds supply, agricultural and hydrological droughts have social and economic implications, resulting in a socioeconomic drought.

Although these more common definitions are based on the consequences of drought, Rao (n.d.) describes two additional drought types that emphasize their cause: political and policy droughts. A political drought is a drought intentionally induced to serve a political purpose. A policy drought, on the other hand, refers to drought-like

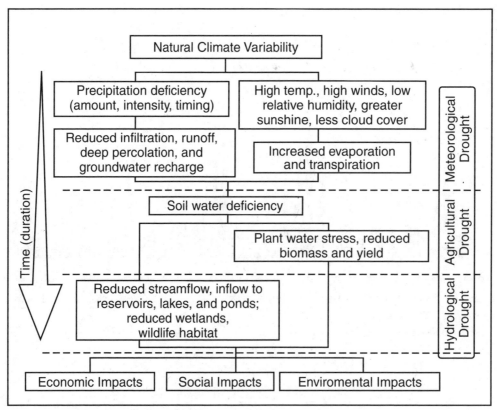

FIGURE 4.2   Drought types typically described in the literature.
Source: National Drought Mitigation Center 2005a.

conditions caused by policies instituted by the government. Policy and political droughts can happen anywhere and at anytime, regardless of hydrologic conditions. Rathore (2005) contends that "[t]his shows that drought is just not the scarcity or absence of rainfall, but is more related to water resource management (or mismanagement)" (p. 1).

Causes of Drought

The causal factors of natural droughts are not well understood and are complicated by multiple feedback loops. In general, the underlying cause of a drought is a change in average atmospheric circulation patterns, which can result from both internal and external forcings. Internal forcings include sea-surface temperature and land-surface characteristics, whereas external forcings include the sun, the Earth's orbit, and volcanoes (Trenberth et al. 2004).

The major climatic phenomenon, known as the El Niño-Southern Oscillation (ENSO), is a primary cause of climatic extremes, including both droughts and floods, in regions throughout the world. ENSO occurs every three to five years and is characterized by warm ocean surface temperatures that result in a change in atmospheric circulation. Changes in atmospheric circulation create dry conditions in some regions but wet conditions in others. La Niña, which follows El Niño, tends to reverse this trend, causing dry periods in areas that were wet during El Niño and vice versa (Trenberth et al. 2004).

Natural and anthropogenic factors form important feedback processes and further aggravate drought conditions. For example, drought-induced soil-moisture deficits can reduce the land cover over large areas, thereby eliminating soil surface shading, increasing soil evaporation, and causing further soil-moisture deficiencies. Anthropogenic activities, such as land management, can elicit a similar ecosystem response. Poor land management can destroy soil structure, leaving it more susceptible to drought. For example, the Dust Bowl has been associated with poor land-management techniques that contributed to severe erosion.

Droughts are often exacerbated by other associated weather conditions, such as high temperatures, low relative humidity, and high winds. Wars, corruption, and other geopolitical events can also exacerbate drought conditions. For example, the Ethiopian-Eritrean border war between 1998 and 2000 intensified an agricultural drought because it resulted in the destruction of essential infrastructure, placement of land mines in agricultural regions, and dedication of resources to fighting the war rather than providing essential services. "For the long term, food security in Eritrea will not depend solely on the weather. Drought is a problem that persists in this semi-arid country, but other factors—like good governance and economic stability—play an important role" (Integrated Regional Information Networks 2005).

Effects of Drought

Because of its temporal and spatial characteristics, as well as its impacts on a broad range of sectors (see later text), many experts describe droughts as the most serious natural hazard. The National Drought Policy Commission (2000), for example, maintains that "[d]rought is perhaps the most obstinate and pernicious of the dramatic events that Nature conjures up. It can last longer and extend across larger areas than hurricanes, tornadoes, floods, and earthquakes." Similarly, the National Aeronautics and Space Administration (NASA n.d.) asserts that "[d]rought is by far the most damaging of all natural disasters. Worldwide, since 1967, drought is responsible for millions of deaths and has cost hundreds of billions of dollars in damage."

Figure 4.3 shows the globally reported deaths due to drought between 1900 and 2005. Since 1900, droughts are estimated to have killed over 10 million people, or about 100,000 people per year on average. Unlike other natural hazards, deaths due to drought occur infrequently but in great numbers. Since 1900, eight droughts accounted for 97 percent of the reported deaths. Most of these deaths occurred in China and India, as a result of agricultural shortfalls that led to starvation.

Figure 4.4 shows the number of people affected by drought between 1900 and 2005.[3] An estimated two billion people have been affected by drought since 1900.[4] As global population continues to rise, the number of people affected by drought each year also seems to be growing, though increases in reported impacts may partly be the result of improvements in monitoring and reporting.

Reliable data on drought-induced economic damages are not available, in part, because drought impacts are predominantly nonstructural and thus difficult to quantify. The U.S. Federal Emergency Management Agency (FEMA) made a recent attempt at estimating economic damages and found that the average cost of drought in

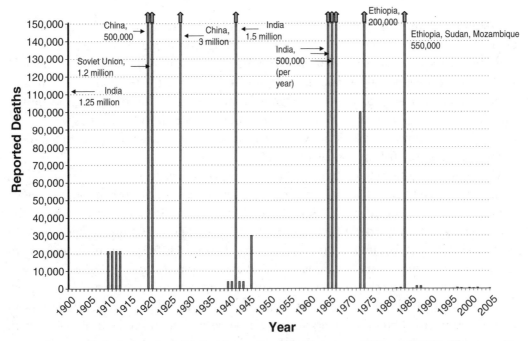

FIGURE 4.3   Global reported deaths due to drought between 1900 and 2005. Some major events far exceeded 150,000 deaths as indicated by arrows at the top of some bars.
Note: Data from EM-DAT 2006.

3. "Affected" is defined as "[p]eople requiring immediate assistance during a period of emergency, i.e. requiring basic survival needs such as food, water, shelter, sanitation and immediate medical assistance" (EM-Dat 2006).

4. Because drought impacts are so pervasive and the historical record before 1964 is incomplete, the actual number of people affected is likely to be significantly higher.

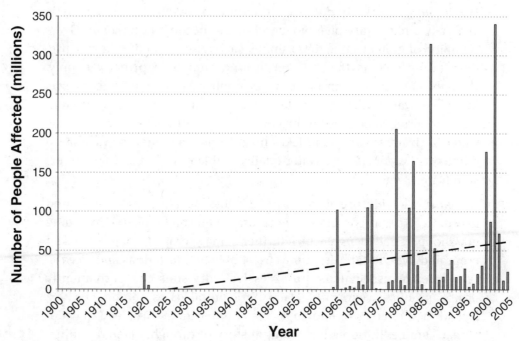

FIGURE 4.4 Global reported number of people affected by drought between 1900 and 2005 Lack of accurate or detailed reporting in early years may underestimate total impacts of drought and change the trend line.
Note: Data from EM-DAT 2006.

the United States alone is $6 to $8 billion annually (1995). Hayes et al. (2004) note that FEMA's estimate is based largely on losses in the agricultural sector and misses other important direct and indirect economic damages. For example, a drought in California from 1987 to 1992 was estimated to have cost $250 million in agricultural losses but over $3 billion overall when other costs, such as lost hydroelectric generation, were included (Gleick and Nash 1991).[5] A standard method to quantify these effects has not been developed, thereby limiting comparisons among events or regions. Hayes et al. (2004) call for more accurate economic data that would allow officials to make better decisions on implementing mitigation strategies based on more consistent comparisons.

Droughts have tremendous environmental, social, and economic implications. Although some conditions can return to normal shortly after the drought has ended, others are permanent. Some effects are described in greater detail (Knutson et al. 1998):

- *Mortality*—A loss of human life, both directly and indirectly. The 1984 drought in Ethiopia, for example, caused nearly one million deaths (UNEP 2002). The 1941–1942 drought in China killed an estimated three million people due to starvation (NOAA n.d.). Humans may die from thirst, heat stress, or malnutrition. Dust-related illnesses, such as suffocation, silica poisoning, and dust pneumonia, can also result in death.

---

5. This analysis also noted that substantial, but unquantified, ecological damages resulted from this long-term drought. Ecological damages of drought are rarely monetized.

- *Water Supply and Quality*—A reduction in water supply, which is generally a temporary phenomenon but can become permanent. Over-pumping of groundwater, for example, can cause an aquifer to collapse, devastating regions that rely on groundwater as a primary supply. Drought can also compromise water quality. Water-quality concerns include elevated salinity, high temperatures, and low oxygen levels. In addition, drought can lead to higher water-pollution levels, because less water in rivers, streams, and lakes means that less water is available to dilute wastewater effluent. Water-quality problems can exacerbate water-supply problems.

- *Agriculture*—Widespread agricultural losses, affecting crops and livestock. Drought conditions are favorable for many insects, including grasshoppers and locusts, which further damage crops. Drought-stricken crops and livestock are also more susceptible to infestations and disease. Wind-erosion associated with excessively dry soils can permanently destroy productive agricultural land. Agricultural losses, combined with a lack of food reserves or limited access to aid, can, in turn, lead to wide-spread famine. Ethiopia has a long history of drought-related famine: "Historical accounts testify that there were seven famines in the 13th century, four in the 14th century, eleven in the 17th century, five in the 18th century, and several more in the 19th and 20th centuries" (Relief and Rehabilitation Commission 1985).

- *Wildfires*—Dry vegetation combined with high temperatures and low humidity often increases the frequency and intensity of fires. Although fire can cause tremendous damage, fire is also a natural disturbance that promotes ecosystem stability and biodiversity. Grassland fires, for example, prevent encroachment from trees and shrubs.

- *Fish and Wildlife*—As water levels in streams, rivers, and lakes decline, fish and wildlife are at risk of dying, and regional extinctions are possible. In addition, stressed fish and wildlife are more susceptible to disease. Some species, however, thrive during droughts, particularly those that are adapted to low oxygen levels and warm water temperatures.

- *Habitat*—Loss of habitat. Wetlands are at tremendous risk. Multiple droughts can contribute to desertification, a process by which once-productive land becomes a barren desert. Land-management decisions can contribute to this problem.

- *Economy*—Job loss, especially for agricultural workers and recreation-oriented workers. In struggling regions, drought can deepen poverty.

- *Energy*—Energy system is strained, particularly in regions dependent on hydropower. The high temperatures associated with drought also increase energy demand due to use of air conditioners and other appliances.

- *Urban Migration*—Rapid migration occurs from rural to periurban environments, particularly where aid is provided at some endpoint, rather than where it is needed. Periurban environments in developing countries often lack adequate infrastructure, leading to heightened risk of contracting water-related diseases.

- *Security*—Regional conflict intensifies. Drought threatens security by exacerbating conflicts over water allocations, especially for rivers that cross political borders. An influx of refugees from drought-stricken regions can also intensify land conflicts, as occurred in parts of Kenya in 2000 (Integrated Regional Information Networks 2002). Drought can also precipitate social unrest. Deaths due to the 1973–1974 drought in Ethiopia, for example, contributed to a revolution to remove the *Ancien Régime* (Relief and Rehabilitation Commission 1985).

Drought Management

Because drought is a pervasive problem that is impossible to prevent, managing drought and drought risk are essential. Historically, drought management has been based on crisis management, a reactionary approach that involves providing relief to those affected by drought. Wilhite (1996) contends that this approach often promotes dependency, encourages poor management practices, and may serve to increase long-term vulnerability: "Unfortunately, the response efforts of many nations have had little, if any, effect on reducing vulnerability, largely because of their emphasis on emergency assistance. In fact, vulnerability to drought has increased in some settings because of relief recipients' expectations for assistance from government or donors. . . . Disincentives to proper management of the natural resource base characterize the provision of relief in most countries." For example, farmers may plant high water-use crops with greater commercial value rather than drought-tolerant crops with lesser commercial value because government relief would provide income in case of crop failure. Similarly, ranchers may stock cattle in excess of what can be sustainably maintained because relief programs may compensate for any losses incurred.

In recent years, the global community has begun to transition away from crisis management in favor of risk management. Risk management seeks to identify risks associated with a particular type of event, to understand the underlying cause of those risks, and to develop appropriate mitigation strategies to avoid (prevent) or limit (mitigate) the impacts associated with those risks. This transition, however, has proven to be difficult because drought and its associated risks are poorly understood, leading to "indecision and/or inaction on the part of managers, policy makers, and others" (Wilhite 1997).

Effective drought management integrates the concept of risk management. It involves both structural and nonstructural measures that must be taken before, during, and after the hazard event. In this chapter, we will adopt a convention established by Wilhite by which drought-management activities are arranged into the following categories: monitoring and early warning; impact and vulnerability assessments; and mitigation and response (Wilhite 2000). An example of actions and policies for each of these categories is described in greater detail.

Monitoring and Early Warning

Because of the complexity and lack of understanding about the driving climatic forces that result in drought, accurately predicting specific droughts is impossible. Monitoring and early warning, however, provide the public and policy makers with information essential for detecting drought conditions and implementing mitigation and response actions. A public education program is integral to any monitoring program. People must understand what the warning means and what actions are appropriate. For example, an early-warning system can inform farmers about what types of crops to grow, when to plant and harvest, and how to fertilize. An early warning system can also prompt farmers to supplement livestock feed or to sell livestock before the animals become too emaciated.

To support this effort, the appropriate institutional capacity must be in place to collect and assess the data and inform the public. This requires a monitoring committee, or some other organization, to collect and maintain data on a range of meteorological and hydrological conditions, including stream and groundwater levels, reservoir levels, soil moisture, snow pack, precipitation, temperature, and humidity (Wilhite 1997). Although this may require establishing an observation network, many regions throughout the world currently operate such systems. In addition, the committee must develop regional and sector-specific definitions of drought and identify and monitor the appropriate drought indices.

Recent technological advances are available to support this effort. For example, environmental satellites can monitor vegetation health, moisture and thermal conditions, and fire risk potential over wide areas. Future work will enable us to detect changes earlier, with greater accuracy and at greater spatial resolution. In a recent discussion of remote sensing and early drought warning, Kogan (2000) voiced optimism about technological breakthroughs: "We begin the 21st century with exciting prospects for the application of operational meteorological satellites in agriculture. . . . drought can be detected 4–6 weeks earlier than before in any corner of the globe and delineated more accurately, and its impact on grain production can be diagnosed long before harvest. This is the most vital step for global food security and trade" (p. 98). In addition, the Internet and other communication systems improve our ability to more rapidly disseminate information to the appropriate parties.

Two centers in Eastern and Southern Africa have developed effective monitoring and early warning systems. In 1989, twenty-four countries in Eastern and Southern Africa established regional drought monitoring centers in Nairobi (DMCN) and Harare (DMCH) with the assistance of the World Meteorological Organization (WMO) and the United Nations Development Programme (UNDP). Although initially pilot projects, these centers have been formally integrated into other institutions and are now a permanent and important element of drought management in Africa. DMCN and DMCH provide information to policy makers, farmers, water-resource scientists, health officials, and environmentalists on the historic, current, and future climate on decadal, monthly, and seasonal time-scales. In addition, they provide information on current and potential socioeconomic conditions related to climate, such as food security. Information is disseminated through various media, including the Internet, high-frequency radio, facsimile broadcasting, and the meteorological data distribution system. Information is also translated into local languages to ensure the greatest distribution possible (Ambenje 2000).

Impact and Vulnerability Assessment

A drought impact assessment identifies those regions, activities, and populations that are affected by drought. As noted previously, drought impacts can include, but are not limited to, energy shortages, crop failure, loss of biodiversity, and loss of human life.

The impact assessment identifies the effects of drought, whereas the vulnerability assessment attempts to determine the underlying cause of drought-related impacts. According to Knutson et al. (1998), "[d]rought may only be one factor along with other adverse social, economic, and environmental conditions that creates vulnerability" (p. A-2). A vulnerability assessment tries to get at the root cause of the vulnerability so that mitigation strategies can be devised to address those vulnerabilities (Wilhite et al. 2005). This approach changes the nature of the discussion and lies at the heart of risk management.

For example, let's say an impact assessment reveals that a drought would reduce crop production. A vulnerability assessment would then address why crop production would be affected. This assessment may indicate that poor crop selection is one of the causes of this reduction. A thorough vulnerability assessment would further question why crop selection was poor in an attempt to elucidate the underlying environmental, social, and economic factors that contributed to the impact. Figure 4.5 shows a simplified tree diagram of a vulnerability assessment.

Conducting impact and vulnerability assessments must include representatives from all affected sectors as well as local community members. This bottom-up approach can be an important component of capacity building. Empowering local people to identify the risks and possible mitigation strategies creates a sense of ownership of the solutions and increases the likelihood of implementation and success of response efforts.

Mitigation and Response

Mitigation and response are core elements of drought management, especially when efforts are made in advance to reduce vulnerabilities. Mitigation refers to actions and programs taken before and in the early stages of a drought to reduce drought-related *risks;* whereas response refers to actions taken immediately before, during, and directly

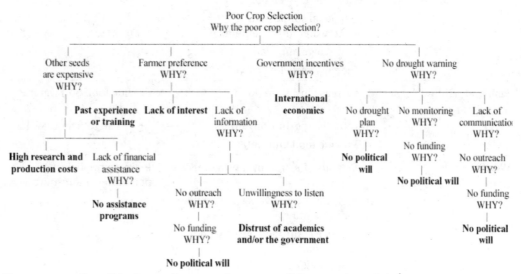

FIGURE 4.5 Simplified tree diagram that exemplifies a vulnerability assessment of the impacts of drought on the agricultural sector.
Source: Knutson et al. 1998.

after drought to reduce drought *impacts* (Knutson et al. 1998). Mitigation is anticipatory and focuses on risk reduction, whereas response is reactive and focuses on impact reduction. Although a response can be critical to relieving suffering and preventing deaths, solely relying on responses can create dependency by encouraging unsustainable practices and may miss efforts that are likely to be especially cost-effective. "A dollar spent on disaster preparedness can prevent $7 in disaster-related economic losses" (Abramovitz 2001). Thus responses should be structured such that they do not increase vulnerability and maximize the benefits of early action.

Table 4.1 highlights mitigation and response strategies for some of the sectors affected by drought. As this table suggests, several mitigation strategies are available,

TABLE 4.1   Mitigation and response strategies for drought-affected sectors

Impact Sector	Mitigation Strategy	Response Strategy
Human Health	Reduce poverty	Provide food, water supplies, and vaccinations
	Empower local population to develop mitigation strategies	Consume wild, edible plants (famine food)
	Food preservation	Establish refugee camp
Water Supply and Quality	Water conservation and efficiency	Truck-in potable water
	Groundwater recharge	Water-use restrictions
	Drought-proof water supplies	Migration
	Construction and proper maintenance of reservoirs	
	Drill emergency well bores	
	Build regional pipelines	
	Establish infrastructure and institutional capacity for water transfers	
	Drill deeper well bores	
	Financial support for wastewater improvements	
	Encourage rain water harvesting	
	Increase recycling capacity	
Fish and Wildlife	Increase instream flows	Provide food and water to wildlife
	Alter fishing/hunting limits	Move fish to deeper water
	Provide food and water	
Wildfires	Establish fuel removal programs	Increase fire fighting personnel and resources
	Identify alternative water supplies for firefighting	
	Practice controlled burns where appropriate	
	Limit public access	

Impact Sector	Mitigation Strategy	Response Strategy
Agriculture	Plant drought-resistant crops	Provide food and water to livestock
	Adjust grazing schedule and intensity	Provide shelter for livestock
	Promote proper soil management techniques	Insurance, grants, and loans
	Reduce herd size	Move herd to watering holes
	Plant alternative forage crop	
	Residue management	
	Weed control	
	Herd diversification	
Economy	Provide job training and placement assistance	Provide emergency funding for businesses
	Extend boat ramps and docks in recreation areas	Insurance, grants, and loans
	Diversify economy	
Habitat	Provide migration corridors	
	Limit tilling during dry periods	
	Promote proper soil management techniques	
Security	Develop regional drought response programs	
	Establish clear water rights and drought protocols	

Sources: Mosley 2001, Najarain 2000, National Drought Mitigation Center 2005b, Wilhite 1993.

and mitigation strategies in one sector can have a positive effect on other sectors. For example, improvements in the efficiency of water use can minimize reliance on the existing supply and reduce wasteful water use, thereby maximizing the current supply and reducing the drought's impact on the other sectors. Likewise, planting drought-resistant crops reduces agricultural losses, with benefits for the economy and habitat.

Post-drought assessments have become more common but must be further encouraged. These assessments are crucial to identifying weakness and developing appropriate mitigation strategies. In addition, they may identify new vulnerabilities that have resulted from changing conditions and encourage an adaptive management approach:

> A plan that is flexible enough to adjust to changing vulnerabilities and adaptive capacities within the state is equally essential. However, it is clear that supporting an 'adaptive' approach to drought planning and preparedness is resource intensive, and it is not clear whether the necessary resources will be available over the long term (Jacobs et al. 2005).

Floods

Definition

Floods are more easily defined and classified than droughts, and their effects are more obvious and visible. A flood is defined as "the rising of a body of water and its overflowing onto normally dry land" (Princeton University 2005). Four primary flood types are categorized according to where they occur. As these categories indicate, floods can occur in a range of environments. These flood types are described in greater detail:

- *Coastal flood.* Inundation of coastal land above normal tide levels due to storm surges or wave action. Strong winds from hurricanes or tropical storms can push water inland. Geologic events, such as earthquakes, landslides, or volcanoes, can cause tsunamis with potentially devastating effects in coastal areas.

- *River flood.* Inundation of land along a riverbank due to a river or stream overflowing natural, or constructed, confines. River floods occur on timescales of hours to weeks, as precipitation occurs in excess of that needed to recharge groundwater. Warm rain on top of snow pack can be particularly dangerous. In coastal areas, river floods can be exacerbated by high tides, which slow drainage.

- *Urban flood.* Flooding that occurs on relatively short time periods when insufficient drainage is available to remove precipitation. Paved surfaces prevent soil infiltration of water, leading to rapid runoff into storm sewers, drains, and local creeks in excess of their ability to transport water. Inadequate storm drainage systems can also induce flooding. Basements and low points are at particular risk.

- *Flash flood.* Typically a highly localized, rapid event that is more common in arid and desert areas. Flash floods are caused by high precipitation levels and are often associated with thunderstorms or other tropical storms. Dam breaks, or even an ice break, can also induce a flash flood. These can occur within minutes to hours of a precipitation event, and the rapid onset can catch people off guard.

Typically, floods are short-term, discrete events. Floods have a distinct beginning and endpoint and last from hours to weeks, depending on the type of flood. As precipitation levels decline or the storm surge subsides, water naturally drains out of the flooded area. In some cases, however, water must be pumped from flooded areas, thereby increasing the duration of the flood; for example, New Orleans, Louisiana, lies below sea level and every drop of water that falls within the basin must evaporate or be manually pumped out. After Hurricane Katrina in late August 2005, many of the pumps were not functioning properly, and the area remained flooded for a substantial period of time.

A flood is often described by its recurrence interval, which refers to the period of time between floods of a particular intensity, for example, a 100-year flood event.

Recurrence interval is based on historic conditions for a given area and is thus site-specific. The terminology used to describe the recurrence interval, however, can be misleading and is often misinterpreted. A 100-year flood does not refer to a flood level that occurs every 100 years. Rather, it refers to a flood that has a 1/100, or 1 percent, chance of occurring in any one year. Over a 30-year period (a typical mortgage period in the United States), a 100-year flood has a 1 in 4 chance of occurring (Box 4.1). It is important to realize that flood frequency is not fixed. Land-use change, such as deforestation or urbanization, and climate change can alter flood frequency. In a watershed in Boston, for example, a 100-year flood became a 20-year flood due to urbanization over the course of 15 years (OTA 1980). A 15-cm sea-level rise is estimated to change the 100-year "highest estimated tide"—a flood planning level for coastal development—into a 1-in-10 year event (Gleick and Maurer 1990).

Box 4.1 Calculating Flood Frequency

What are the chances that a 100-year flood will occur during a 30-year period?

To make this determination, we must apply basic probability theory. Flooding is a random event, that is, the odds of it occurring in any year are independent of past conditions. Thus, the odds of a storm not occurring over a 30-year period can be calculated, using the following methodology:

Suppose that an event has an X percent chance of occurring in a given year, then the odds that the event will *not* occur in a given year are

$$1 - X$$

The odds that an event will not occur in two successive years is

$$(1-X)(1-X) = (1-X)^2.$$

The odds of an event not occurring over y number of years is

$$(1-X)^y.$$

Now calculate the odds that a 100-year flood event will not occur over 30 years.

In this case,

$$X = 1/100 = 0.01 \text{ and } y = 30$$
$$(1-X)^y = (1-0.01)^{30} = 0.74.$$

Thus, there is a 74 percent chance that a 100-year storm will *not* occur over a 30-year period and a 26 percent, or 1 in 4, chance that it will occur.

Causes of Floods

Floods are caused by an input of water in excess of the infiltration or runoff capacity. Heavy rains, particularly on a snow pack, can induce all types of flooding (coastal, urban, river, and flash floods). In addition, storm surges and high tides can cause flooding in coastal areas. Further, mechanical blockage of rivers, that is, from logs or ice, can temporarily dam a river channel; when these obstacles are cleared, severe downstream flooding can occur.

Like droughts, floods are largely climatological phenomena induced by weather that diverges from normal conditions due to random variability or cyclical weather patterns. Flood events are usually confined to a particular time of year, such as during the rainy, hurricane, or monsoon season, although there are notable exceptions (described later). As described previously, a multi-year cyclical pattern, ENSO, is associated with floods in regions throughout the world.

Floods may also be intensified by human activities, such as soil compaction or urbanization. Paving over the land surface, which is associated with urbanization, inhibits soil infiltration and leads to water accumulation in low-lying areas or rapid discharge into local streams and waterways. Both conditions increase the likelihood of floods. In addition, poorly planned urbanization leads to the construction of buildings and other developments in areas prone to periodic inundation.

Unlike droughts, geologic or anthropogenic activities can also cause floods. For example, a magnitude 9.1 to 9.3 earthquake off the coast of Sumatra initiated a series of powerful tsunamis that slammed into countries surrounding the Indian Ocean, causing extensive damage in Sri Lanka, India, Bangladesh, Thailand, Somalia, Malaysia, and the Maldives on December 26, 2004. Although a precise death count is not known, it is estimated that as many as 300,000 people were killed, making it among the worst natural disasters in history (Wikipedia 2006a). Dam breaks are also responsible for some of the most devastating floods ever recorded.

Effects of Floods

Figure 4.6 shows global reported deaths due to flood between 1900 and 2005.[6] Since 1900, nearly seven million people have died from floods. Over 90 percent of the deaths, however, are from five floods, all of which occurred in China before 1960. Deaths from floods occur annually throughout all parts of the world, although people in developing countries are typically at greater risk for several reasons, which include (1) poor early-warning systems and less access to transportation, thereby reducing the opportunity to flee threatened areas; (2) houses that are not structurally sound and more likely to be destroyed during a flood event; and (3) less access to adequate water supply and sanitation, and thus a greater likelihood of contracting infectious diseases that are commonly associated with floods.

Substantial numbers of people are affected by flood each year, and this number appears to be growing. Figure 4.7 shows the number of people affected by floods between 1900 and 2005.[7] An estimated three billion people have been affected by flood

---

6. Note that these figures do not include storm and wave surges.

7. *Affected* is defined as "[p]eople requiring immediate assistance during a period of emergency, i.e. requiring basic survival needs such as food, water, shelter, sanitation and immediate medical assistance" (EM-Dat 2006).

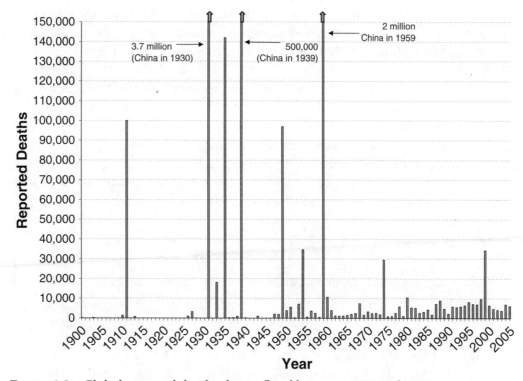

FIGURE 4.6 Global reported deaths due to flood between 1900 and 2005.
Note: Data from EM-DAT 2006.

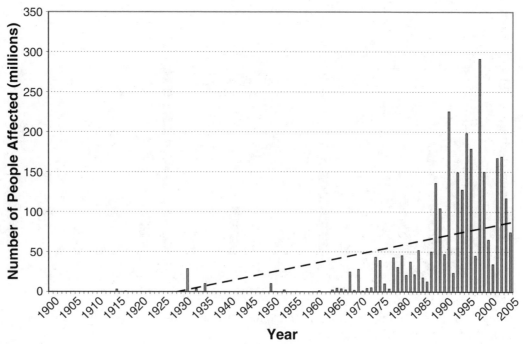

FIGURE 4.7 Global reported number of people affected by floods between 1900 and 2005. Data are less reliable in earlier years.
Note: Data from EM-DAT 2006.

since 1900. A linear trend fitted to the data suggests that this number is growing due to a greater incidence of flooding, as well as greater vulnerability to this hazard, though, as noted earlier for droughts, determining a clear trend is complicated by inadequate data for earlier years.

Figure 4.8 shows reported global flood damage between 1985 and 2005 for extreme flood events.[8] The average annual flood damage was over \$40 billion (in 2000 dollars) but was over \$250 billion in 1998 alone, suggesting that there is a high degree of annual variability. The estimates shown in the figure are due to direct damage, such as property loss, and would be substantially higher if indirect damages were included. In addition, these estimates do not include social and environmental losses, because there is no standard method for capturing these losses.

Like droughts, floods have tremendous environmental, social, and economic implications. Although some effects are temporary, others are permanent. Some effects are described in greater detail:

- *Mortality.* Floods result in a tremendous loss of human life, both directly and indirectly. Large floods can sweep villages into a fast-moving river in a short period of time. Coastal floods can also level entire villages in seconds, as was evidenced by the tsunami that struck Southeast Asia in December 2004, killing at least 300,000 people. Floods can destroy essential infrastructure, such as wastewater treatment and water-supply systems, increasing the risk of contracting water-related and vector-borne diseases (WHO 2005).

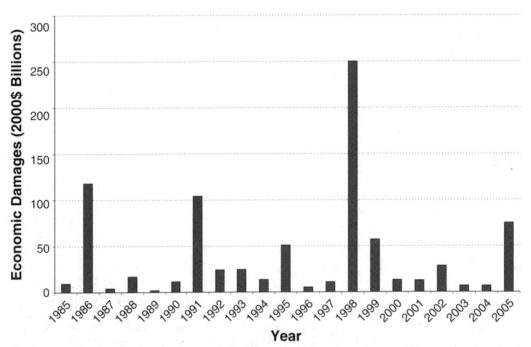

FIGURE 4.8 Reported global flood damage for extreme floods between 1985 and 2005.
Source: Brakenridge et al. 2006.

---

8. An extreme flood event is defined as a flood causing substantial damage to infrastructure or agriculture, fatalities, and/or has a decades-long recurrence interval.

- *Agriculture.* Floods can hurt or benefit agricultural production. Floodwaters deposit nutrient-rich sediment on the floodplains, thereby creating fertile soil. Early and modern civilizations realized the bounty that floods can provide and often settled in floodplains to reap the benefits. The early Egyptians realized the importance of the floods and attributed this annual event to the coming of Hapi. They would throw offerings to Hapi at specific points along the river because they knew that if the floodwaters were not sufficient, famine would follow (Seawright n.d.). These benefits are sometimes offset by the vulnerability of agricultural production to floods that destroy farms and crops.

- *Water supply and quality.* Flooding can cause toxic spills and leaks that contaminate water bodies. Floods can also expose buried contaminates and redistribute them along the river. Storm surges or levee breaks can induce saltwater intrusion in coastal areas, contaminating freshwater ecosystems. In some areas, such as the Sacramento-San Joaquin Delta in California, saltwater intrusion can contaminate the water supply for a large segment of the population. Floods, however, can also recharge groundwater levels.

- *Human health.* In addition to direct death and injury from floods, toxic substances released into the environment from household and industrial chemicals can cause skin irritations, vomiting, and other acute reactions (Olson 2005). In addition, flooding can lead to outbreaks of microorganisms in homes, causing severe respiratory and skin problems.

- *Economy.* Floods cause both direct and indirect economic damage, including tremendous structural damage. After the 2004 Southeast Asia tsunami, for example, many tourists were reluctant to visit the affected areas, causing greater economic damage than reported. Local fisheries and farmers can also incur damage. Local economies in New Orleans and the Gulf Coast were devastated by the flooding caused by Hurricane Katrina (see Box 2).

- *Fish and wildlife.* Debris, high water velocities, and contact with contaminated water can kill fish and wildlife. Floods can both reduce and rejuvenate spawning habitat. Some species require floods for survival. For example, floodwaters carry diadromous fish eggs out to sea, where they hatch, grow, and return to the river as adults (Jowett 1997). Floods can rejuvenate fish spawning beds by depositing large rocks and boulders and removing silts that can suffocate fish eggs. The construction of Glen Canyon Dam on the Colorado River greatly reduced high flows and contributed to severe impacts on many Colorado River fish species. In 1996 and again in 2004, test releases of flood flows were performed in an effort to restore damaged habitat previously maintained by periodic natural flood events.

- *Habitat.* Floods also destroy habitats, particularly streamside vegetation. Saltwater can destroy marshes (Olson 2005). Inundation of forested land can also kill trees. In subsequent years, large fires may result from the dead material. Floods can also scour new pools and deposit debris that provide essential habitat (McSwain et al. 1996).

Box 4.2   Hurricane Katrina

Hurricane Katrina, the eleventh tropical storm of the 2005 Atlantic hurricane season, originated in the Bahamas on August 23, 2005. As the hurricane moved over warm waters in the Gulf of Mexico, it strengthened to a Category 5 storm. On August 29, 2005, the hurricane, which had weakened to a Category 3 storm, made landfall on the Gulf coast of the United States. Katrina caused extensive wind damage along the coast, as well as a storm surge roughly 10 meters high that extended from 200 meters to 1 kilometer into the interior of Louisiana, Alabama, and Mississippi (Kron 2005). The most serious damage occurred when levees separating Lake Pontchartrain and New Orleans were breached by the storm surge, causing extensive flooding throughout New Orleans. Because of a lack of preparation, poor response, and extreme poverty in the region, over 1,400 people lost their lives. Families were separated during the flood, with limited means of reconnecting with loved ones. Damages are estimated to exceed $100 billion, making it the most expensive flood catastrophe in history (Kron 2005).

In addition to the factors described previously, poor land-use decisions and environmental degradation also contributed to the devastation caused by Hurricane Katrina. In the early nineteenth century, levees were built along the Mississippi River as flood control measures, allowing development along the river's edge. Because they were designed to increase river flow and transport sediment into the Gulf, the levees prevented deposition along the river. Soils in the surrounding area began to oxidize and many areas, including New Orleans, began to sink. Because sediment was not being deposited along the coast, coastal erosion rates were high. In addition, the wetlands were destroyed, partly due to the coastal erosion, but also by deliberate removal to promote development. These conditions combined to create a devastating flood with significant long-term impacts.

Flood Management

The histories of flood and drought management are fundamentally different, in part, due to differences in the characteristics of the hazards and their impacts. Drought management has been hampered by unclear and imprecise definitions. Floods, on the other hand, are relatively clearly understood. A flood is a distinct event with a quantifiable beginning and endpoint. It causes structural damage that is visible and is therefore more likely to gain public attention and provoke political response. As a result, flood management has included more proactive mitigation strategies. These mitigation strategies, however, typically relied upon controlling rivers through large-scale, structural measures, such as dams, levees, and diversions.

Levees are among the oldest and most common flood-control methods. Artificial levees are earthen embankments that are oriented parallel to the river and whose primary purpose is to provide flood protection. The ancient Egyptians built the first levees over 3,000 years ago. These levees ran along the left bank of the Nile River and extended nearly 1,000 km (Wikipedia 2006b). Today, extensive levee systems are found on rivers throughout the world, including the Mississippi River and Sacramento River in the United States, the Yellow River in China, and the Po, Rhine, and Danube Rivers in Europe.

Levees and other structural methods, however, have several environmental and social disadvantages. First, structural methods tend to isolate the river from the flood-plain, eliminating the important ecosystem functions and processes that rivers provide (Natural Heritage Institute 2002). They can also increase vulnerability to the hazard by encouraging development in flood-prone areas and give those who live behind the structure a false sense of security. According to the United Nations (2004), "[p]rotective works have a tendency to increase the level of development in floodprone areas, as the assumption is made that it is now safe to build and invest in areas that are protected. However, it must be recognized that at some point in the future the design event will likely be exceeded and catastrophic damages will result" (p. 31). In addition, structural measures require regular maintenance, a task that is often overlooked due to budgetary constraints. Failure to maintain levees can lead to structural failures and catastrophic damage, even during moderate weather conditions.

In addition to structural mitigation measures, traditional flood management, like drought management, has relied on disaster response, such as evacuations and relief aid. As is the case with droughts, this approach encourages unsustainable practices and often exacerbates the problem: "Disaster relief encourages continued occupancy of unsafe locations" (OTA 1980). Thus, although this approach can reduce the short-term *impacts* of floods, it does not reduce the *risk* of floods and may actually increase long-term consequences.

In recent years, flood management has begun to encompass nonstructural mitigation measures to reduce vulnerability. Land-use management is among the most effective mitigation measure available. Land-use management for flood-risk reduction consists of locating appropriate land uses, such as parks, wildlife, and recreation areas, in flood-prone areas. Although parks and recreation areas may sustain damage during a flood, the damage and potential loss of life are small when compared to flooding in an urbanized region. In addition, proper land-use management can increase the benefits of floods; floodwaters, and the sediments they contain, provide an important resource for maintaining agricultural productivity. For example, the Yolo Bypass was established as a flood conveyance channel around communities in the Sacramento River watershed in California. Although the Bypass is an effective flood-control method, it also provides several other benefits, including essential upland and wetland habitat for wildlife, as well as productive agricultural land for various farm uses. Rice, safflower, tomatoes, and corn are among the late spring and summer crops produced (Natural Heritage Institute 2002).

Although not typically considered an element of land-use management, leaving certain elements of the natural environment in place may help mitigate flood risk. Wetlands, for example, can absorb large volumes of water and release the water after the flood peak has receded (U.S. EPA 1995). Wetlands, as well as mangroves, can also absorb some of the energy associated with storms and minimize coastal inundation.

Effective land-use management is often politically charged, particularly in developed or urban areas. When private property is at stake, it may be necessary for a governmental authority to seize or purchase private land through some legal means, such as eminent domain or expropriation. The U.S. Office of Technology Assistance suggested that tension is related to the time lag between costs and benefits as well as the difficulty in quantifying the benefits: "Land use management is the most effective tool for mitigating flood hazards in the long term. Its costs, however, are incurred in the short term and its benefits are deferred and difficult to evaluate. Therefore, it is politically the most difficult measure to implement" (OTA 1980).

The Future of Droughts and Floods

Some statistics indicate that the number of people affected by natural disasters, including floods and droughts, and economic losses associated with these events are on the rise.[9] As described previously, Figures 4.4 and 4.7 show an upward trend in the number of people affected by both floods and droughts. Likewise, a recent report by the WMO (2006) shows an increase in the number of natural disasters, the number of people affected by natural disasters, and the economic losses associated with disasters over the past thirty years, but a decline in the number of deaths. Although death toll reductions may be due, in part, to better early-warning systems and disaster preparedness and improved sanitation in many regions, increases in economic losses and the number of people affected by natural disasters suggest that overall vulnerability and exposure to these events are on the rise (Abramovitz 2001).

As weather- and climate-related events, floods and droughts are subject to a certain degree of natural variability. In addition to natural variability, however, a growing body of evidence indicates that increases in greenhouse gases are causing a larger, systemic change to our climate, with implications for the intensity and frequency of hydrologic extremes. Atmospheric climate models indicate that global warming will induce significant changes to global water resources and coastal ocean conditions during the next century; average surface air temperature is projected to rise 1.4°C to 5.8°C; sea level is expected to rise 10 to 90 cm; and average precipitation is expected to increase due to higher evaporation rates caused by warmer temperatures (IPCC 2001a).

As noted in the U.S. National Water Assessment (Gleick et al. 2000), the majority of studies on climate change have emphasized changes in average conditions and "[w]hile many factors of concern are affected by such average conditions, some of the most important impacts will result, not from changes in averages, but from changes in local extremes." Increasingly, scientists are investigating the impacts of climate change on hydrologic extremes. These studies suggest that climate change will increase the frequency and intensity of extreme events, such as floods and droughts. The IPCC (2001b) concluded that climate change is likely to have already caused more intense precipitation events and greater incidence of inland drying and associated drought risk in a few areas and will very likely cause these changes in most areas during the twenty-first century. In addition, climate change will likely intensify flood and drought risk

---

9. It is important to note there is neither a single agency responsible for collecting disaster data nor a single method for measuring impacts. In addition, historical data on impacts is likely incomplete.

associated with El Niño events in many parts of the world.[10] In some regions, the likelihood of both floods and droughts will increase: Warmer temperatures would cause more precipitation to fall as rain and an earlier snowmelt, thereby increasing the likelihood of floods; warmer temperatures will also increase the evapotranspiration rates, which when combined with a reduced snow pack and less spring runoff, would increase the likelihood of drought.

In addition to changing the frequency and intensity of floods and droughts, human activities are increasing vulnerability and exposure to these events. Population growth and continued development in drought- and flood-prone areas put greater numbers of people at risk. In addition, poverty plays a major role in increasing vulnerability to floods and droughts; it prevents people from evacuating threatened areas or implementing preventative actions or mitigation strategies; the poor are also less likely to have access to adequate water supply and sanitation and are more likely to contract infectious diseases that are commonly associated with floods and droughts. The WMO (2004) maintains that heightened vulnerability will increase the number of people affected by floods and droughts more than natural climate variability or climate change:

> The extent of future changes cannot be predicted with certainty, as these changes may be random (e.g., climate variability), systemic (e.g., climate change), or cyclical (e.g., El Niño). However, hydrological uncertainty is perhaps subordinate to social, economic, and political uncertainties. For example, the biggest and unpredictable changes are expected to result from population growth and economic activity (WMO 2004).

Conclusion

Floods and droughts are natural processes that serve important ecosystem functions. They are perceived as disasters or hazards only when they affect humans and their institutions. Ironically, humans have been attracted to both floodplains and drought-prone areas throughout their history because many of their characteristics make them attractive for development. Thus, human activities have created and perpetuated these hazards.

The number of people affected by floods and droughts and economic losses associated with these events are on the rise, suggesting that overall vulnerability and exposure to these events are also on the rise. Human activities are largely responsible for this trend, though a growing body of evidence indicates that increases in greenhouse gases are causing a larger, systemic change to our climate that will increase the intensity and frequency of hydrological extremes. Further, population growth and development in flood- and drought-prone areas, and poverty, increase vulnerability to these events. The question of which of these activities—climate change or social factors—will increase vulnerability to floods and droughts to a greater degree is debatable. However, human activities are responsible for both factors, and it is critical that we begin to evaluate our actions and make serious effort to correct those that

---

10. Likely refers to an event with a 66-90% change of occurring. Very likely refers to an event with a 90-99% chance of occurring.

increase our vulnerability to floods and droughts, including reducing greenhouse gas emissions, making more informed land-use decisions, and fighting poverty that both puts more people at risk and makes it harder to reduce those risks.

RERERENCES

Abramovitz, J. 2001. *Unnatural disasters.* Worldwatch Paper 158. Washington, DC: Worldwatch Institute.

Ambenje, P. G. 2000. Regional drought monitoring centres—The case of Eastern and Southern Africa. In *Early warning systems for drought preparedness and drought management.* Wilhite D. A. et al. (editors). World Meteorological Organization. Proceedings of an expert group meeting, September 5–7, 2000, Lisbon, Portugal. http://www.drought.unl.edu/monitor/EWS/EWS_WMO.html.

Brakenridge, G. R., Anderson, E., Caquard, S. 2006. Global active record of large flood events. Dartmouth Flood Observatory, Hanover, USA, digital media. http://www.dartmouth.edu/%7Efloods/Archives/index.html.

EM-DAT 2006. The OFDA/CRED International disaster database. Université Catholique de Louvain, Brussels, Belgium. http://www.em-dat.net.

Federal Emergency Management Agency. 1995. National mitigation strategy: Partnerships for building safer communities. Mitigation Directorate, p. 2. Washington, DC: Federal Emergency Management Agency.

Gleick, P. H., and others. 2000. *Water: The potential consequences of climate variability and change for the water resources of the United States.* Report of the Water Sector Assessment Team of the National Assessment of the Potential Consequences of Climate Variability and Change. Oakland, CA: Pacific Institute for Studies in Development, Environment, and Security.

Gleick, P. H., Maurer, E. P. 1990. *Assessing the costs of adapting to sea-level rise: A case study of San Francisco Bay.* Berkeley, CA: Pacific Institute for Studies in Development, Environment, and Security; Stockholm, Sweden: Stockholm Environment Institute.

Gleick, P. H., Nash, L. 1991. *The societal and environmental costs of the continuing California drought.* Berkeley, CA: Pacific Institute for Studies in Development, Environment, and Security.

Hayes, M. n.d. *Drought indices.* National Drought Mitigation Center White Paper. http://drought.unl.edu/whatis/indices.htm.

Hayes, M. J., Svoboda, M. D., Knutson, C. L., Wilhite, D. A. 2004. *Estimating the economic impacts of drought.* Extended abstract. The 84th AMS annual meeting. Seattle, WA. http://ams.confex.com/ams/pdfpapers/73004.pdf.

Integrated Regional Information Networks. 2002. *Land competition in Garissa.* First referred to in Environment and Development Challenges News 2001–02. http://www.edcnews.se/Research/KenyaLand.html.

Integrated Regional Information Networks. 2005. *Eritrea: Drought a major cause of hardship.* UN Office for the Coordination of Human Affairs. http://www.irinnews.org/webspecials/DR/47339.asp.

Intergovernmental Panel on Climate Change (IPCC). 2001a. *Climate change 2001: Synthesis report.* 398 pp. A contribution of working groups I, II, and III to the third assessment report of the Intergovernmental Panel on Climate Change. Watson, R. T., and the Core Writing Team (editors). Cambridge, UK: Cambridge University Press; New York.

Intergovernmental Panel on Climate Change (IPCC). 2001b. *Climate change 2001: Impacts, adaptation, and vulnerability.* Cambridge, UK: Cambridge University Press; New York.

Jacobs, K. L., Garfin, G. M., Morehouse, B. J. 2005. Climate science and drought planning: The Arizona experience. *Journal of the American Water Resources Association,* 41(2):437–445. http://www.ag.arizona.edu/AZWATER/presentations/JAWRApublished.pdf.

Jowett, I. G. 1997. Instream flow methods: A comparison of approaches. *Regulated Rivers: Research and Management,* 13:115–127.

Knutson, C., Hayes, M., Phillips, T. 1998. *How to reduce drought risk.* Preparedness and Mitigation Working Group. Western Drought Coordination Council. http://www.drought.unl.edu/plan/handbook/risk.pdf.

Kogan, F. N. 2000. Contribution of remote sensing to drought early warning. In *Early warning systems for drought preparedness and drought management.* Wilhite D. A. et al. (editors). World Meteorological Organization. Proceedings of an Expert Group Meeting, September 5–7, 2000, Lisbon, Portugal. http://www.drought.unl.edu/monitor/EWS/EWS_WMO.html.

Kron, I. W. 2005. *Storm surges, river floods, flash floods—Losses and prevention strategies.* Munich, March 2005.

McSwain, M., Bickford, D., Capurso, J. 1996. *Flood of '96 will help rejuvenate McKenzie River.* Eugene-Register Guard. http://www.4j.lane.edu/partners/eweb/ttr/mckenzie/resources/flood96.html.

Meteorological Hazards and Seasonal Forecasting Group. 2005. *Global Drought Monitor: Standard Precipitation Index.* London: Benfield Hazard Research Centre at University College. http://drought.mssl.ucl.ac.uk/spi.html.

Mosley, J. 2001. Grazing management during and after extended drought. *Beef Question and Answer Newsletter,* 6(3):1, 6–7. http://animalrangeextension.montana.edu/articles/Beef/q&a2001/beef6-3.pdf.

Najarian, P. A. 2000. An analysis of state drought plans: A model drought plan proposal. Masters thesis. Lincoln: University of Nebraska.

National Aeronautics and Space Administration (NASA). n.d. *Drought.* http://earthobservatory.nasa.gov/Drought/.

National Drought Mitigation Center. n.d. *Understanding your risk: A comparison of droughts, floods, and hurricanes in the United States.* http://drought.unl.edu/risk/us/compare.htm.

National Drought Mitigation Center. 2005a. *What is drought?: Understanding and defining drought.* http://drought.unl.edu/whatis/concept.htm.

National Drought Mitigation Center. 2005b. *Mitigating drought: Mitigation tools for states.* http://www.drought.unl.edu/mitigate/tools.htm.

National Drought Policy Commission. 2000. *Preparing for drought in the 21st century.*

National Oceanographic and Atmospheric Administration (NOAA). n.d. Climate TimeLine. http://www.ngdc.noaa.gov/paleo/ctl/cliihis.html.

National Weather Service (NWS). 2006. *What is meant by the term drought?* http://www.wrh.noaa.gov/fgz/science/drought.php?wfo=fgz.

Natural Heritage Institute. 2002. *Habitat improvement for native fish in the Yolo Bypass.* http://www.n-h-i.org/Projects/RestorationBiodiversity/BayDelta/YoloBypass/Final%20Yolo%20Report.pdf.

Office of Technology Assistance (OTA). 1980. *Issues and options in flood hazards management.* Congress of the United States. http://govinfo.library.unt.edu/ota/Ota%5f5/DATA/1980/8012.PDF.

Olson, E. D. 2005. *The environmental effects of Hurricane Katrina.* Testimony before the Committee on Environment and Public Works of the United States Senate. Natural Resources Defense Council. http://www.nrdc.org/legislation/katrina/leg_05100601A.pdf.

Princeton University. 2005. WordNet 2.1. Princeton, NJ. http://wordnet.princeton.edu/.

Rao, N. n.d. *Drought and agrarian crisis in Andhra Pradesh, India.* African Drought Risk and Development Network Web Portal. http://www.unisdrafrica.org/droughtnet/issIndianagri.htm.

Rathore, M. S. 2005. *State level analysis of drought policies and impacts in Rajasthan, India.* International Water Management Institute. Working Paper 93: Drought Series Paper No. 6. Colombo, Sri Lanka. http://www.iwmi.cgiar.org/pubs/working/WOR93.pdf.

Relief and Rehabilitation Commission. 1985. *Tackling the ravages of drought.* Addis Adaba, Ethiopia.

Seawright, C. n.d. *The inundation.* Tour Egypt feature story. http://www.touregypt.net/featurestories/nile.htm.

Trenberth, K., Overpeck, J., Solomon, S. 2004. Exploring drought and its implications for the future. *EOS,* 85(3):27.

United Nations Environment Programme (UNEP) and the Government of Kenya. 2000. *Devastating drought in Kenya: Environmental impacts and responses.* Nairobi, Kenya.

UNEP. 2002. *Africa environment outlook: Past, present, and future perspectives.* http://www.unep.org/dewa/Africa/publications/AEO-1/index.htm.

United Nations. 2004. *Guidelines for reducing flood losses.* United Nations. http://www.unisdr.org/eng/library/isdr-publication/flood-guidelines/Guidelines-for-reducing-floods-losses.pdf.

United States Environmental Protection Agency (U.S. EPA). 1995. *America's wetlands: Our vital link between land and water.* Office of Water, Office of Wetlands, Oceans and Watersheds.

Wikipedia. 2006a. *2004 Indian Ocean earthquake.* http://en.wikipedia.org/wiki/Asian_tsunami.

Wikipedia. 2006b. *Levees.* http://en.wikipedia.org/wiki/Levee.

Wilhite, D. A. 1993. *Drought mitigation technologies in the United States: With future policy recommendations.* Final report of a cooperative agreement between the Soil Conservation Service, U.S. Department of Agriculture; and the International Drought Information Center, Lincoln: University of Nebraska. IDIC Technical Report Series 93–1, International Drought Information Center. Lincoln: University of Nebraska.

Wilhite, D. A. 1996. A methodology for drought preparedness. *Natural Hazards,* 13:229–252.

Wilhite, D. A. 1997. *Improving drought management in the West: The role of mitigation and preparedness.* National Drought Mitigation Center. Submitted to Western Water Policy Review Advisory Commission.

Wilhite, D. A. 2000. Drought planning and risk assessment: Status and future directions. *Annals of Arid Zone,* 39(3):211–230.

Wilhite, D. A., Hayes, M. J., Knutson, C. L. 2005. Drought preparedness planning: Building institutional capacity. 93–135. In *Drought and water crises: Science, technology, and management issues.* Wilhite, D. A., editor. CRC Press.

World Health Organization (WHO). 2005. *Flooding and communicable diseases fact sheet.* http://www.who.int/hac/techguidance/ems/flood_cds/en/.

World Meteorological Organization (WMO). 2004. *Integrated flood management.* The Associated Programme on Flood Management. APFM Technical Document No. 1.

World Meteorological Organization (WMO). 2006. *Preventing and mitigating natural disasters.* Geneva, Switzerland.

Environmental Justice and Water

Meena Palaniappan, Emily Lee, Andrea Samulon

Introduction

Water makes up 60 to 75 percent of the average person and is required for nearly every biological process that sustains life. Water flows through the Earth's lands in the same way it flows through our veins, bringing life and sustenance. Yet not all people have equal access to water. Like most natural resources, water flows to those who have power. Those with political and economic power are able to direct or stop the flow of water through cities and rural areas, drying up or flooding areas along the way, effectively denying the poor and powerless the health and sustenance that clean, safe, and affordable water provides. In the past decade, these challenges have increasingly been considered and addressed in the field of environmental justice. Despite the fact that almost all countries are naturally endowed with enough water to meet basic water needs, billions of people still lack access to reliable and safe drinking water and sanitation. Absolute water availability is not the problem; globally, only twelve countries have less than 1000 liters of freshwater available per person per day, and even this quantity is more than adequate for fundamental needs (Gleick 1999). The failure to satisfy these needs appears to be one of political will and priorities.

In November 2002, the United Nations Committee on Economic, Social, and Cultural Rights declared access to water a fundamental human right, entitling everyone to affordable, safe, and accessible water supplies for domestic uses. The UN Committee's General Comment on Water states that "Water is fundamental for life and health. The human right to water is indispensable for leading a healthy life in human dignity. It is a pre-requisite to the realization of all other human rights" (UN CESCR 2002).

Despite the clear link between access to safe water and health, food, livelihood, and other basic human needs, the population without sanitation and water coverage remains unacceptably large. Over two and a half billion people worldwide lack access to adequate sanitation, and over a billion do not have access to a safe and affordable supply of water (Data Table 5). Unsafe water continues to be one of the leading causes of disease and death in the developing world. The United Nations and World Health Organization (WHO) estimate that more than two million people in developing countries, most of them children, die every year from water-related diseases (WHO and UNICEF 2000). Mega-water projects flood huge swaths of land, dislocating people from their lands, livelihoods, and cultural homelands. Climate change is changing the very nature of water problems, fundamentally shifting the hydrogeological cycle, and altering where water is located and how and when it falls.

These and other water problems are not borne equally by humans worldwide. The majority of people who lack safe water and sanitation, and whose livelihoods are threatened by polluted water or overextraction, are predominantly poor, people of color (Box 5.1), and indigenous people. This condition is part of a larger pattern of "environmental discrimination."

Box 5.1   People of Color: International Context

The U.S. environmental justice movement developed partly in response to inequitable and disproportionate burdens borne by low-income people of color, including higher levels of workplace hazards, greater exposure to pollution from chemical plants, and closer proximity to municipal landfills, smelters, incinerators, and abandoned toxic waste dumps than their white counterparts.

When discussing environmental justice in an international context, and framing issues of disproportionate burden, or unequal access to natural resources as an environmental injustice, where does the term *people of color* (used widely in the environmental justice field in the United States) fit in? This question becomes particularly important in discussions over environmental justice in the context of societies that are predominantly, or almost exclusively, nonwhite. Does the term *people of color* have significance in this instance? Do those who suffer environmental injustices outside the United States identify in any way with the term *people of color* or correlate their struggle with racial, ethnic, and cultural marginalization?

While we will try to minimize our use of this term in the international context, we note that even within nation-states that could be characterized as predominantly nonwhite, political and economic power, social capital, and access to natural resources are still very much distributed and acquired consistent with racial, ethnic, and cultural categories. In Latin America, for example, both the indigenous population and the Afro-Latino population are considered "nonwhite" and have experienced social, political, and economic exclusion and oppression. In Latin America, the indigenous population has been marginalized, oppressed, and socially excluded since the arrival of the Spaniards on the continent. In addition, Afro-Latinos, who constitute 30 percent of the region's population, also bear a high degree of social exclusion throughout Latin America (Hooker 2005). According to Nobles (2005), "The claim that inequalities in Latin America are borne disproportionately by indigenous groups and blacks has been significantly boosted by census data, when such data are collected and tabulated against other socioeconomic indicators." Nobles also reports that, in Brazil, people who have identified themselves as "black" or "brown" on the census continue to earn less than whites with comparable levels of education.

The concept of "environmental discrimination" is typically applied when the poor and the politically powerless are denied access to the resources that they need to protect their health and their livelihoods or are disproportionately exposed to environmental ills and suffer the associated health and livelihood impacts. Civil society organizations in the United States sometimes refer also to "environmental racism" or "environmental injustice," which is discussed at the international level in the context of inequities between the "North" and the "South" or between developed and developing countries. In the United States, the environmental justice movement arose to address the disproportionate impacts of environmental pollution on communities of color and low-income communities.

Environmental justice in the water context requires that every one has access to affordable and safe water, that people are not dislocated as a result of water projects without consultation and complete, prompt, and appropriate compensation, and that water decisions are made with adequate public participation. Some examples of water issues with environmental justice implications include access, quality, exchanges and sales, privatization, governance, consequences of dams and other infrastructure, and climate change. In this chapter, we consider the distributional inequities of water resources and water projects, as well as the disproportionate impacts of water-related pollution, and poor governance on the world's most marginalized communities. Who has access to water and water services? Who is affected by water pollution and water projects? How can these distributional conflicts be resolved?

A Brief History of Environmental Justice in the United States

The roots of the environmental justice movement in the United States can be traced to community protests against the siting of toxic waste dumps or hazardous and polluting industries in areas inhabited by predominantly low-income communities of color. An early Government Accounting Office (GAO) (1983) report indicated that African Americans comprised the majority population in three-fourths of all communities in the Southeastern United States, where hazardous waste landfills were located. The landmark study, *Toxic Wastes and Race in the United States,* demonstrated that race was the most significant variable in the national distribution of hazardous waste facilities (United Church of Christ Commission for Racial Justice 1987). From this work arose what has become the traditional definition of environmental justice—that low-income communities and communities of color live with a disproportionate share of environmental hazards or pollutants due to their lack of access to decision-making and policy-making processes, and they suffer the resulting health problems and decreased quality of life.

The environmental justice movement in the United States has been shaped and influenced by various social movements and sectors, such as the civil rights movement of the 1950s to 1970s, the diverse ethnic minority liberation movements (e.g., African American, Chicano/Chicana, Asian American, and Native American liberation struggles), the grassroots anti-toxics movement, academic research, and the labor movement. As a result of these diverse influences, the environmental justice movement has developed a much broader effort to achieve racial and social justice in the United States and to achieve access to safe and clean environments for low-income communities of color.

The environmental justice movement differentiates itself from the mainstream environmental movement in several ways. The mainstream environmental movement in the United States is deeply rooted in conservationist ideologies and wilderness preservation. It has been criticized as elitist, with little regard for or attention to the environmental and health concerns of low-income communities of color. The environmental justice movement, however, is concerned with the basic civil rights of marginalized communities and immediate public health dangers resulting from exposure to toxic pollutants. Whereas the mainstream movement uses strategies based on litigation, lobbying, and technical evaluation, the environmental justice movement has added the strategies of community organization, education, and capacity building for community members, with a goal of promoting democratic decision making, community empowerment, and social and racial justice.

As environmental justice activists struggled to define the movement beyond the boundaries drawn by traditional environmentalists, seventeen guiding principles were created to guide the environmental justice movement at the First National People of Color Environmental Leadership Summit in 1991. The principles united communities throughout the United States in forming a cohesive, powerful voice for the environmental justice movement (Box 5.2.)

Box 5.2   The Principles of Environmental Justice

WE THE PEOPLE OF COLOR, gathered together at this multinational People of Color Environmental Leadership Summit, to begin to build a national and international movement of all peoples of color to fight the destruction and taking of our lands and communities, do hereby re-establish our spiritual interdependence to the sacredness of our Mother Earth; to respect and celebrate each of our cultures, languages, and beliefs about the natural world and our roles in healing ourselves; to insure environmental justice; to promote economic alternatives which would contribute to the development of environmentally safe livelihoods; and, to secure our political, economic, and cultural liberation that has been denied for over 500 years of colonization and oppression, resulting in the poisoning of our communities and land and the genocide of our peoples, do affirm and adopt these Principles of Environmental Justice:

1. Environmental justice affirms the sacredness of Mother Earth, ecological unity and the interdependence of all species, and the right to be free from ecological destruction.

2. Environmental justice demands that public policy be based on mutual respect and justice for all peoples, free from any form of discrimination or bias.

Box 5.2

3. Environmental justice mandates the right to ethical, balanced, and responsible uses of land and renewable resources in the interest of a sustainable planet for humans and other living things.

4. Environmental justice calls for universal protection from nuclear testing, extraction, production and disposal of toxic/hazardous wastes and poisons and nuclear testing that threaten the fundamental right to clean air, land, water, and food.

5. Environmental justice affirms the fundamental right to political, economic, cultural, and environmental self-determination of all peoples.

6. Environmental justice demands the cessation of the production of all toxins, hazardous wastes, and radioactive materials, and that all past and current producers be held strictly accountable to the people for detoxification and the containment at the point of production.

7. Environmental justice demands the right to participate as equal partners at every level of decision making including needs assessment, planning, implementation, enforcement, and evaluation.

8. Environmental justice affirms the right of all workers to a safe and healthy work environment, without being forced to choose between an unsafe livelihood and unemployment. It also affirms the right of those who work at home to be free from environmental hazards.

9. Environmental justice protects the right of victims of environmental injustice to receive full compensation and reparations for damages as well as quality health care.

10. Environmental justice considers governmental acts of environmental injustice a violation of international law, the Universal Declaration on Human Rights, and the United Nations Convention on Genocide.

11. Environmental justice must recognize a special legal and natural relationship of Native Peoples to the U.S. government through treaties, agreements, compacts, and covenants affirming sovereignty and self-determination.

continues

Box 5.2 *continued*

12. Environmental justice affirms the need for urban and rural ecological policies to clean up and rebuild our cities and rural areas in balance with nature, honoring the cultural integrity of all our communities, and providing fair access for all to the full range of resources. Environmental justice calls for the strict enforcement of principles of informed consent, and a halt to the testing of experimental reproductive and medical procedures and vaccinations on people of color.

13. Environmental justice opposes the destructive operations of multi-national corporations.

14. Environmental justice opposes military occupation, repression, and exploitation of lands, peoples and cultures, and other life forms.

15. Environmental justice calls for the education of present and future generations which emphasizes social and environmental issues, based on our experience and an appreciation of our diverse cultural perspectives.

16. Environmental justice requires that we, as individuals, make personal and consumer choices to consume as little of Mother Earth's resources and to produce as little waste as possible; and make the conscious decision to challenge and reprioritize our lifestyles to insure the health of the natural world for present and future generations.

Adopted October 27, 1991
The First People of Color Environmental Leadership Summit
Washington Court on Capitol Hill, Washington, DC, October 24–27, 1991.

The U.S. environmental justice movement has identified a broad range of water-related environmental and health issues that affect low-income communities of color. They include concerns over dams and indigenous rights, contaminated drinking water and groundwater, water privatization and rising costs of water service, subsistence fishing and toxins, and lack of access to waterfront space. In California, the Environmental Justice Coalition for Water (EJCW) was created to try to integrate and help coordinate assistance to low-income communities of color who are fighting to protect their water rights. In its report, *Thirsty for Justice*, which documents community-based water struggles and California water policy, EJCW (2005) argues that deep-rooted issues of racism and injustice in state water policy and infrastructure have resulted in stark water inequalities between poor and rich communities, communities of color, and predominantly white communities.

EJCW points out that at least 80,000 of California's 10.4 million households "may have a vulnerable source of water," according to the 1990 U.S. Census Bureau. A California Department of Water Resources analysis of state health data revealed that about 250,000 Californians sometimes go without water due to insufficient supply. The study showed that over 4 million of the state's 35.5 million residents may be drinking unfiltered surface or well water that is potentially contaminated with fecal matter or *Escherichia coli* (also known as *E. coli*), and almost a million residents contend with sewage contamination of their water supply. The report documents that many of these Californians "reside in rural, economically disadvantaged communities" and in metropolitan areas with overcrowded housing conditions.

In addition to health impacts and economic costs, the EJCW finds that poor people, people of color, and indigenous people also suffer from cultural erosion. The Winnemem Wintu, a band of the Native American Wintu tribe, and their ancestral lands along the McCloud River near Mount Shasta, are threatened by the United States Bureau of Reclamation's plans to raise the Shasta Dam, which impounds the largest reservoir in California. In 1945, the dam's gates were closed and tribal lands, homes, sacred sites, and burial grounds were flooded. The loss of Winnemem lands threatened cultural practices, left many homeless, and resulted in further impoverishment of the tribe. The current plan to again raise the dam would result in flooding more Winnemem burial grounds and sacred sites, such as Puberty Rock, where young women participate in a "coming of age" ceremony. Yet tribal members have been excluded from decisions over this project.

In many ways, environmental injustices that face indigenous, poor populations around the world are similar to those in the United States. As we shall see, the impacts of large dams and hydroelectric projects in the global South disproportionately affect indigenous and tribal peoples and ethnic minorities. In the United States, a quarter of the North Dakota reservation of the Three Affiliated Tribes and almost all of their productive land was flooded by Garrison Dam in 1953, displacing 80 percent of the reservation's population (Parker 1989). As the following section demonstrates, the U.S. environmental justice movement is deeply connected globally to the struggles of marginalized communities and the poor, and, ultimately, the exchange of lessons learned, strategies, and principles among these global "peoples movements" will be critical to the success of economic and social justice objectives.

Environmental Justice in the International Water Context

In their book *Varieties of Environmentalism*, Ramachandra Guha and Martinez-Alier place the U.S. environmental justice movement within a broader global movement that they call the "environmentalism of the poor." The environmentalism of the poor is a struggle for livelihoods, health, and survival itself. It is a response to "ecological distribution conflicts": disputes that stem from "social, spatial and temporal asymmetries or inequalities in the use by humans of environmental resources and services, i.e., in the depletion of natural resources (including the loss of biodiversity) and in the burdens of pollution" (Guha and Martinez-Alier 1997, p. 31).

Examples include conflicts over the dumping of toxic wastes in poor countries, mining that destroys livelihoods and pollutes the environment, inadequate health and safety protections for workers, and the appropriation of genetic resources without proper compensation or recognition. Specific cases include struggles against foreign oil companies by the indigenous population of Amazonia in Ecuador and the Ogoni people of Nigeria; efforts by the black, poor population in Esmeraldas to stop the shrimp farming industry from destroying mangroves; and the Chipko movement in India against deforestation of community forests.

Worldwide, environmental justice requires equal sharing of resources and the burdens of pollution. In the United States, environmental injustice affects the poor and people of color. In a global context, environmental justice issues expand to include ecological distribution conflicts between the global North and South, as well as those between the "haves" and the "have-nots." Within countries, resources are distributed unequally between the rich and the poor, those with political power and those without, men and women, and among different ethnic groups and indigenous people.

Internationally, numerous struggles over water have environmental justice implications. Broad trends that influence the availability and affordability of water affect marginalized communities most seriously. In the following section we explore a range of international water issues with environmental justice implications, including access, quality, privatization, dams and other infrastructure, and climate change. We then consider the fundamental elements of equitable water policy.

Environmental Justice and International Water Issues

Water Access

One of the world's most serious injustices is the failure to meet basic water needs for all. The WHO and UNICEF (2000; see also Data Table 5) estimated that in 2000 and again in 2002, one-sixth of the global population, or 1.1 billion people, did not have access to any form of improved water supply within 1 km of their residences. Most of these people live in Asia and Africa, with two-fifths of Africans lacking access to improved water. More than 2.6 billion people lack access to improved sanitation, including more than half of all Asians. Rural areas have far less service than urban areas. Eighty percent of the people without safe sanitation live in rural areas, including about 1.3 billion in China and India alone.

Numerous people have attempted to estimate the amount of water needed to sustain life. Peter Gleick estimated in 1996 that about 25 liters per day (L/d) are needed for basic human consumption, cooking, bathing, and washing, and an additional 25 L/d are needed for sanitation. This means that about 50 liters per capita daily (Lpcd) are required to maintain minimum domestic needs for human health and well-being, excluding water for food production or energy. At that time, hundreds of millions of the poorest people on Earth lacked even this minimum (Gleick 1996).

Increasing and competing demands for water exist. Household demand for water is growing rapidly among wealthy consumers in urban areas. The increased use of water in household appliances and garden irrigation contributes to greater stresses in regions with limited water supplies. Industrial water use is also expanding significantly, further aggravating shortages. In China, rapid industrialization is projected to require a fivefold increase in industrial water use by 2030 (Brown and Halweil 1998).

One-third of the world's population, living primarily in rural areas, relies on groundwater supplies for drinking water and irrigation. Overextraction of groundwater beyond natural recharge rates is widespread in many parts of the Arabian Peninsula, China, India, Mexico, Russia, and the United States. Groundwater, which was formerly a "free resource," is increasingly used unsustainably, and the costs of groundwater use are rising as both quality and groundwater levels decline (Moench 2004).

Who Is Affected?

Lack of access to water is a factor in ongoing poverty for numerous reasons. First, the poor are often forced to purchase water from more expensive sources, including water tankers. Cost disparities are dramatic, not only when measured as a proportion of income, but in absolute terms as well. The Asian Development Bank reports that the urban poor in Asia may pay twenty to forty times more per liter than those connected to municipal water systems. In Manila, connected users pay only $5 per month, whereas the poor who are not connected to the system pay $15 a month (ADB 2006). The poor are also forced to spend their time and energy collecting water from distant locations, reducing their time and ability to earn additional income. Furthermore, the economic costs of poor health and disease due to lack of water are borne by individuals as well as governments. The United Nations includes access to water services as a key component in their Human Poverty Index (UNDP 2004). Women, in particular, are burdened by lack of access to water, as described in Box 5.3.

Quickly diminishing groundwater supply contributes to greater social inequity in many parts of the world. For example, in the state of Gujarat, India, overextraction of groundwater has caused the water level in aquifers to fall by as much as 40 meters (UNEP 1996; Moench 2004). This has led to a lack of freshwater supply for many poor farmers, because they cannot afford to deepen their wells. Larger and wealthier farmers, on the other hand, can purchase land with better water supply. The use of groundwater in bottling operations has also been the subject of controversy. Box 5.4 describes the role of the Coca-Cola company in India.

That billions of people remain without access to water, despite nearly three decades of international efforts and publicity about the issue, demonstrates not only the intractability of the problem, but also a lack of funding priority and failure to focus on the right solutions.[1]

Seventeen of twenty-two industrial nations have not yet met the very modest target set by the United Nations of spending 0.7 percent of a nation's Gross National Income on overseas development assistance (Shah 2006). Although the health and human impacts of the water crisis are well known, only a small portion of all international aid from industrialized nations is targeted to water and sanitation projects (Gleick 2004).

Even the limited funding provided for water issues—mostly in the form of capital and resource-intensive technologies from the global North to the global South—has not resulted in adequate progress in addressing water access problems. Much of the funding from multilateral aid and development agencies has traditionally gone to

---

1. We note that we are rapidly approaching the thirtieth anniversary of the 1977 Mar del Plata water conference where the lack of access to basic water resources was first brought to the world's attention.

Box 5.3   Women and Water

Women play a critical role in the provision of water and are also most affected by disparities in water access and availability of adequate sanitation. In most places, women are chiefly responsible for collecting and transporting water for all household uses and are the primary users and managers of in-home water. These facts have increasingly been recognized and acknowledged by the water community. For example, in 1992, the International Conference on Water and the Environment in Dublin, Ireland, released what have become known as the "Dublin Principles." Principle 3 states: "Women play a central part in the provision, management and safeguarding of water." At the 2000 Water conference at The Hague, the first challenge of the Hague Declaration refers specifically to the need to "empower people, especially women, through a participatory process of water management."

The FAO (2005) estimates that in some parts of Africa, women and children spend eight hours a day collecting water. Because of the amount of time women spend in collecting and transporting water, water availability closer to the home can significantly reduce their workloads, allowing them to engage in economic activities, and for girls and young women, allowing them to go to school.

Safe and private methods of sanitation are also important for women. For reasons of safety and modesty, women often do not use the same open locations as men to perform their daily activities. The availability of clean, private, and accessible locations to defecate is significant. As the primary caretakers of children, women are most affected by waterborne diseases caused by unsafe sanitation. As the primary educators of children, they can most effectively teach safe hygiene to the next generation.

Despite their central role, women are the least involved in decisions about water in their community. In many cases, women know best where water is located, how to access it, and its quality and reliability. Yet women's traditional economic and political roles often prevent them from acting as stakeholders in water resource decisions and benefiting from these decisions.

Irrigation decisions, failing to account for the imbalance in ownership rights between men and women, and labor and income inequities, often benefit men to the detriment of women (FAO 2005). As described later in the chapter, decisions around dams can also fail to account for women's lost income and property, further damaging women's livelihoods and futures. Despite the deep knowledge women have and the key role they play in this sector, they continue to be marginalized. Involving women in decisions around water is vital to ensuring environmental justice, and far more progress needs to be made in protecting women's rights to water and ensuring their direct and active participation in water policy and decision making.

Box 5.4 The Coca-Cola Company in India

When community members in Plachimada, Kerala, began to experience severe water shortages around Coca-Cola's bottling plant, they decided to organize a grassroots campaign to hold the company accountable. In December 2003, the Kerala High Court determined that Coca-Cola was illegally and indiscriminately using water resources around the Plachimada plant for its beverage production and was ordered to immediately find alternative sources. The plant in Plachimada has remained closed since March 2004. In a case that is being watched internationally, the state government of Kerala has now joined the community in challenging Coca-Cola's right to extract water. The Supreme Court of India is expected to hear the case shortly. In another community, a similar challenge to Coca-Cola is emerging. In Mehdiganj, the village council has revoked Coca-Cola's license to operate because of water scarcity and pollution, and community leaders are initiating a legal challenge to The Coca-Cola Company.

At the heart of the issue is access to, and local control over, resources. Although beverage bottling plants are often located in rural and low-income areas, consumers of commercial beverages in India are typically the urban middle- and upper-classes. Many of the community residents who bear a disproportionate burden cannot even afford to buy a can of Coca-Cola. One of the central demands of the community-based movement against Coca-Cola in India is to assert control over local natural resources—in this case water—by communities. Challenging Coca-Cola's right to exploit local groundwater resources is central to the effort by affected communities to protect their livelihoods and their health.

support centralized, large-scale water systems that often depend upon exogenous expertise and resources at the expense of traditional and small-scale systems (Data Tables 6–8). There is strong evidence that smaller-scale projects that involve the community and work to conserve resources while providing basic water and sanitation services can be more effective (Gleick 2000).

Water Quality

Poor water quality poses a major threat to human health. Water can be degraded by various contaminants, in various ways, many of them caused by human activity: poor sanitation and lack of wastewater treatment, industrial pollutants, agricultural runoff, naturally occurring contaminants, and saltwater intrusion.

One of the main sources of surface water pollution in developing countries is direct sewage discharge and ineffective sewage treatment. As noted earlier, more than 2.6 billion people lack access to improved sanitation facilities, and entire watersheds and rivers are contaminated by untreated human waste.

Nearly 70 percent of the available water resources in India are polluted, and less than 7 percent of India's 3,000 cities have any kind of sewage treatment facilities (Sampat 1996). Of the 300 largest cites in India, 30 percent have little or no sewerage system or sewage treatment. And of the total wastewater generated in metropolitan areas, the great majority, about 70 percent, goes untreated into water systems (Planning Commission 2002).

In Chennai, India, researchers estimate that the sewerage system serves 31 percent of the metropolitan population, and raw sewage flows freely into the metropolitan area's natural watercourses at many points (Cairncross et al. 1990). This is common in many cities in developing countries, where urban waterways are little more than open sewers, serving as vectors for disease. The amount of fecal coliforms in some of Asia's rivers is fifty times the WHO guidelines (UNEP 2000). Sewage also contaminates groundwater in many areas, affecting the quality of water procured from wells and boreholes.

Pesticide and fertilizer runoff from agriculture also contributes to freshwater pollution. Nitrate pollution from fertilizer runoff is now considered one of the most serious water quality problems. Nitrates can lead to brain damage and even death in infants (OECD 1994).

Industrial wastes are another growing and important source of water pollution. The chemicals released by industrial processes are often not treatable in conventional wastewater treatment plants, leading to heavy metal and persistent organics contamination (including lead, mercury, arsenic, and cadmium) in water bodies and groundwater. Chlorinated solvents were found in 30 percent of groundwater supplies in fifteen Japanese cities, sometimes traveling as far as 10 kilometers (km) from the source of pollution (UNEP 1996).

Additionally, groundwater overextraction leads to saltwater intrusion in coastal urban groundwater and agricultural land. In Chennai, India, overextraction of groundwater has resulted in saline groundwater nearly 10 km inland of the sea (UNEP 1996).

Who Is Affected?

Worldwide, infectious diseases, such as waterborne diseases, are the number one killer of children under five years old, and more people die annually from unsafe water than from all forms of violence, including war. According to the United Nations, 80 percent of diseases in developing countries result from unsafe water. Unsafe or inadequate water, sanitation, and hygiene cause approximately 3.1 percent of all deaths worldwide or 1.7 million, and 3.7 percent of DALYs (disability adjusted life years) worldwide (WHO 2002).

Unsafe water causes about four billion cases of diarrhea and approximately two million deaths annually, mostly of children under five years of age. Fifteen percent of child deaths each year are attributable to diarrhea, and this represents a child dying every fifteen seconds (WHO and UNICEF 2000). In India alone, the single largest cause of ill health and death among children is diarrhea, which kills nearly half a million children each year.

Those who are primarily affected by poor water quality are the people who live near unsafe water courses streaming with untreated human waste and those who do not have the resources to pay for safe water. Slums often develop along the banks of waterways in urban areas, because this land is usually public. Lacking other options, slum dwellers use the waterways to directly discharge sewage, sullage, and other garbage. Slum dwellers and those without adequate access to water are forced to use polluted surface water for washing, bathing, or drinking and are thus at high risk for waterborne diseases.

Sanitary Infrastructure

While the developed world further refines its advanced methods of wastewater treatment, inhabitants of developing countries continue to die of diseases that could easily be prevented with existing technologies. Why are sophisticated sanitation systems a fixture in many parts of the world and absent in so many others?

Part of the disparity can be traced to the colonial period and policies and priorities during this time. While there are now vast differences between sanitary conditions in the global North and South, 150 years ago most urban citizens—rich *and* poor—lived amidst excrement and sewage. Nineteenth century London was described as a "cesspool city" by the London journalist George Goodwin in 1854:

> The entire excrementation of the Metropolis . . . shall sooner or later be mingled in the stream of the river, there to be rolled backward and forward around the population (Goodwin, cited in Wilson 1991, p. 37).

At the time, conditions were not much different in Madras, India. Major Cunningham, the Director of the King Institute at Guindy, Madras, wrote:

> The large mass of people, living under the most primitive and unsanitary surroundings afforded an almost unbounded field for the spread of every kind of epidemic disease (Balfour and Scott 1924, p. 126).

The sanitation systems of the colonial capitals and the colonies began to diverge in the mid 1800s, when the project of Public Health arrived in Great Britain. Sanitation was improved both in England and across the Atlantic in North America.

Similar measures, only fitfully enacted and rarely enforced, were less successful in India. As time went on, the difference between Britain and India grew more palpable. By the 1890s, Britain, apart from a few exceptional areas like Glasgow, was decisively set apart from its Indian dependency in health and sanitation. Calcutta, one might say, became filthy only as London became clean (Metcalf 1995, p. 173).

Who Is Affected?

It would be simplistic to argue that colonialism is entirely responsible for the current condition of sanitation in formerly colonial cities. Nevertheless, it is true that the sanitary histories of London, Paris, and Madras diverged in the latter half of the nineteenth century when the colonial powers set different standards for sanitation at home than in the colonial port cities. The underdevelopment of sanitation infrastructure in these early days undeniably influenced the current state of public health in many post-colonial cities.

Even when efforts were made to expand sanitation in colonial cities, that effort was limited to places where colonial rulers and the upper class lived and focused on protecting the colonial armies from disease (one of the primary causes of death among British soldiers in India). Colonial cities were often spatially bifurcated, with the area for the colonists separated from areas for the indigenous population (Prashad 1994) via buffer zones of railway lines, industrial/commercial premises, or other landscaped features (Lowder 1986). The term *cordon sanitaire* in French means literally "quarantine line" and was implemented in many colonial cities. These separate cities within cities developed separate systems of sanitation, one for the colonists and another one for the native population.

Colonial governments also prioritized the infrastructure needed to move natural resources from the colonies' interior to the ports over municipal infrastructure and sanitation systems. Improvements in public works infrastructure are some of the most expensive undertakings of a government. Documenting the financing of waterworks in Imperial Rome, nineteenth-century England, and twentieth-century America, Gunnerson (1991) posits:

> The implication that large scale sewerage systems can only be financed with exogenous wealth is inescapable (p. 1).

The Romans used revenues from North Africa, and the British financed sewers with revenues from the East India Company. In discussing the financing of sanitary improvements, it is impossible to separate the financial resources available to colonial city municipal governments from the larger global economy and the nature of imperial extraction of wealth from colonies occurring in the nineteenth century.

> The British in Bengal administered an economy that systematically exported wealth to England, a flow of tribute aptly titled, 'the great drain,' amounting to tens of millions of pounds by the late eighteenth century (Home 1997, p. 63).

The uneven distribution of sanitation that resulted is an injustice noted by many colonial historians:

> It is difficult to forgive the deliberate rejection of recommendations made by knowledgeable medical authorities of the Crown, which are now available for all to see in the official colonial records. Further, natural justice is offended by policies which charged or taxed residents for services which were only supplied to a few, or which reached only designated areas from which the indigenes were excluded (Lowder 1986, p. 94).

The results of this historical discrimination are most apparent when one compares the current situation of sanitation in cities throughout the world. Post-colonial governments throughout the Global South inherited networks of underdeveloped urban infrastructure that were materially consistent with the colonial project. Struggling with an inconsistent foundation, these post-colonial governments also failed to comprehensively meet the sanitary needs of their populations. In Madras, India, less than one third of the metropolitan population has adequate sanitation (Cairncross et al. 1990), while Chicago runs the largest water and sewage treatment plants in the world, managing the waste water for the population equivalent to 10.1 million people. While Londoners celebrate the rebirth of the Thames River, many developing country waterways are little more than open sewers.

The current value of existing sewerage assets in the United Kingdom alone is some $200 billion, and it is several times larger than that in the United States. The history of underinvestment that began in the latter half of the nineteenth century led to drastically lower levels of investment in the cities of developing nations. Camdessus (2003) estimated that sanitation wastewater treatment annual investment needs over the next two decades could be as high as $87 billion per year. The lack of exogenous wealth with which to fund this enormous need continues to plague the cities and rural areas of the Global South.

Privatization

Privatization is a growing and controversial trend in the water and wastewater service sector. We define privatization as "the transfer of the production, distribution, or management of water or water services from public entities into private hands" (Gleick et al. 2002, p. 21). Privatization has been proposed as the solution to many of the woes facing water utilities, including inadequate service coverage, corruption, inefficiency, and large projected capital needs. In practice, the extent to which privatization will improve water management is not yet clear. However, private companies and investors are not the panacea that some advocates of privatization claimed they would be, just five years ago. Prematurely terminated contracts in Manila, Philippines, Atlanta, Georgia, and elsewhere, after only a few years of operation under long-term concessions, demonstrate how hard it is to forge successful public-private partnerships, even in a regulated market economy such as the United States.

Public water companies still provide most water and wastewater services worldwide, nearly 95 percent by some estimates. But the number of people served by private companies has grown from 51 million people in 1990 to nearly 300 million in 2002. Six water companies alone expanded operations from twelve countries in 1990 to over fifty-six countries by 2002 (CPI 2003). Pinset Masons, a global water sector consulting firm, estimates that 560 million people are currently served by some aspect of the private sector for their water and wastewater (Pinset Masons 2005), though sometimes this includes modest private consulting, billing, or maintenance functions rather than comprehensive private services or system ownership.

Who Is Affected?

How does this new privatization trend affect the poor and politically powerless? Private involvement in water supply has a long history. In some places, including the United States, private ownership and provision of water was initially very common. In the latter half of the nineteenth century, private water systems in the United States began to be municipalized because private operators were not equitably providing access and service to all citizens or making necessary infrastructure investments. In the Southern United States, at the beginning of the twentieth century, typhoid rates among African Americans, which were twice as high as for Caucasians, dropped significantly after water systems became public (Troesken 2001). On the other hand, recent water privatizations may have improved public health in some places. Galiani et al. (2002) report that infant mortality declined 5 to 7 percent in parts of Argentina, where water services were privatized.

In a WaterAid and Tear Fund report, the authors seek to investigate the role of privatization in providing services to the poor (Gutierrez et al. 2003). The authors attribute the failure to deliver affordable and sustainable water and sanitation to the poor from either public or private systems to capacity issues, including lack of funding, lack of management ability, corruption, poor financial management, lack of independent regulation, and lack of civil society participation. Through analysis of rural and urban case studies from around the world, they find that the private sector participation does not comprehensively address the underlying issues that result in poor water and waste-water services for the poor.

If private-sector water-service provision provides improved efficiency, and the underlying problem of political corruption is not addressed, then privatization will not lead to greater service for those who are currently underserved. Political corruption will continue to force water-service providers, public or private, to line the pockets of the politically powerful at the expense of the health of communities. One solution may be to specifically require that priority be placed on serving the poor in privatization contracts; otherwise, there is a risk that only those who are already connected to the water-supply system will benefit from improved service.

The authors identify four concerns regarding the role of water privatization and its effects on the poor. First, privatization can contribute to the irreversible loss of public sector skills and capacity for water and sanitation provision. Because privatization contracts can be as long as twenty years, privatization may undermine the ability of governments to take back water and sanitation provision if privatization contracts fail (as they have in numerous cases). Weak governments are also unable to effectively regulate private sector operators who are providing what is ultimately a public responsibility.

Second, privatization can also reduce transparency and community involvement in water and sanitation provision. Third, the drive to "cost recovery" often forces the poor to pay far more than they can afford for water. Cost recovery is often an underlying driver of the push to private sector management, and, although it is essential for the sustainable operation of a water utility, it can have adverse impacts on equity. The economists determine that if there is a willingness to pay, cost recovery must be pursued.

> The problem here is that 'willingness' is often measured in ways that do not capture the complexities of poverty. For example, the urban poor in Accra pay as much as 5 times more than other users per liter to fetch water from distant sources. Paying 5 times more should not be seen as an indicator of willingness, but rather, that there is no alternative. The poor's willingness to pay should be measured against the proportion of the income spent on water and not against prices paid (Gutierrez et al. 2003, p. 20).

The authors provide the example of an Accra family of six who pays 22.4 percent of their household income for water expenses versus a London family of four who pays 0.22 percent of their income for water. These disparities often force the poor to compromise on other essentials like health, food, and education to pay for water. Last, the authors argue that institutional reform is needed to combat corruption and inefficiency.

The authors argue that a *context-sensitive approach* is needed to address the problem of water and sanitation and the role of privatization. Privatization may work in some cases and not in others, but the focus needs to be on improving coverage and service to the poor and involving the public in decision making about the best strategies to meet their water and wastewater needs. In the report "The New Economy of Water" (Gleick et al. 2002), the Pacific Institute identified a core set of principles that are critical to the success of private sector participation in the water sector.[2] These principles include meeting basic human needs for water as the top priority, subsidiz-

---

2. This subject was also addressed in Chapter 3 of *The World's Water 2002-2003* (Washington, D.C.: Island Press).

ing water rates when necessary for reasons of poverty, ensuring that governments retain control over the water resource itself, and ensuring that negotiations over privatization contracts be open, transparent, and include all affected stakeholders. When these principles are integrated into planning for privatization, better contracts that serve the poor can result.

Dams

Large dam projects have received considerable attention worldwide due to massive protests from environmental and grassroots community groups over their impacts on local populations, the environment, and cultural and social resources. The three main functions of dams are to protect against the impacts of floods and droughts, to produce hydroelectricity, and to provide water supply. Other benefits can include providing recreation, supporting fisheries, and improving navigation. From an environmental justice perspective, many of the benefits of large dams never reach the local, often poor and rural, communities most affected by their impacts.

Although dams have existed since 3000 B.C., large dams are a fairly modern development. During the late 1800s and into the twentieth century, U.S. water and land policies and legislation encouraged the construction of larger dams, not only to supply electricity, but also to conquer the wilderness of the West, open lands to irrigation, and symbolize technological advances (McCully 1996). By the end of the twentieth century, development was slowing in the United States because of economic, environmental, and social concerns, while China and India's governments seemed to embrace the idea that bigger dams were critical to faster development for their country. Both the Three Gorges Dam in China's Yangtze River Basin and the Sardar Sarovar Dam on India's Narmada River are examples of gargantuan projects with serious consequences for local populations. Yet both China and India have pushed forward with such projects, despite national and international pressure and loss of support from various international financial institutions.

However, dam critics are increasingly gaining in the argument against large dams, by proposing alternative solutions to meet the needs that are supposedly satisfied by dams. Small-scale water schemes, using traditional and new techniques, can provide water quickly and cheaply for drought-prone areas. Increasing water efficiency and the use of reclaimed or recycled water can reduce the need for new water supply. Renewable energy, such as solar and wind power, is another option to generate electricity without fossil fuels. The most compelling reason to wean ourselves off large dams is the human destruction that dams cause during their entire process, from conception to completion.

Who Is Affected?

How do dams affect the poor and those without power? In countries throughout the world, dams have resulted in population displacements, unjust or uncompensated resettlement, desecration of sacred lands, and loss of access to resources for food, water, fish, game, grazing land, timber, and fuel wood. According to the World Commission on Dams report in 2000, "Little or no meaningful participation of affected people in the planning and implementation of dam projects—including resettlement and rehabilitation—has taken place."

Although the exact number of people worldwide that have been displaced by dams has never been accurately reported, indigenous and tribal peoples and other marginalized ethnic minorities make up a disproportionately large percentage of those who are displaced and negatively impacted by dams. As Wong (1999) ironically notes, "Areas with people who are well off and well connected do not make good reservoir sites."

The Chinese government estimates that 10.2 million people have been displaced by dams in China between 1950 and 1989 (World Bank 1993). Dam critics argue that the true number of displaced people in China is between 40 and 60 million, and the recent Three Gorges Dam may, by itself, have displaced as many as 1.9 million people. Worldwide estimates of the total number of people physically displaced due to dams range from forty to over eighty million (McCully 1996; WCD 2000).

According to estimates by the government of India (1985), 40 percent of all those displaced by dams in India are *adivasis,* the indigenous people, who represent less than 6 percent of the country's population. The majority of the large dam schemes built and proposed in the Philippines are in the territory of the country's six to seven million indigenous people (WCD 2000). In the case of Vietnam's largest dam, Hoa Binh, the majority of the 58,000 people evicted were ethnic minorities, as are most of the 112,000 to be displaced by the even bigger Ta Bu Dam planned further upstream (Hirsch 1992). The Waimiri-Atroari tribe of Northern Brazil numbered 6,000 in 1905. Eighty years later, massacres and disease left only 374 Waimiri-Atroari alive. In 1987, the Balbina Dam flooded two of their villages, displacing 107 of the remaining members of the tribe (WCD 2000). Similar experiences of displacement of indigenous, tribal, and ethnic minorities are recorded in Indonesia, Malaysia, Thailand, Brazil, Argentina, Mexico, Panama, Colombia, Guatemala, United States, Canada, and Siberia (WCD 2000). Table 5.1, taken from the World Commission on Dams report, describes a few of the many examples of groups negatively impacted by large dams (see also Gleick 1998; Data Table 15, p. 281).

Women are also disproportionately affected by the negative impacts of dams, which can further exacerbate existing gender inequities or create new inequities, while receiving few of the potential benefits from dam projects. The Kariba Dam built by the British colonial authorities displaced the egalitarian Gwembe Tonga community in Northern Rhodesia (now Zambia) and completely disregarded women's recognized right to own land. As a result, women lost their land and received no compensation during displacement and resettlement (Colson, cited in Mehta and Srinivasan 1999). Loss of access to traditional natural resources has also disproportionately affected women. For example, at Thailand's Pak Mun Dam the loss of local edible plants due to submergence resulted in loss of income and a source of subsistence for local women, who are responsible for collecting and processing these plants (WCD 2000).

In addition to bearing a burden of the costs of dams, women are less likely to receive the potential benefits, such as jobs resulting from the construction of the dam, which are primarily filled by men. The allocation of the irrigated land made available by dams is also often done in a manner that exacerbates gender inequalities. In the Mahaweli irrigation scheme in Sri Lanka, 86 percent of the land allocations were made to men, and only two local female-headed households were granted land. The state's land allocation policies recognized only one family heir to be given land (usually a son), which overrode the prevalent inheritance rule that allowed women to own and control land (Agarwal, cited in Mehta and Srinivasan, 1999).

TABLE 5.1   Some Examples of People Affected by Large Dams
Profile of Groups Adversely Affected by Large Dams: Illustrations from WCD Case Studies

Dam Project	Displaced People	Profile of Displaced People	Others Adversely Affected
Glomma and Laagen 1945–75	None displaced		None documented in case study
Grand Coulee 1934–75	5,000 to 6,500 people	1,300 to 2,000 from Colville and Spokane tribes not compensated until 1990s; rest are settlers.	Several thousand members of First Nation groups (Colville, Spokane, Nez Perce, and Canadian tribes) located throughout the upstream basin, due to flooding of fishing falls and blockage of salmon
Kariba 1955–59	57,000 people	Tonga subsistence farmers; ethnic minority in Zimbabwe; most were resettled in resource-depleted areas	Thousands of downstream people lost floodplain livelihoods, and lakeside inhabitants experienced increased prevalence of schistosomiasis
Tarbela 1968–76	96,000 people	Composed of 93% farmers; 5% artisans or semiskilled workers; and 2% boatmen	Pastoralists, landless people, low cast groups of fisher folk, boatmen, basket makers, and weavers suffered from loss of wetlands, forests, and grazing
Gariep and Vanderkloof 1963 to late 1970s	1,380 people; families of 40 white farmers and 180 black farm workers	Most white farmers felt they were fairly compensated. Black farm-workers were not eligible for compensation and eventually moved to other farms and urban areas	Farmers living along the river, where 1 million sheep suffered from proliferation of biting black-fly, leading to livestock losses
Aslantas 1975–85	1,000 families	Mostly former immigrants from Eastern European countries; composed of small and medium-size landowners and landless families	None documented in case study, although it is claimed that customary users of forests that were not recognized were the most adversely affected
Tucurui 1975–85	25,000–35,000 people plus indigenous peoples	Subsistence farmers, fisher folk, pastoralists, and riverbank cultivators (126 of 2,247 disputes are not yet settled)	100,000 people (subsistence farmers, pastoralists, fisher folk, and other natural-resource-dependent communities) affected by reduced water quality, loss of riverbed cultivation, and decreased downstream fish populations
Pak Mun 1991–94	1,700 families	Rural families dependent on rice farming and fisheries income; cash compensation failed to provide livelihood regeneration	More than 6,000 families of subsistence farmers and fisher folk suffered loss of livelihood from fisheries reduction

135

There are exceptions to these impacts, if care is taken in design and implementation of projects. For example, benefits from dam projects may serve to reduce gender disparities in some cases when social services are provided as part of transparent and comprehensive resettlement programs. For example, 80,000 resettled people from the Akosombo Dam in Southeastern Ghana benefited from services, such as schools, markets, public latrines, boreholes, wells, and water stand pipes (Tamakloe, cited in WCD 2000). But policies requiring equitable and balanced programs need to be guaranteed as part of any dam proposal in order to ensure environmental justice.

Climate Change

Climate change is affecting water resources in diverse and complex ways, as previously noted in earlier volumes of *The World's Water* (e.g., the 1998–1999 report). The Intergovernmental Panel on Climate Change's *Synthesis Report* (2001) documents that climate change will exacerbate water shortages in many water-scarce areas of the world by substantially reducing available fresh water, changing the timing and frequency of floods and droughts, greatly reducing snowfall and snow pack in mountainous regions, and more. Water quality is also expected to decrease in many regions as water temperatures increase and streamflows change.

Who Is Affected?

Nations in the Global South are increasingly faced with the reality that global climate change is affecting them immediately and disproportionately. Although the United States and other industrialized nations in the North contribute to the majority of global warming emissions, island nations and countries of the global South will be forced to deal with the ensuing consequences by either spending limited resources to adapt to the worst impacts, or, more likely, suffering the human, environmental, health, and financial consequences of unavoidable impacts. In his book, *The Heat Is On*, Gelbspan (1997) identifies one bloc of nations who are taking climate change seriously—those most vulnerable to its effects, including island nations and poorer nations without resources for climate adaptation investments. In contrast, he describes developed nations as having drowned their concerns by "international jockeying, diplomatic posturing, conflicting domestic political agendas, and intense obstructionism by the oil-producing nations and their industrial allies."

One immediate concern is the impact of climate change on the water resources in vulnerable small island nations, such as the Pacific Island Developing Countries (PIDCs). These countries are responsible for only 0.03 percent of the world's carbon dioxide emissions yet are expected to experience some of the earliest and most severe impacts of climate change (Burns 2002). The independent island nation of Tuvalu is an example of a particularly vulnerable country. The nine, small coral atolls that make up Tuvalu lie, on average, only 2 meters above sea level, and the government is worried about more frequent tropical cyclones, more severe droughts, accelerating coastal erosion, frequent saltwater intrusion into soils and groundwater, and increasing sea-level rise. In 2004, the island nation began a "migration plan" for residents to resettle in New Zealand due to rising waters and floods.

Many other small island nations throughout the world, such as Maldives, Samoa, and the Marshall Islands, share Tuvalu's plight. However, due to their small population size and weak economic power, their troubles seem to be of limited concern to countries so far unwilling to participate in international agreements to reduce green-house gas emissions. Now that many countries have signed the Kyoto Protocol into action, special attention must be given to the issue of winners and losers, and, ulti-mately, equity.

Recommendations

The growing complexity of water issues requires more explicit attention to equity if we are to ensure that water policies are socially just. National governments, in partnership with civil society and the business community, must recognize and implement a human right to water; ensure meaningful public participation in water decisions and transparency in water governance; address the consequences of climate change; improve water quality; and reduce waterborne disease and mortality, to name a handful of the most pressing challenges. A truly equitable system would fairly distrib-ute both water resources and the costs and benefits of decisions about the design and operation of water infrastructure.

While realizing a just system of water management will require substantial reforms, here we focus on two fundamental elements of equitable water resource planning: implementation of a human right to water and good governance.

Recognizing and Implementing a Human Right to Water

Water has both social and economic values, but a growing number of laws, statements, and agreements set the public good above private financial gain. In November 2002, the UN Committee on Economic, Social and Cultural Rights issued General Comment 15, which states that access to water is a human right and that water is a public good fundamental to life and health. The UN Committee's declaration is an interpretation of the provisions of the Covenant on Economic, Social and Cultural Rights (ratified by over 140 member countries) and a range of other international treaties, agreements, and declarations on other acknowledged rights and on water. The Committee affirmed that water, like health, is essential to achieving other human rights, including the right to adequate nutrition, housing, and education. It also acknowledged the role of economics but described it as a secondary one. "Water should be treated as a social and cultural good, and not primarily as an economic commodity" (UN CESCR 2002, para. 11).

The 1992 Earth Summit in Rio de Janeiro in Agenda 21 also describes these two roles:

> In developing and using water resources, priority has to be given to the satisfaction of basic needs and the safeguarding of ecosystems. Beyond these requirements, however, water users should be charged appropri-ately (United Nations 1992).

These two widely acknowledged statements plainly assert that water's economic value is secondary to its social and ecological value (Varghese 2003).

UN General Comment 15 is the clearest interpretation of the human right to water in customary international law. It notes that a basic minimum amount of water must be provided for all people. It also calls for this basic water requirement to be affordable and available to all citizens, which may require subsidies or even a decision to offer free water, as has been explored in South Africa. The countries that have ratified the International Covenant now have a "constant and continuing duty under the Covenant to move as expeditiously and effectively as possible towards the full realization of the right to water" (UN CESCR 2002, para. 18).

The ratifying countries of the International Covenant are also required to provide individuals with the necessary institutional, economic, and social capacity to achieve this right. This means that states must prioritize the human right to water over other allocations. The costs of meeting basic human needs for water is far lower than the economic and social costs of failing to meet these needs. Pearce and Warford (1993) estimate that water-related disease alone cost about $125 billion dollars per year (in 1970 dollars)—a number that would be far larger in today's dollars. Moreover, this estimate only includes medical costs and lost work time, and not other social costs of disablement, missed school days, or social disruption. Yet the investment required to meet these needs has been estimated by some to be about $25 to $50 billion per year (Jolly 1998, cited in Gleick 2000; Rogers et al. 2000). This estimate considers the costs of new infrastructure needed for all major urban areas. Because this does not include the rural needs, this is a low estimate but strongly suggests that the benefits of providing a human right to water clearly outweigh the costs.

Implementing a human right to water will require that the international community and governments spend more than they currently do on water issues. The average rate of spending on water projects in the 1980s and 1990s, and in more recent years, has been a fraction of the money needed (United Nations 1997). According to Gleick (2000), even these investments were not focused on the right priorities:

> The WSSCC estimates that 80 percent of water investments in the 1980s represented expenditures to meet the needs of a relatively small number of affluent urban dwellers. Studies on investment alternatives reveal that 80 percent of the unserved can be reached for only 30 percent of the costs of providing the highest level of service to all. The WSSCC, for example, estimates that 35,000 rural people could be provided with basic sanitation services for the same cost of providing 1,000 urban residents with a centralized sewerage system (pp. 13–14).

Meeting the human right to water will require a change in international and national priorities. A shift from large, centralized systems (which, so far, have not solved the problem of safe water for all) to community-driven, locally appropriate solutions is required. Wolff and Gleick (2002) have described this paradigm shift as part of the move to a Soft Path for water, which includes a focus on community-scale infrastructure designed, built, and maintained by the people who use it. International funding efforts need to focus on these solutions and support national governments in implementing locally appropriate community-designed solutions.

Good Governance in the Water Sector

Many argue that the world water crisis is a crisis of governance (GWP 2000). Poor governance in the water sector has led to constrained livelihoods, increased social inequality, and inadequate and unsafe water and sanitation for billions of people. Characteristics of poor governance include corruption, political interference, weak management, inadequate legal structures, lack of transparency in water decisions and contracts, and the low priority placed by governments on water financing, regulation, and oversight.

The failure to meet basic human needs for water is, in good measure, a failure of governments at many levels. Yet there is growing recognition that "governance" is more than simply the formal processes of governments. Political, economic, social, and environmental issues are becoming more entwined and a broader range of actors are getting involved in water management. Thus, governance entails bringing together public institutions, private actors, NGOs, and the public to pursue social goals related to water.

As traditionally defined, governance involves the "traditions, institutions and processes that determine how power is exercised, how citizens are given a voice, and how decisions are made on issues of public concern."[3] It also involves political accountability, an effective judicial system, public access to information, freedom of expression, and the capacity of all people to affect decisions. Good governance is characterized by transparency and accountability, stakeholder participation, public participation, and a proper definition of the role of government agencies (Manzungu and Mabiza 2004). Good water governance can also be a tool of democracy, by institutionalizing the democratic right of every citizen to participate in and influence the water policy and management decisions that will affect him or her (Mostert 2003).

Good governance in water issues has been widely advocated by governments, civil society, and international bodies. The second principle of the 1992 Dublin Statement maintains that "water development and management should be based on a participatory approach, involving users, planners and policy-makers at all levels" (ACC/ISGWR 1992). The 2000 Hague Declaration lists several challenges for achieving water security, including, "Govern water wisely: to ensure good governance, so that the involvement of the public and the interests of all stakeholders are included in the management of water resources." The first challenge of the Hague Declaration refers specifically to the need to "empower people, especially women, through a participatory process of water management." Yet, despite a number of official recognitions, decrees, and statements, tangible evidence of good water governance is still rare, and there is no consensus of what constitutes good water governance in the practical sense (Mostert 2003). More progress is being made in this area, and organizations like the Third World Centre for Water Management and the United Nations have begun to systematize the idea of good governance through case examples and histories. It is clear that eradicating corruption and political interference, and ensuring the participation of all stakeholders, will be critical to the successful governance of water.

Democratizing Decision-Making

A key aspect of good governance is the issue of democratizing decision making or involving all affected stakeholders in decisions in the water sector. Development and environmental professionals have learned over the last few decades that the problems of water supply and sanitation cannot be solved by technology alone. In the past, an engineering-driven approach to water focused almost exclusively on centralized infrastructure and decision making. Communities and users were regularly left out of decisions that determined how their drinking water and wastewater needs were met.

---

3. From the Public Health Agency of Canada's website http://www.phac-aspc.gc.ca/vs-sb/voluntarysector/glossary.html

As a result, technologically brilliant solutions to the water problem often failed in their implementation because either the community was not using the technology for its intended purpose or the system fell into disrepair and lacked regular maintenance. Examples ranged from toilets built in a rural Indian village that a year later were covered in dirt as residents continued to perform their morning ablutions in the fields, to mega-wastewater treatment plants that did not have a reliable energy source to operate or local expertise to maintain. These well-intentioned and even well-designed projects failed because they did not involve the receiving community in the decision-making process.

Ensuring access to water and sanitation throughout the world requires that the voices of affected communities be part of decisions about water management. Social, economic, and political factors have turned out to be just as important as technological ones and must be considered at the beginning of any potential project. By devolving decision making to a local level, people can become directly engaged in the decisions that affect their lives, and ensure that potential water projects are culturally and economically appropriate, sustainable, and address their needs.

A major challenge is identifying and integrating stakeholders into water management. But who are the "stakeholders"? Stakeholders should include all those who are affected by a particular decision, whether positively or negatively. Often, disenfranchised stakeholders are excluded from the process and marginalized, which results in further powerlessness. For example, stakeholder participation in water management has been adopted as a policy in Zimbabwe's catchment and subcatchment councils, which prepare outline plans, determine permits, regulate the exercise of water rights, and perform actual day-to-day water management functions. The expectation was that improved governance through stakeholder participation would reduce the racial disparities found in Zimbabwe's water access.

In practice, however, urban residents in Harare are often represented by city employees, rather than by local groups. Disparities also exist in representation of the majority black population in decisions about commercial farming.

> A white commercial farming sector representing less than 1% of the population used 85% of the developed water resources to the disadvantage of the majority black population. This situation was blamed for food and water insecurity among the black population. Low crop production in the black-tilled [land] was blamed on poor access to water (Manzungu and Mabiza 2004).

When we consider the principle of empowering people, as The Hague Declaration challenges us to do, the principles of environmental justice would argue for direct democracy and direct participation from affected community members. As described in Box 5.3, women in particular play a significant role in how water is collected, managed, and used (Singh et al. 2003). In considering the needs and interests of women as water managers, it is necessary to consider women's sociocultural position within society and not within a vacuum (Upadhyay 2003). Many traditional assumptions about the role of men in using and allocating labor and household resources lead to water policies and programs detrimental to women's water rights and their sustainable management and use of water. Good water governance will help protect women's water rights by also addressing imbalances between men's and women's ownership rights, division of labor, and incomes.

At the local level, until we can ensure that affected stakeholders—particularly those most disenfranchised in society, including indigenous peoples, ethnic minorities, poor people, and women—are given the opportunity to engage in meaningful participation, good water governance will remain empty words on international declarations. And until governments as a whole commit to broader reforms in water planning, management, and development, efforts to meet basic human needs for water will fail.

The four key elements for effective community water-management efforts are

1. *Involve communities from project inception.* Water projects should involve community organizations, local leaders, and residents as early as possible to ensure that decisions about the scope of the project, the technology used, and work plan are decided with full community input. The planning phase needs to be community-driven and led to set the groundwork for a successful implementation (Brooks 2002).

2. *Involve communities in decision making.* Community residents and organizations need to make key decisions about the design, management, and ongoing maintenance of a system based on their needs and cultural, social, demographic, or economic factors. Key to a community-driven effort is that communities feel ownership over the infrastructure and are invested in its ongoing maintenance and upkeep (Lammerink et al. n.d.).

3. *Involve all stakeholders in the process.* A process that involves the community must be truly representative of all stakeholders, including those typically left out of decisions. Often women and children, who are key providers of water, sanitation, and health needs of their families, are left out of community-driven processes because they are not recognized as important stakeholders (McIvor 2000). Evidence shows that the most successful water projects are those that include affected stakeholders, even if they are not traditional or elected leaders.

4. *Partner with governments and NGOs.* A community-driven or controlled process should not be seen as an excuse for the government to give up providing water and wastewater services. Although a community-driven process is critical in ensuring the success of a water project, the government needs to provide financial and political support to ensure that the project is sustained over time (Lammerink et al. n.d.).

Ultimately, what is most exciting about community-driven water management is not only that the resulting projects are usually more successful and sustained over time, but also that community efforts in one sector lead to a flowering of civic engagement that, in turn, benefits other important community needs, such as education, health, and economic development.

Conclusion

The water sector is poised for significant changes in the coming years. New ideas about who owns and controls water resources, growing water scarcity and stress, climate change, and other critical issues are transforming the sector in important ways. At the same time, lack of access to water and sanitation, the health effects of waterborne diseases, and impacts of new water infrastructure disproportionately affect the poor, women, and people of color. Achieving more equitable distributions of the benefits and costs of water and water services is critical to the success of efforts to reform the water sector. As the water sector responds to these challenges, environmental justice principles must inform the development of new decision making and governance structures that involve all affected stakeholders in meaningful ways. Equitable access to water and water services is a cornerstone in the global effort to eradicate poverty and create a more just society.

REFERENCES

ACC/ISGWR. 1992. *The Dublin statement and report of the conference.* International Conference on Water and Environment: Development issues for the 21st century. 26–31. January 1992, Dublin, Ireland. Geneva: United Nations Administrative Committee on Co-ordination Inter-Secretariat Group for Water Resources.

ADB. 2006. Water brief: Should Asia's urban poor pay for water? http://www.adb.org/Water/Water-Briefs/should-poor-pay.asp.

Balfour, A., Scott, H. H. 1924. *Health problems of the empire: Past, present and future.* London: W. Collins Sons & Co. Ltd.

Brooks, D. B. 2002. *Water: Local level management.* Canada: International Development Research Center.

Brown, L., Halweil, B. 1998. *China's water shortage.* Worldwatch press release. April 22. Washington, DC: Worldwatch Institute.

Burns, W. C. G. 2002. The potential impacts of climate change on the freshwater resources of Pacific Island developing countries. 113–131. In *The world's water 2000–2001: The biennial report on freshwater resources.* Gleick, P. H. (editor). Washington, DC: Island Press.

Cairncross, S., Hardoy, J. E., Satterthwaite, D. 1990. *The poor die young: Housing and health in third world cities.* London: Earthscan Publications.

Camdessus report. 2003. *Financing water for all: Report of the world panel on financing water infrastructure.* World Water Council, Global Water Partnership. (This report, written by J. Winpenny, is often called the *Camdessus report* after its chair, Michael Camdessus.) March 2004. Marseilles, France.

Center for Public Integrity (CPI). 2003. *The water barons.* The Center for Public Integrity, International Consortium of Investigative Journalists. http://www.icij.org/dtaweb/water/.

Environmental Justice Coalition for Water. 2005. *Thirsty for justice: A people's blueprint for California water.* Oakland, CA: Environmental Justice Coalition for Water.

Food and Agriculture Organization of the United Nations (FAO). 2005. Women and water resources. In *FAO focus: Women and food security.* http://www.fao.org/FOCUS/E/Women/Water-e.htm.

Galiani, S., Gertler, P., Schargrodsky, E. 2002. *Water for life: The impact of the privatization of water services on child mortality.* Center for Research on Economic Development and Policy Reform, Working Paper No. 154. August.

Gelbspan, R. 1997. *The heat is on.* Cambridge, MA: Perseus Books.

General Accounting Office (GAO). 1983. *Siting of hazardous waste landfills and their correlation with racial and economic status of surrounding communities.* Washington, DC: Government Printing Office.

Gleick, P. H. 1996. Basic water requirements for human activities: Meeting basic needs. *Water International,* 21(2):83–92.

Gleick, P. H. 1998. *The world's water 1998–1999: The biennial report on freshwater resources.* Washington, DC: Island Press. (Table 15, p. 281, Populations displaced as a consequence of dam construction, 1930 to 1996.)

Gleick, P. H. 1999. The human right to water. *Water Policy,* 1(5):487–503.

Gleick, P. H. 2000. The human right to water. 1–18. In *The world's water 2000–2001: The biennial report on freshwater resources.* Gleick, P. H. (editor). Washington, DC: Island Press.

Gleick, P. H. (editor). 2004. *The world's water 2004–2005: The biennial report on freshwater resources.* Washington, DC: Island Press.

Gleick, P. H., Wolff, G., Chalecki, E. L., Reyes, R. 2002. *The new economy of water*. Oakland, CA: Pacific Institute for Studies in Development, Environment, and Security. ISBN No. 1-893790-07-X. http://www.pacinst.org/reports/new_economy.htm.

Global Water Partnership (GWP). 2000. *Towards water security: A framework for action*. Stockholm and London: Global Water Partnership.

Government of India. 1985. *Report of the Committee on Rehabilitation of Displaced Tribals due to development projects*. Ministry of Home Affairs. New Delhi.

Guha, R., Martinez-Alier, J. 1997. *Varieties of environmentalism: Essays North and South*. London: Earthscan Publications Ltd.

Gunnerson, C. G. 1991. Costs of water supply and wastewater disposal: Forging the missing link. In *Water and the city*. Chicago: Public Works Historical Society.

Gutierrez, E., Calaguas, B., Green, J., Roaf, V. 2003. *New rules, new roles: Does PSP benefit the poor? The synthesis report*. London: WaterAid and Tearfund.

Hirsch, P. 1992. Social and environmental implications of resource development in Vietnam: The case of Hoa Binh reservoir. RIAP Occasional Paper, University of Sydney.

Hooker, J. 2005. Indigenous inclusion/black exclusion: Race, ethnicity and multicultural citizenship in Latin America. *Journal of Latin American Studies*, 37:285–310.

Home, R. 1997. *Of planting and planning: The making of British colonial cities*. London: Alden Press.

Lammerink, M., Bolt, E., Jong, D., Schouten, T. n.d. Strengthening community water management. International Water and Sanitation Centre. http://www.irc.nl/products/planotes35/pnts5.htm.

Lowder, S. 1986. *The geography of world cities*. New Jersey: Barnes & Noble.

Manzungu, E., Mabiza, C. 2004. Status of water governance in urban areas in Zimbabwe: Some preliminary observations from the city of Harare. *Physics and Chemistry of the Earth*, 29:1167–1172.

McCully, P. 1996. *Silenced rivers: The ecology and politics of large dams*. London and New Jersey: Zed Books.

McIvor, C. 2000. Community participation in water management experiences from Zimbabwe. *Development and Cooperation*, 1:22–24.

Mehta, L., Srinivasan, B. 1999. *Balancing pains and gains: A perspective paper on gender and large dams*. Report for the World Commission on Dams thematic review: Social impacts. http://www.dams.org.

Metcalf, T. R. 1995. *Ideologies of the Raj*. Cambridge: University Press.

Moench, M. 2004. Groundwater. 79–100. In *The world's water 2004–2005: The biennial report on freshwater resources*. Gleick, P. H. (editor). Washington, DC: Island Press.

Mostert, E. 2003. The challenge of public participation. *Water Policy*, 5(2):179–197.

Nobles, M. 2005. The myth of Latin American multiracialism. *Daedalus*, Winter.

OECD. 1994. *Towards sustainable agricultural production_Cleaner technologies*. Paris, France: OECD.

Parker, L. S. 1989. *Native American estate: The struggle over Indian and Hawaiian land*. Honolulu, HI: University of Hawaii Press.

Pearce, D. W., Warford, J. J. 1993. *World without end: Economics, environment, and sustainable development*. New York: Oxford University Press.

Pinset Masons. 2005. *Pinset Masons water yearbook 2005–2006*. London, England.

Planning Commission, Government of India. 2002. India assessment 2002: Water supply and sanitation. http://planningcommission.nic.in/reports/genrep/wtrsani.pdf.

Prashad, V. 1994. Native dirt/imperial ordure: The cholera of 1832 and the morbid resolutions of modernity. *Journal of Historical Sociology*, 7(3):243–260.

Rogers, P., H. Bouhia and J. Kalbermatten. 2000. "Water for big cities: Big problems, easy solutions?" In C. Rosan, Blair A. Ruble, J S. Tulchin, eds., *Urbanization, Population, Environment, and Security: A Report of the Comparative Urban Studies Project*, Woodrow Wilson International Center for Scholars, Washington, DC, 2000.

Sampat, P. A. 1996. River's long decline. *World Watch*, July/August, 25–32.

Shah, A. 2006. The U.S. and foreign aid assistance. http://www.globalissues.org/TradeRelated/Debt/USAid.asp?so=p2002#Almostallrichnationsfailthisobligation.

Singh, N., Bhattacharya, P., Jacks, G., Gustafsson, J. E. 2003. Women and water: A policy assessment. *Water Policy*, 5:289–304.

Troesken, W. 2001. Race, disease, and the provision of water in American cities, 1889–1921. *Journal of Economic History*, 61(3):750–777.

United Church of Christ Commission for Racial Justice. 1987. *Toxic wastes and race in the United States*. New York: United Church of Christ Commission for Racial Justice.

United Nations Development Programme (UNDP). 2004. Human development report. http://hdr.undp.org/statistics/data/.

United Nations Economic and Social Council, Committee on Economic, Social, and Cultural Rights (UN CESCR). 2002. General Comment No. 15. *Substantive issues arising in the implementation of the International Covenant on Economic, Social and Cultural Rights. The Right to Water*. E/C.12/2002/11 Twenty-ninth session, Geneva. November 26.

United Nations Environment Program (UNEP). 1996. *Groundwater: A threatened resource*. UNEP Environment Library No. 15, UNEP. Nairobi, Kenya.

UNEP. 2000. Global environmental outlook 2000. UNEP. http://www.unep.org/geo/geo2000/.

United Nations. 1997. *Comprehensive assessment of the freshwater resources of the world*. Commission on Sustainable Development. United Nations, New York. Printed by the World Meteorological Organization for the Stockholm Environment Institute.

United Nations. 1992. Protection of the quality and supply of freshwater resources: Application of integrated approaches to the development, management and use of water resources. *Agenda 21*, Chapter 18, United Nations Publications, New York.

Upadhyay, B. 2003. Water, poverty and gender: Review of evidences from Nepal, India and South Africa. *Water Policy*, 5:503–511.

Varghese, S. 2003. Transnational led privatization and the new regime for the global governance of water. *WaterNepal*, 9/10(1/2).

World Bank. 1993. China: Involuntary resettlement. 72. Washington, DC. June 8.

World Health Organization. 2002. World health report: Reducing risks, promoting healthy life. France. http://www.who.int/whr/2002/en/whr02_en.pdf.

WHO and UNICEF. 2000. *Global water supply and sanitation assessment 2000 report*. WHO and UNICEF Joint Monitoring Programme for Water Supply and Sanitation (JMP).

Wilson, E. 1991. *The sphinx in the city: Urban life, the control of disorder, and women*. Berkeley, Los Angeles, Oxford: University of California Press.

Wolff, G., Gleick, P. H. 2002. The soft path for water. In *The world's water 2002–2003: The biennial report on freshwater resources*. Gleick, P. H. (editor). Washington, DC: Island Press.

Wong, S. 1999. The impacts of large dams and the international anti-dam movement. Environmental NGOs' International Symposium on Dams. International Rivers Network. Berkeley, CA.

World Commission on Dams 2000. Dams and development: A new framework for decision-making. http://www.dams.org.

CHAPTER 6

Water Risks that Face Business and Industry

Peter H. Gleick, Jason Morrison

Introduction

During most of the twentieth century, commercial or industrial access to water was taken for granted, even in regions with scarce supplies. Industry has always been able to pay premium prices for water, compared to agricultural users, and water is often a relatively minor cost of production. As a result, the focus of most industrial managers and planners has been on ensuring access to labor, energy, and capital, rather than to water resources. This is changing.

In the coming decades, the availability of clean, adequate, and consistently reliable water may become crucial to economic development for some sectors of local and national economies. Corporations and institutional investors face new risks and challenges related to water, including changing allotments, more stringent water-quality regulations, growing community activism and control over local resources, and increased public scrutiny of water-related private sector activities. These factors are already affecting factory site selection, license to operate, productivity, costs, revenues, and, ultimately, profits and corporate viability.

In recent years, multinational corporations have had to close major factories or change operations because of water problems. At the same time, poor, high-risk investments have been pursued, and key investment opportunities have been overlooked. Opportunities for businesses and investors include access to multibillion dollar markets for water supply and treatment equipment, improved public perception and goodwill, and reduction in supply-chain risks for forward-thinking companies. This chapter summarizes key trends in water use and availability for businesses, describes water-related risks, and identifies strategies that the financial community can take to assess investment opportunities and dangers.[1]

Understanding these issues is vital if investors, companies, and policy makers are to successfully address the growing business, social, and environmental risks posed by water problems. Neglecting water factors is not prudent. Indeed, evaluating risks

---

1. More detail on these issues can be found in two reports prepared by the Pacific Institute (Morrison and Gleick 2004 and Gleick et al. 2005). The authors acknowledge the contribution of Jim Newcomb and Todd Harrington, our co-authors on the second of these reports.

145

related to water is vital when water plays an important role in production and operations, or in the extended supply chain feeding industrial activities. Understanding opportunities may offer unusual or unexpected rates of return. It is increasingly critical that investors, managers, and directors work toward a better understanding of the business sectors with the greatest exposure to water-related problems and of approaches for reducing their exposure. Ultimately, organizations that fail to think strategically about water will find themselves embroiled in highly public and emotionally charged disputes over a resource considered by many to be a basic human right.

Water Risks for Business

Water supply and quality problems are likely to pose increasingly significant and direct threats to businesses in the decades ahead. This section describes selected areas of water risk that businesses face. These areas are likely to become increasingly important, especially in regions where water supplies are under the greatest stress.

Decreasing Water Availability and Reliability of Supply

Water shortages are increasing as demand exceeds the available supply in some regions due to either natural events or other factors, such as population growth, increased industrial activity, or expanding irrigated acreage. As a result of these shortages, some regions are limiting water availability for new industry, and existing industries are facing growing public opposition to their activities from communities dependent on the same limited resources. In Beijing, authorities have announced plans to severely limit development of new water-intensive businesses in the region, with explicit constraints on the location of new textile, leather, metal smelting, and chemical industries. Chinese makers of beverages, plastics, and pharmaceuticals may have to meet water conservation restrictions to gain approval (*U.S. Water News* 2004). Box 6.1 summarizes some of the significant industrial water challenges in China.

In Kerala, India, both Pepsi and The Coca-Cola Company lost their licenses to use groundwater at bottling plants, after a drought increased competition for local aquifers. Coca Cola announced that it might permanently shut down its Kerala facility—its largest bottling operation in India—after continuing difficulties regaining a groundwater-pumping permit from local authorities (*The Hindu* 2004). Similarly, the city of Bangalore, India, is losing information technology firms because of worries about water scarcity and reliability (*Economic Times* 2004a). Textile plants in India have been forced to shut down due to water shortages and conflicts with local farmers over water allocations (*Economic Times* 2004b). Box 6.2 summarizes some of the significant industrial water challenges in India.

Declining Water Quality

Although some industries face shortages and constraints due to competition for limited supplies, others are affected by declining water quality. Poor water quality can threaten industrial production in two ways: some industries require high-quality water for production, and contamination of local sources may force industries to invest in costly,

Box 6.1 China's Industrial Water Challenges

China is widely regarded as one of the most highly challenging areas for companies that either require substantial amounts of water or provide water technologies and services. With relatively limited water resources, severe distribution problems, extensive contamination, and soaring demand from industrial users and rapidly growing cities, China's water infrastructure needs are vast. Although China has 22 percent of the world's population, it has only 7 percent of the world's fresh water resources, and the water is unevenly distributed around the country. The rapid growth of China's cities is putting serious strains on the ability of water and waste-water infrastructures to keep up. As one example of the water problems now emerging, heavy reliance on groundwater to meet soaring water demand in and around Beijing has led to water tables in the area falling at a rate of nearly 2 meters a year. This, in turn, is already constraining corporate development in the Beijing region. Business opportunities in China's water services sector have shifted significantly because the Chinese government has opened the door to build-operate-transfer (BOT) and transfer-operate-transfer (TOT) deals with private sector companies.

Another major driving force in China, along with economic growth and population shifts to the cities, is new environmental regulation. China's water pollution control laws are forcing cities to build sewage-treatment plants and requiring industrial companies to restrict water use and comply with new water-pollution laws. In January 2005, China's Ministry of Water Resources released a report stating that more than 53 percent of the water in major river systems is undrinkable, with half the water in fifty-two lakes surveyed and 35 percent of groundwater deemed too polluted to drink (Red Herring 2005). According to China's Ministry of Construction, China will need to double its urban sewage disposal capacity in the next five years, requiring investment of RMB 300 billion (China Knowledge Press 2005). The official governmental goal is for 90 percent of all Chinese cities to have sewage treatment plants capable of processing 60 percent of all waste by 2010—a huge leap from current capacity.

front-end water treatment systems or even to move operations; and in some regions, government efforts to impose water-quality regulations to stop uncontrolled or unregulated industrial water discharges can lead to dramatic changes in industrial activities.

Specific industries, such as semiconductor manufacturing, biotechnology, pharmaceutical, and food processing firms, often require high-quality water for their processes and products. Such industries can be particularly vulnerable to degradation or contamination of source water, which can necessitate costly capital expenditures for

Box 6.2   India's Industrial Water Challenges

Like China, India has an enormous population, a rapidly growing industrial sector, and serious water challenges. Although the World Bank recently pledged to increase its water-sector loans to India from US$200 to $900 million per year over the next four years, India will be under severe pressure to revamp its water policies, improve water management, and renew ageing infrastructure (Kuber 2005; Parsai 2005). Water-sector opportunities in India are most likely to fall in areas of metering, desalination, improving irrigation efficiency, and sewage treatment.

Nearly 80 percent of India's water resources are used in irrigation, largely from groundwater aquifers. Farmers and entrepreneurs have invested an estimated US$12 billion in groundwater pumps and nearly one in four of India's 100 million farmers own a pump set and well. This investment has, however, greatly accelerated the unsustainable pumping of groundwater in some parts of India, suggesting a serious risk of growing costs to farmers from energy needs to pump from greater depths, as well as major losses in agricultural production if wells go out of service. According to a recent World Bank study, India produces 15 percent of its food and meets 80 percent of its household needs by "mining" rapidly depleting groundwater resources. The Bank estimates that by 2050 water demand in India will exceed all available supplies (Mukerjee 2005). Upgrades to improve the energy and water efficiency of this widely dispersed rural infrastructure will come slowly, constrained by limited capital available to farmers and inappropriate government policies that subsidize electricity used for groundwater pumping.

Finally, India's government is under pressure to find ways to fund sewage-treatment projects for increasing volumes of sewage generated by rapidly growing cities. The country has the capacity to treat only about 10 percent of the 30 billion liters of sewage generated every day by urban areas. Similarly, only a few of India's major cities, including Bangalore and Mumbai, have achieved comprehensive water metering coverage. As a result, water usage in many areas is inefficient and is rationed physically rather than by price. Changes in India's water economy could bring stronger pressures to accelerate the installation of water-metering systems in other major cities.

treatment technology. The risks and costs associated with these kinds of water-quality problems, at present, are poorly assessed and managed by businesses.

   Although most industrialized countries have managed to curtail concentrated "point source" pollution emitted from factories and sewage-treatment plants, an estimated 90 percent of wastewater in developing countries is still discharged directly to rivers and streams without any waste processing or treatment (WHO/SEI 1997). As economic development continues in these countries and per capita income rises, companies will likely have to absorb the costs associated with meeting new water-treatment requirements.

As countries try to improve water quality, some governments are restricting the type, size, or location of specific industrial investments that contaminate water. The government of Victoria, Australia, is considering plant closures to help eliminate the discharge of untreated industrial wastewater in the pulp and paper industry (Koutsoukis 2004). Five of the seven major river systems in China are considered to be severely polluted, with the Hai He River near the Beijing and Tianjin industrial centers in the worst condition. The severity of this pollution is already affecting local industrial development plans (UPI 2005).

Poor water quality can also affect national strategies for economic development. Some national governments already impose strict water-quality standards for water supply or wastewater discharge; some impose both. Other governments have yet to develop, impose, and enforce water-quality standards. Growing concerns about water quality may lead to the imposition of new and costly requirements on company's wastewater discharges. China's new 5-year plan, announced in October 2005, calls for a departure from the "old, growth-at-any-cost model" that has led to many rivers being polluted (*China Securities Journal* 2005). These regulatory requirements often lead to significant changes in the cost of water for industry, or in the required investment for water-quality treatment systems.

Supply-Chain Vulnerability

Understanding the full extent of water-related risks to a company involves assessing factors outside of the company's immediate operations. For industries as diverse as apparel, forest products, and beverages, water is required to produce key upstream inputs that many companies use in production (Box 6.3). Indeed, it can take more than a thousand times as much water to produce some inputs than is used in all on-site activities. However, businesses' traditional water-use estimates fail to address the water risks throughout the supply chain and entire production cycle.

No corporate water-risk management approach can be considered complete unless it addresses these broader supply-chain issues. As an example, Unilever has been analyzing its water impact, taking into account water used by suppliers in growing raw materials and water used by consumers in using Unilever products. Since its initial efforts to systematically measure water use in the mid-1990s, the company has reportedly cut its water use in half, and they have initiated efforts to address water use in their supply chain (Unilever 2006a). For example, Unilever provides financial and technical support to help tomato farmers in Brazil switch to drip irrigation, which can reduce water consumption by up to 30 percent while increasing crop yield and cutting the need for pesticides and fungicides (Unilever 2006b). At the consumer end of its industry, the company estimates that a reformulated version of laundry detergent that requires less rinsing could have a considerable impact on water use in water-stressed areas of India, where washing clothes accounts for more than 20 percent of water consumption. Other companies are also beginning to address supply-chain issues associated with their production and companies that do not do so in the future risk challenges to competitiveness, public perceptions, and profitability.

Box 6.3   Anheuser-Busch, Water, and their Supply Chain

In 2001, Anheuser-Busch, the world's largest brewer of beer, experienced business impacts from unexpected water shortages that affected its supply chain through several key inputs: specifically barley and aluminum. An unusually dry winter in the U.S. Pacific Northwest created intense competition for freshwater. At the same time, a turbulent West Coast electricity market, highly dependent on water for power generation, led to upward pressure on electricity prices. Higher electricity costs led to a reduction in aluminum production, whereas reduced allocations of water for irrigation in Idaho resulted in reduced acreages of barley, a key brewery ingredient. This combination of events led Anheuser-Busch to take a more comprehensive and strategic approach to water issues (GEMI 2002).

Failure to Meet Basic Water Needs

Businesses that operate in developing countries face a broader set of risks than those in richer nations. Many people in poorer countries often lack basic, affordable clean water supply and sanitation systems. As a result, significant populations suffer from debilitating water-related diseases, and governments suffer from inadequate water expertise and institutional capacity. They also often lack the capital needed for significant water-system investments, or they choose to spend their limited capital elsewhere.

The failure of governments to provide 100 percent coverage for water services means that international and local businesses will increasingly find themselves with operations in regions where people lack some of the basic resources either used or produced by the company.

Tensions can also arise between businesses and local communities in developing countries, when the relative cost of a unit of water for a commercial facility is negotiated to be below what local residents pay. This can lead to resentment and community opposition over perceived and real inequities. Chapter 5 addresses in more detail some water-related concerns over equity and environmental justice. Balancing these public and private benefits is a challenge increasingly facing corporations with facilities (and extensive water use) in poor communities.

Some New Water Trends: Looking Ahead

Emerging Role of the Public in Water Policy

A dramatic shift is under way in the role of the public in setting water policy. In the twentieth century, water-policy decisions were typically made by a small number of technical or engineering experts who were responsible for water systems. By the end of the century, however, many countries had witnessed a transformation in the way that water decisions are made. In support of this trend, numerous major international water conferences in recent years have called attention to the importance of public participation in water decision making.

In part, this change has come about because of some spectacular and highly publicized water project failures or controversies, in which decisions were made that affected large numbers of people, without adequate local consultation. A high-profile example of this is the displacement of between 1.3 and 1.9 million Chinese by the construction of the Three Gorges Dam. This project led to numerous public protests and opposition in a country where open public debate over government projects is rare. Similarly, opposition has arisen to large-scale water projects in India, Southern Africa, and much of the developed world. This public interest and activism is increasingly also targeting the private sector. Corporations that fail to think strategically about water typically find themselves embroiled in highly public and emotionally charged disputes over a resource considered by many to be a basic human right.

Not surprisingly, as the public's interest has grown, so too has the attention of mainstream media. It is increasingly common to see headlines on disputes among users in water-scarce regions, protests against water-infrastructure proposals, or unusually severe droughts and floods that affect local water conditions and industrial activities. Growing public opposition to "globalization" has also brought renewed attention to water, especially bulk transfers of water from one region to another, and corporate control and use. As one example, the media has widely covered protests and controversy surrounding a Perrier plant in Michigan, which was perceived to be pumping substantial amounts of groundwater in the Great Lakes basin (*U.S. Water News* 2001). This increased attention has direct consequences for businesses, brand image, and public goodwill and will affect companies' long-term strategic plans, markets, and public affairs.

The Water-Energy Factor

Strong links exist between energy and water. As the world increasingly tries to address energy challenges, there will be a need and an opportunity to address the links between these two vital resources. Water is required to produce and use energy, and energy is used to clean, transport, and use water (Gleick 1994; Cohen et al. 2005). Some parts of the world are heavily dependent upon hydropower as the primary means of fulfilling their energy needs. For example, Brazil, a favorite recipient of foreign direct investment, generates over 90% of its electricity from hydropower. Areas that disproportionately rely upon hydroelectricity for energy (or lack energy diversity, in general) can present particular risks. In other regions, substantial amounts of energy are required to move, clean, and heat water. California's water infrastructure accounts for nearly 20 percent of the state's electricity consumption (CEC 2005).

The case of the state of Sao Paulo, Brazil, in 2001 illustrates how severe drought cycles, coupled with water-policy decisions, can disrupt economic productivity. Brazilian energy production in 2001 was highly constrained, as a result of both severe drought and government energy tariff policies that favored the development of hydro-electric systems over thermal plants. To prevent blackouts, the government imposed quotas aimed at reducing energy consumption by 10 to 35 percent based on the value added of particular industries and the number of jobs affected. Private electric companies were hard-hit by the reduction quotas, including the hydroelectric company AES Tiete, which had recently completed and closed a fifteen-year, U.S. $300 million bond offering. The effects of the energy rationing were so severe that the bond payment schedule had to be postponed and, ultimately, renegotiated. Many other industries based in the Brazilian Southeast region (which accounts for almost 60 percent of the country's GDP) were plagued by reductions in operational capacity, production delays, or increases in the costs of production. The effects of the drought-induced energy rationing extended to the national economy, with an estimated reduction of 2 percent of the country's GDP or a loss of around US$20 billion (UNEP 2005).

Climate Change

Compelling scientific evidence reveals that climate change will impact water supplies and pose formidable challenges to water systems (IPCC 2001; Gleick et al. 2000). Global warming threatens to disrupt traditional rainfall and runoff patterns and could increase the frequency and severity of both drought and floods. Changes in natural water availability will affect water management, allocations, prices, and reliability. Changing climate may also degrade water quality by changing water temperatures, flows, and runoff rates and timing, with significant impacts on water users. By increasing temperatures in lakes and streams, melting permafrost, and reducing water clarity, climate change can also seriously threaten fish and other aquatic organisms, as well as harm critical habitat like wetlands. In countries with robust biodiversity protection laws, this, in turn, could lead to reductions in available water supply, or changes in the reliability of current supplies. In addition, rising sea levels will threaten coastal aquifers and water supplies, with implications for businesses in coastal metropolitan areas that depend on groundwater resources. These effects will vary regionally, but, at the very least, climate change is expected to add more complexity and unpredictability to sustainable water-management efforts for the industry.

Public Opposition to Water Privatization

One of the most important—and controversial—trends in the global water arena is the transfer of the production, distribution, or management of water or water services from public entities into private hands—a process loosely called *privatization*. Over the past decade, private companies have been invited to take over the management, operation, and sometimes even the ownership of former public water systems. International development agencies that used to work with governments to improve water services are also encouraging outright privatization efforts, or more frequently, partnerships between public and private entities. At the same time, there is growing public awareness of, and opposition to, these efforts (Gleick et al. 2002).

The economic and political implications of privatization for industrial water users are complex and regionally specific. Among the possible impacts for businesses (both positive and negative) are changes in the cost of water supply and/or wastewater services, quality of water, and reliability of supply. Furthermore, privatization may lead to real or perceived inequities between commercial and residential water users, controversies over rates, and greater public scrutiny of large water users. Public feelings over water are highly volatile, and much opposition to water privatization is based on emotional, yet very real, connections to water. Companies would be well advised to be aware of them in any region where they operate.

Managing Water Risks

A small but growing number of companies around the world are taking steps to strategically address water-related risks in ways that protect long-term value. At many other companies, though, issues of water risk are not addressed or are addressed only in an *ad hoc*, piecemeal fashion. Although companies' water risks vary significantly, depending on their business sector and areas of operations, many businesses would benefit from preparing a strategic assessment of their current water-related business risks and plans to mitigate them. This is particularly true for companies that are dependent on high quality, reliable supplies, or on large volumes of water; companies with key operations in arid areas; and companies that rely on inputs that are themselves highly water-dependent. Table 6.1 lists a wide range of major international corporations and reviews their environmental reports for elements that address water resources. As noted, few companies consistently measure water use, report that use, set standards for future operations, engage stakeholders, or address other important water issues.

For companies that are concerned about water-related risks, or financial analysts who are interested in understanding corporate exposure to water risks, comprehensive water-risk management activities should include the following ten elements.

Measure Current Water Use and Wastewater Discharges

Companies need to understand and measure the water use and wastewater discharges associated with their operations and production. Although harder to track and quantify, they should also assess water uses and discharges associated with key suppliers and inputs. This will provide the baseline for assessing risks, prioritizing efforts, and measuring progress. In its first environmental report, for example, The Coca-Cola Company reported that its bottling plants used an average of approximately three liters of water for ever liter of beverage produced, identifying both a benchmark of current use and a way to measure improvements in water-use efficiency over time. Unilever and many other companies are now regularly assessing and reporting water use. Table 6.1 lists a wide range of companies that report, or fail to report, water and wastewater information in their annual or environmental reports.

Assess Water Landscape and Water Risks

For key areas of operation, companies should assess local hydrological, social, economic, and political factors related to water. This includes identifying shortage risks, rapidly water growing demands, problems with local institutional and political "water governance" capacity, and disparities or inequities in local and regional water access and pricing. Such an analysis of relative water conditions around a company's facilities can permit advance warning of places where tensions with the local community may appear during dry periods. In addition, renegotiating allocations and prices may help reduce these tensions without much added expense. For regions of

TABLE 6.1 Water-Related Activities and Efforts of Selected Companies[1]

Company	Industry	Online Annual or Environmental Report Available/Reviewed[2]	Measure and Report Current Water Use[3]	Water Stakeholder Consultation Reported	Engage the Supply Chain	Water Targets/Goals[4]
Anheuser-Busch	Beverage	Yes	Yes		Yes	
The Coca-Cola Company	Beverage	Yes	Yes	Yes	Yes	Yes
Pepsi Co.	Beverage	Yes				Yes
Kirin	Beverage	Yes	Yes	Yes		
McDonalds	Food	Yes			Yes	Yes
ConAgra	Ag/food processing	Yes				
Kraft	Ag/food processing	Yes	Yes			
Smithfield Foods	Ag/food processing	Yes	Yes			
Nestlé	Ag/food processing	Yes	Yes	Yes	Yes	Yes
Unilever	Food/health product	Yes	Yes		Yes	Yes
PG&E	Utility	Yes				
Duke Energy	Utility	Yes				
IBM	Electronics/high tech	Yes	Yes			Yes
HP	Electronics/high tech	Yes	Yes			
Intel	Electronics/high tech	Yes	Yes			Yes
LG	Electronics/high tech	Yes	Yes			
Sony	Electronics/high tech	Yes	Yes			Yes
Samsung	Electronics/high tech	Yes	Yes			
Haier	Electronics	No				
Lenovo	Electronics (PC maker)	No				
TCL	Electronics (TV maker)	No				
GAP	Apparel	Yes			Yes	
Levi Strauss	Apparel	No				
Burlington	Apparel	No				
VF Corporation	Apparel	No				
Bayer	Chemical/biotech	No				

Company	Industry					
Monsanto	Chemical/biotech	Yes				
Amgen	Chemical/biotech	No				
Genentech	Chemical/biotech	No				
Pfizer	Chemical/biotech	No				
P & G	Chemical/biotech	Yes	Yes		Yes	
Johnson & Johnson	Chemical/biotech	Yes				Yes
Abbott	Biotech/health	Yes	Yes	Yes		Yes
ICI	Chemical/biotech	Yes	Yes			
ExxonMobil	Refining	Yes				Yes
BP	Refining	Yes	Yes	Yes		
Chevron	Refining	Yes			Yes	
Shell	Refining	Yes	Yes			
International Paper	Forest product	Yes	Yes			
Georgia-Pacific	Forest product	Yes	Yes			
Kimberly-Clark	Forest product	Yes	Yes			Yes
Weyerhaeuser	Forest product	Yes	Yes			
POSCO (Korea)	Steel	Yes	Yes			
Nippon Steel	Steel	Yes	Yes			Yes
Alcan	Metal/aluminum	Yes	Yes			
Shanghai Baosteel Group Corp.	Steel/metal	No				
Volkswagen	Automobile	Yes	Yes			Yes
DaimlerChrysler	Automobile	Yes	Yes			
Hyundai	Automobile	Yes	Yes			
Toyota	Automobile	Yes	Yes			Yes
Honda	Automobile	Yes	Yes			
General Motors	Automobile	Yes	Yes	Yes		Yes

Notes:

1. Based on an informal online review of major corporate environmental reports. Reporting efforts are changing rapidly every year, and new reviews will no doubt find new information from a wider range of companies.

2. As of late 2005, the most recent online annual environmental reports were reviewed. These represent different years, from 2003 to 2005.

3. Some companies report total use; some report use per unit of production or employee; some report single years; some report time-series data. No standard reporting is accepted by all companies.

4. Most companies that report any water-related targets set water-use reductions over time. No standard of reduction is accepted by all companies.

Source: Morikawa 2005.

high risk, companies should develop plans to minimize their water use and impacts, and they should establish contingency plans to respond to water supply and related risks, such as decreasing water quality, higher water prices, extreme hydrological events, and local economic development.

Consult Stakeholders

Communities often feel very strongly about the use of local water resources. As a result, transparent discussions with local communities are vital to good business planning. Although public participation in local water policy in the past has been limited, civil society representatives and nongovernmental organizations now play increasingly important roles in water policy. There is growing recognition that communities should be part of water resource decision making. In instances where a company plays a large role in a community, developing early and ongoing ties with local groups can prevent or reduce the risks of future water-related disputes. In addition, proactive efforts by the company to improve water quality or water availability can help build positive relations with regional stakeholders. These efforts can include direct participation in developing local water systems, provision of funds or appropriate technology, education, or water resource planning.

Like many policy debates, questions over water are often seriously polarized. Local conditions vary from region to region, making most generalizations difficult and inappropriate. In some water-scarce regions, community backlash about corporate use of local water resources can occur, particularly during prolonged dry periods. Potential local problems should be identified early and efforts made to include the public in decisions over water, to improve company practices and to work with local groups on education and outreach. Experience has shown that early identification of local actors and their water-related issues, coupled with a policy of open communication, can reduce risks of controversy that, in extreme cases, can lead to loss of license to operate.

Engage Supply Chain

Many companies' most significant water impacts and risks may be embedded in their supply chain. To address this, companies should assess and evaluate water use in their supply chain and work collaboratively with suppliers to reduce water use and minimize risks of supply chain disruptions from water-related problems. A few companies are starting to do this (Table 6.1). As an example of one supply-chain initiative yielding tangible business benefits, Levi Strauss, Gap, and Nike joined with Business for Social Responsibility in the 1990s to develop a set of minimum water-quality guidelines to be met by apparel factories that manufacture their products. The guidelines help protect human health and water quality in countries and regions without strong clean water standards, or where local standards are poorly enforced.

Establish a Water Policy and Set Corollary Goals and Targets

Top management, particularly in water-intensive industry sectors, should clearly articulate the organization's policy regarding water-resource issues (Table 6.1). In addition, companies should establish supporting quantifiable goals and targets for water-use efficiency, conservation, and minimizing water impacts (and associated water-related risks). Efficiency programs can have multiple benefits, including cost savings, reduced energy use, and reduced regulation. As an example of the latter, "closed-loop" cooling systems in certain circumstances can reduce regulatory costs by eliminating the need for water-discharge permits. Aggressive water conservation programs developed within formal agreements with water-service providers or local governments can also reduce reliability risks during periodic drought periods.

As one example, Intel established the Corporate Industrial Water Management Group to develop and implement programs to improve water-use efficiency at its major manufacturing sites, which use large amounts of highly treated water for chip cleaning. The group includes representatives from fabrication sites, corporate technology development experts, and regulatory compliance staff. For 2003 and beyond, Intel set a goal to offset at least 25 percent of its total incoming fresh water supply needs with reclaimed water and more efficient systems. In 2002, the company exceeded this goal for the year by achieving 35 percent water savings through reclaimed water and efficiency gains.

Implement Best Available Technology

Numerous technologies can reduce water use and improve water quality, reclaim and reuse process water, filter out pollutants, and reduce cooling water requirements. Some of these require significant capital outlays; others do not. In either case, companies have often found that such technology investments can have very short payback periods and generate high returns on investment, which is likely to be increasingly true, as water scarcity becomes more severe. Companies should assess best available technology for reducing water use and wastewater discharges and commit to using such technology in new facilities and retrofitting existing facilities in areas of significant water stress.

Factor Water Risk into Relevant Business Decisions

Given its growing importance, companies should consider water scarcity and water-related risks as an important factor when making a range of strategic business decisions, from factory siting to new product development. Procter & Gamble estimates that nearly 85 percent of its product sales are associated in some way with household water use and has thus focused product research and development on addressing water use efficiency. The company has directed its product development team, "As you improve current products, or develop new-to-the-world products and services, think about how you could apply our technologies to use less water, use water differently, or use no water at all" (BSR n.d.).

Measure and Report Performance

To meet increased expectations and demands for transparency, companies should publicly report key metrics on their water use and impacts and track how their performance changes over time. This information can help investors, customers, local communities, and other key stakeholders assess how companies are managing their water risks, and is often a useful tool for engaging employees across the enterprise in supporting water programs. In February 2003, the Global Reporting Initiative (GRI) produced its "Water Protocol" to provide resource-specific guidance for organizations implementing the GRI's 2002 Sustainability Reporting Guidelines. More recently, the Facility Reporting Project is in the process of developing facility-level reporting metrics (including for water) that are based on the GRI framework (GRI).[2] In a recent Pacific Institute internal review of corporate environmental/sustainability reporting, almost all the surveyed companies measure and report at least one metric for water, most often total water consumption (Morikawa 2005).

Form Strategic Partnerships

Many water-related issues cannot be addressed in isolation, by single companies working alone. Often, water challenges are regional, affecting multiple sectors and stakeholders, and requiring broad and comprehensive action. As a result, some companies are working together through organizations like the World Council on Sustainable Development to promote watershed protection and to improve access to water for impoverished communities. Sustainable Silicon Valley, a multistakeholder initiative in the electronics and telecommunications center of California, is developing a "regional environmental management system." This initiative has identified fresh water as one of the main environmental challenges that face the region and has brought companies in the Silicon Valley together with other stakeholders to establish long-term objectives and implementation plans to work toward regional water sustainability.[3]

Commit to Continuous Improvement

Despite increased efforts from companies and other sectors of society, water scarcity and water-related business risks are likely to grow in the future. A commitment to continuous improvement in assessing and managing these risks and lessening impacts of the company's water use on local communities and the environment can help protect its operations (and long-term shareholder value) from unexpected water-related business disruptions. Such a commitment should be in written form and can be a stand-alone statement, or part of an organization's overall environmental policy, such as the one required in ISO 14001 (ISO 1996).

An Overview of the "Water Industry"

In addition to risks that face all water-using companies, many companies are involved in the manufacture or sale of a diverse range of goods and services related to water. Altogether, annual revenue from these sectors totals more than US$400 billion (Gleick et al. 2006), including over $50 billion in sales of bottled water alone. Significant growth in some developing regions, on the order of 10 to 15 percent annually, will occur in

---

2. More information on the Facility Reporting Project can be found at: http://www.facilityreporting.org/.

3 More information on Sustainable Silicon Valley can be found at: http://www.calepa.ca.gov/EMS/SiliconEMS/.

some segments of the water industry. Somewhat slower growth rates, on the order of 5 to 10 percent annually, are more likely in developed countries, though some segments of the industry may have substantially higher rates of growth.

Efforts to describe and categorize the complex water industry face problems of overlap. Below we identify major categories in the water industry and briefly profile areas of investment interest, with a focus on those sectors with emerging technologies or growth prospects that could be drivers for future opportunities. We also highlight particular risks associated with these sectors.

Disinfection/Purification of Drinking Water

A key objective of the United Nations and international nongovernmental organizations is to provide high-quality, affordable drinking water for all people. Many technologies and approaches exist for meeting these needs. For large urban areas, centralized treatment of municipal water supply is a common approach, whereas in rural areas, community and village-scale systems are more common. In some regions of the world, a second market is developing in the sales of "point of use" systems that homeowners purchase and install individually. Centralized disinfection of drinking water is a US$18 billion per year market growing at an annual rate of 10 to 15 percent. Chlorination has been the dominant technology in this sector, accounting for more than 80 percent of the market, but rapid technological changes are under way. New disinfection technologies, including ozone treatment, membrane filtration, and ultraviolet (UV) treatment are spreading and could account for half or more of the new sales of disinfection systems in the next decade. UV water disinfection is a US$500 million global market today that could see rapid growth as a result of tightening environmental regulations and concerns about *Cryptosporidium*, other microorganisms, or chemicals not adequately removed by chlorination and other traditional techniques.

In 1993, Milwaukee battled a *Cryptosporidium* outbreak that resulted in the deaths of an estimated 100 residents and nearly 400,000 illnesses. Because of the worry about new water-quality problems, combined methods of disinfection that involve UV or ultrafiltration (UF) in conjunction with other methods could become increasingly popular approaches for protecting water quality. In September 2005, GE Infrastructure, Water & Process Technologies, announced the opening of one of the world's largest potable water UF plants, designed to provide up to 78 million gallons per day (g/d) to over half a million residents of the city of Minneapolis (Business Wire 2005).

Major drivers for this sector include the growing demand for water purification treatment, rapidly growing urban centers in developing countries, new and tighter drinking water standards in Europe, especially Eastern Europe, which has seen less previous investment in these systems; and new water-quality standards for new pollutants in North America. Major risks include rapidly changing standards that may affect project designs and development and inconsistent standards from one region to another that may increase costs in systems that cannot be easily standardized.

Wastewater Treatment

The market for wastewater treatment for human and industrial waste is far larger than that for drinking water purification. This market currently totals about US$140 billion per year, with prospects for significant growth in some rapidly developing countries,

where water-quality problems are rapidly increasing. The Chinese Ministry of Construction, for example, estimates that only 45 percent of current urban wastewater
volumes are treated to any standard at all (and typically to very low standards), and
they have set a target of increasing the level of urban wastewater treatment to 60 to 70
percent of wastewater flows by the end of the Eleventh Five Year Plan (2006–2010), with
a planned expenditure of $50 billion. In a bid to attract foreign capital, advanced tech-
nology and management expertise, the central government has begun introducing
more market-based mechanisms to the urban water industry (UPI 2005).

Conventional wastewater treatment systems vary in their approach to treating
wastes but typically include centralized settling and filtration systems followed by
some kind of chemical or biological treatment. These conventional water-treatment
technologies are now being supplemented by combinations of mechanical and natural
systems that are intended to more closely mimic natural biological purification
processes, and by advanced systems capable of processing wastewater to very high
standards for reuse in certain applications. In Kuwait, GE recently completed the
world's largest membrane filtration water-purification plant, which will treat up to 100
million g/d of wastewater for reuse by agriculture and industry. A sophisticated water-
treatment system in Singapore now produces high-quality "NeWater" water for direct
potable reuse. Major drivers for this sector include severe river contamination in major
developing countries like China and India, leading to pressures to implement new and
strict wastewater discharge and treatment standards.

Industrial Water Treatment

Some industrial processes require high-quality water, including semiconductor manufacturing, pharmaceuticals, and certain chemical processing. Specialized industrial water
treatment systems to meet these needs is a highly differentiated, high-value market of
around $80 to $85 billion per year, including equipment, services, and chemicals.
Stringent water-quality requirements drive markets for water-treatment services that are
tailored to the needs of these customers. For example, a 300-mm diameter silicon wafer
chip produced by the high-tech industry may require as much as 8,600 liters of deionized
ultrapure freshwater (UNEP 2005). Large-scale water users in other sectors, such as pulp
and paper, chemicals, food, and petrochemicals, require water-treatment services for
various needs, including prevention of corrosion and contamination.

Major drivers in this sector include growing water contamination problems in
countries with weak water-quality regulations or limited enforcement, the expansion of
industries with special water requirements for either quality or reliability, the high value
of such water (compared with its relatively low cost of production), and technological
advances that open new markets. Major risks include competition from other regions
where high-quality water may be more readily available and regional restrictions on
water-intensive industries. Regions with severe water-quality problems may be passed
over for investment, if additional processing and treatment costs are lower elsewhere.

Infrastructure for Water Distribution

Equipment and services are required to build and maintain water distribution and
sewage systems for residential, commercial, and industrial markets, including pumps,
valves, water testing and monitoring, and related engineering services. Altogether,

these sectors represent a global market with total annual revenues of approximately $50 billion, spread among thousands of manufacturing companies and service providers. In richer countries, much of this market supports maintenance of old infrastructure, as opposed to the initial construction of distribution systems needed in poorer countries. Unmet needs in many developing countries are driving rapid expansion of these markets; a shift from large-scale centralized infrastructure to community-scale systems will facilitate market expansion. Another component of this sector is a newly developing focus on instrumentation and metering to recover revenues from water sales, reduce losses and leaks, and monitor use. Major risks that face this sector include inability or unwillingness of countries to spend scarce capital in this area and the presence of many players and many diverse technologies.

Desalination

Desalination has long been considered the holy grail of water supply because of its potential to convert the vast stocks of salt water in the oceans into usable freshwater (see Chapter 3; Cooley et al. 2006). At present, desalination provides only around three one-thousandths of global freshwater supplies, and recent growth rates have been around 7 percent per year. Nevertheless, some believe that the desalination market will grow more rapidly in coming years, as major cities in some regions come up against freshwater supply constraints or are affected by drought conditions, and as technological advances reduce the high price of product water. GE projects that the desalination market will grow at a compound annual growth rate of 9 to 14 percent over the next decade, growing from $4.3 billion in 2005 to $9.2 to $14.1 billion by 2014 (*The Wall Street Transcript* 2005). German consultancy Helmut Kaiser (2005) sees the market growing even faster, reaching $30 billion by 2015.

The biggest challenge that faces the desalination industry remains its high cost compared to most alternatives. The minimum cost of desalination in large-scale plants is still above $1 per cubic meter,[4] typically much higher than typical urban water costs, though optimistic observers suggest that desalination costs are dropping rapidly. Until they do, ocean desalination will be affordable in only coastal regions with serious water scarcity or with significant public subsidies (Cooley et al. 2006).

Demand-Side Efficiency

Growing water shortages and higher prices for water are driving investments and interest in improving the efficiency of water use. This sector is underdeveloped but rapidly growing, comprising companies that provide technology and services to improve "demand-side" or "end-use" efficiency for water users. Effective management of existing water resources can reduce the need for expensive, capital-intensive, new water-supply systems. In many developed countries, actual demand for water is no longer growing as fast as population and economic growth, as improvements are made in the efficient use of water. In the United States, for example, less water is now used for all purposes in 2000 than in 1980. Water use has also leveled off in places like Hong Kong, even while total economic productivity has continued to grow (Gleick 2002).

---

4. Excluding subsidies that governments often offer desalination plants, such as low-cost energy, free land, or outright reduced rates.

Companies that provide high-efficiency equipment, including showers, faucets, toilets, washing machines, and other residential and commercial appliances, may benefit as water prices rise, limits to economic supply are reached, new regulations are adopted, and awareness of efficiency potential increases in certain markets. New technologies that reduce or eliminate water use are appearing regularly on the market, such as digital X-ray machines for hospitals, efficient dishwashers and washing machines, and improved irrigation technologies. Providers of integrated water-efficiency technology and consulting services may see increased opportunity in helping address water supply crises faced by several major cities around the world.

Irrigation

The greatest human use of water worldwide is water for irrigation for food and fiber production. Growing populations, increased competition for water from urban centers, and rising prices are all putting pressure on agricultural water allocations. As a result, efforts are expanding rapidly to improve the efficiency of irrigation water use to permit more food to be grown with less water.

Providers of high-efficiency irrigation systems, such as precision sprinklers (including low-energy, precision application systems), drip irrigation, and advanced monitoring and control systems for irrigation management, are seeing growth opportunities. Drip irrigation can reduce water use by 30 to 70 percent, reduce risks of increased soil salinity, and increase agricultural yields. Although drip irrigation currently accounts for only a small fraction of the irrigation market, it is growing rapidly globally and is regionally significant. In 1970, no vineyards in California were watered with drip/precision sprinklers. By 2005 nearly 70 percent of these lands were watered with drip systems (Gleick et al. 2005). Markets could also expand rapidly if low-cost small-scale drip systems become widely available in developing countries where drip irrigation has been too expensive. Progress is being made in developing and marketing such systems (Polak and Sivanappan, n.d.).

Water Utilities

Thousands of large and small water utilities around the world provide water and wastewater services. In many cases, these utilities are public entities, owned and operated by municipalities and local governments. In some cases, private water utilities provide these services. In Switzerland alone there are still approximately 3,000 municipal water utilities; in Germany, there are 6,000 water utilities and 10,000 wastewater utilities. The U.S. water industry is also highly fragmented, with more than 50,000 municipal water utilities, of which 84 percent serve fewer than 3,300 customers. In the United States, approximately 85 percent of water utilities are publicly owned. In other countries, like Great Britain and France, private ownership dominates.

The three largest private water utility companies in the world are Veolia Environnement, Suez SA, and RWE/Thames. The first two are headquartered in France and have operations in water and energy. RWE/Thames is headquartered in Germany. All three companies have relatively high debt loads because of extensive acquisition efforts in the 1990s and early 2000s, when they purchased many smaller private water companies and pushed aggressively to expand operations globally, especially in Asia, Eastern Europe, North America, and Africa. Some high-profile project failures and strong public opposition to privatization have taken a toll on this sector and slowed the push toward more private involvement in water management. At present, all the companies are reevaluating their operations and considering splitting off or selling parts of their water business.

Future performance in this sector will depend significantly on public opinion, governmental regulatory efforts, water rates, and the form and implementation of contracts for public/private partnerships. Regulatory and legislative changes can shift the performance outlook for water utilities, although the sector generally offers stable returns.

Bottled Water

Sales and consumption of bottled water have skyrocketed in recent years. From 1988 to 2004, the global sales of bottled water have more than quadrupled to over an annual 154 million cubic meters (Gleick 2004; see also the Water Brief in this volume). Bottled water sales worldwide are increasing at 10 percent per year, whereas the volume of fruit drinks consumed is growing less than 2 percent annually, and beer and soft drink sales are growing at less than 1 percent per year, though there was a slight slowdown in the rate of growth in 2004. More than half of Americans drink bottled water occasionally or as their major source of drinking water. Total consumer expenditures for bottled water could be as high as $100 billion per year.

The slowest growth is occurring in countries of Europe, where bottled water has long had a commercial foothold, yet even there, growth rates of 5 to 10 percent per year are common in some countries. The highest growth rates are occurring in Asia and South America, with annual increases in sales of 15 percent or more in places as diverse as Egypt, Kuwait, the United States, and Vietnam. Figure 6.1 shows the trend in regional bottled water sales showing the especially rapid growth in Asia. Figure 6.2 shows the 2004 regional breakdown in sales. Table 6.2 shows annual global consumption from 1996 to 2004 (preliminary), along with the annual percent increase.

Many companies produce and sell bottled water, but a few major players dominate the market, including Nestlé, The Coca-Cola Company, and Pepsi Co. One notable trend that may have long-term ramifications is the entry of a few companies that are choosing to devote a portion of profits from bottled water sales to water development projects in poor countries. Ethos Water in the United States has committed $0.05 per bottle to such projects and was recently acquired by Starbucks Company, which now distributes Ethos Water in over 5,000 coffee shops. A Canadian bottled water company, Earth Water, has committed 100% of net profits to water programs around the world, and they have partnered with the UN Refugee Agency (UNHCR) to distribute these profits.

Conclusion

Water is a key resource for nearly all industries. Although in the past, water was a relatively low-cost factor in production, growing problems with water allocations, availability, and quality are beginning to pose serious risks to some industrial sectors. Understanding and managing these risks can save companies huge amounts of money as well as their reputations; failing to do so can lead to loss of licenses to operate,

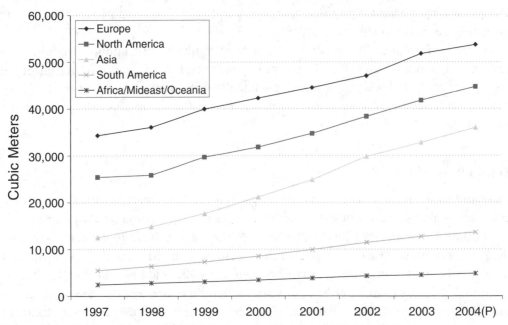

FIGURE 6.1 Regional growth in bottled water sales from 1997 to 2004.
Source: BMC 2005.

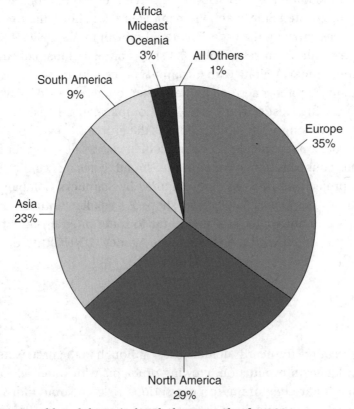

FIGURE 6.2 Regional breakdown in bottled water sales for 2004.
Source: BMC 2005.

TABLE 6.2 Total Global Consumption of Bottled Water, 1996 to 2004 (Preliminary)

Year	Thousands of Cubic Meters	Change (%)
1996	72,675	—
1997	80,649	11.0
1998	87,838	8.9
1999	97,848	11.4
2000	107,381	9.7
2001	117,876	9.8
2002	132,499	11.4
2003	144,925	9.4
2004 (P)	154,381	6.5

Source: Beverage Marketing Corporation 2005.
(P), Preliminary.

reduced productivity, dropping revenues, and threats to profits and corporate reputation. Some recent high-profile cases have highlighted these risks. Major companies in India, China, and elsewhere have been forced to shut down, relocate, or invest in expensive modifications to their operations.

Among the most significant risks are decreasing water availability and reliability, declining water quality, the vulnerability of supply-chain production to water problems, and a wide range of community concerns about access, equity, and transparency. In the coming years, the failure to think proactively about the water implications of industrial production and operation will carry with it higher and higher unexpected costs.

At the same time, these risks present possible opportunities for the multibillion dollar water "industry" that produces technologies and provides services for meeting water needs. New investments are being made by governments and corporations in water distribution, collection, treatment, monitoring, and processing technologies, and some sectors of this industry are seeing double-digit growth rates, especially in developing countries where past investments have been minimal and inadequate. Careful analysis of opportunities and needs may prove valuable to corporations and investors who are seeking to capitalize on smart and effective solutions to water problems.

REFERENCES

Beverage Marketing Corporation. 2005. (Personal communication of data to Peter H. Gleick.)
Business for Social Responsibility. n. d. *Issue brief—Water issues.*
    http://www.bsr.org/CSRResources/IssueBriefDetail.cfm?DocumentID=49620.
Business Wire. 2005. *GE technology plays critical role in opening of world's largest potable ultra-filtration plant; GE technology aids Minneapolis in meeting today's stringent regulatory standards & water safety concerns.* http://www.findarticles.com/p/articles/mi_m0EIN/is_2005_Sept_6/ai_n15345766.
Budds, J., McGranahan, G. 2003. Are the debates on water privatization missing the point? Experiences from Africa, Asia and Latin America. *Environment & Urbanization,* 15(2). October 2003. http://www.iied.org/human/eandu/sample_pubs/budds_mcgranahan.pdf.
California Energy Commission (CEC). 2005. *2005 Integrated energy policy report.* California Energy Commission. CEC-100-2005-007-ES. Sacramento, CA.

China Knowledge Press. 2005. *China welcomes investment in water sector.* China Knowledge Press, November 1.

China Securities Journal. 2005. *Dangers in economic growth model.* September 29. http://www.cs.com.cn/english/opinion/t20051129_810642.htm.

Cohen, R., Nelson, B., Wolff, G. 2005. *Energy down the drain: The hidden costs of California's water supply.* A report of the Natural Resources Defense Council, San Francisco, and the Pacific Institute. Oakland, CA.

Cooley, H., Gleick, P. H., Wolff, G. H. 2006. *With a grain of salt: A review of seawater desalination.* A report of the Pacific Institute for Studies in Development, Environment, and Security. Oakland, CA.

The Economic Times. 2004a. *Why are IT firms fleeing Bangalore?* July 29. http://economictimes.indiatimes.com/articleshow/793907.cms.

The Economic Times. 2004b. *Grasim's VSF plant closure hits textile units in MP.* July 16. http://economictimes.indiatimes.com/articleshow/901725.cms.

Gleick, P. H. 1994. Water and energy. *Annual Review of Energy and Environment* 19:267–299. Annual Reviews, Inc. Palo Alto, CA.

Gleick, P. H. 2002. *The world's water 2002–2003.* Figure 1.4, p. 25. Washington, DC: Island Press.

Gleick, P. H., and others. 2000. *Water sector report of the U.S. national assessment, water: The potential consequences of climate variability and change.* Washington, DC: U.S. Global Change Research Program. http://www.gcrio.org/NationalAssessment/water/water.pdf.

Gleick, P. H., Wolff, G., Chalecki, E. L., Reyes, R. 2002. *The new economy of water: The risks and benefits of globalization and privatization of fresh water.* A report of the Pacific Institute for Studies in Development, Environment, and Security. Oakland, CA.

Gleick, P. H. 2004. The myth and reality of bottled water. In *The world's water 2004–2005,* ed. Gleick, P. H, 17–43. Washington, DC: Island Press.

Gleick, P. H., Cooley, H., Groves, D. 2005. *California water 2030: An efficient future.* A report of the Pacific Institute for Studies in Development, Environment, and Security. Oakland, CA. http://www.pacinst.org//reports/california_water_2030/index.htm.

Gleick, P. H., Morrison, J., Newcomb, J., Harrington, T. 2006. *Remaining drops: Freshwater resources—A global issue.* Hong Kong: CLSA Blue Books.

Global Environmental Management Initiative. 2002. *Connecting the drops toward creative water strategies: A water sustainability tool.* http://www.gemi.org/water/resources.htm.

Global Reporting Initiative Water Protocol (for use with the GRI 2002 Sustainability Reporting Guidelines). 2003. February. http://www.globalreporting.org/guidelines/protocols/WaterProtocol030501.pdf.

The Hindu. 2004. *Government undecided on Coke plant's future.* July 21. http://www.hindu.com/2004/07/21/stories/2004072108870400.htm.

Intergovernmental Panel on Climate Change (IPCC). 2001. *Climate change 2001: Impacts, adaptation and vulnerability.* Cambridge, UK: Cambridge University Press. http://www.grida.no/climate/ipcc_tar/wg2/index.htm.

ISO. 1996. *ISO 14001 environmental management systems—Specification with guidance for use.* Geneva, Switzerland: International Organization on Standards.

Kaiser, H. 2005. *Study: Water desalination worldwide for sea water and brackish water 2005–2010–2015.* http://www.hkc22.com/waterdesalination.html.

Koutsoukis, J. 2004. Gunnamatta priority in push to end ocean outfalls. *The Age,* April 13. http://www.theage.com.au/articles/2004/08/12/1092102596047.html?oneclick=true.

Kuber, G. 2005. India's on the brink of a water crisis. *The Economic Times Online,* October 12.

Morikawa, M. 2005. *Corporate water reporting industry summary.* Internal report. September 14. Oakland, CA: Pacific Institute.

Morrison, J., Gleick, P. H. 2004. *Freshwater resources: Managing the risks facing the private sector.* A report of the Pacific Institute for Studies in Development, Environment, and Security. Oakland, CA.

Mukerjee, A. 2005. Forget oil: India's bigger problem is water. *Bloomberg,* October 27.

Parsai, G. 2005. Manage water resources: Expert. *The Hindu,* October 6.

Polak, P., Sivanappan, R. K. n.d. *The potential contribution of low cost drip irrigation to the improvement of irrigation productivity in India.* International Development Enterprises. http://www.ide-international.org/Files/The%20Potential%20Contribution%20of%20Low%20Cost%20Drip%20Irrigation.pdf.

Red Herring. 2005. Water: The new oil—China. *Red Herring,* August 15.

Unilever. 2006a. *Water.* http://www.unilever.com/ourvalues/environmentandsociety/shortstories/Water/.

Unilever. 2006b. *Brazil: Drip irrigation cuts water and pesticides on tomato farms.* http://www.unilever.com/ourvalues/environmentandsociety/casestudies/agriculture/brazil. asp.

United Nations Environment Programme. 2005. *Challenges of water scarcity: A business case for financial institutions.* 9. Stockholm, Sweden: UNEP Finance Initiative and Stockholm International Water Institute. http://www.unepfi.org/fileadmin/documents/challenges_ water_scarcity_2005.pdf.

United Press International (UPI). 2005. *China's water crisis said to be severe and urgent.* November 1. http://www.wpherald.com/storyview.php?StoryID=20051101-110404-6078r.

U.S. Water News Online. 2001. *Group doesn't want Perrier bottling Michigan water.* March. http://www.uswaternews.com/archives/arcsupply/1grodoe3.html.

U.S. Water News Online. 2004. *Dry Beijing to shun water-intensive industry.* March. http://www.uswaternews.com/archives/arcglobal/4dryxbeij3.html.

The Wall Street Transcript. 2005. GE's water financing strategy. *The Wall Street Transcript,* June 7.

WHO/SEI. 1997. *Comprehensive assessment of the freshwater resources of the world.* Stockholm, Sweden: World Meteorological Organization and Stockholm Environment Institute.

Bottled Water: An Update

Peter H. Gleick

The 2004 volume of *The World's Water* discussed the growing phenomenon of bottled use around the world, particularly in regions where high-quality tap water is available, as in most of North America and Western Europe (Gleick 2004). This "In Brief" updates recent events and provides new data on bottled water use.

Bottled-water sales continue to grow rapidly, as does controversy over its use (Water Technology News 2006a; Arnold 2006). Total annual sales are now on the order of $50 to $100 billion dollars, for over 150 million cubic meters (m^3). Growth in sales has been particularly rapid in Asia and South America, where sales have nearly tripled since 1997. Figure WB 1.1 and Table WB 1.1 show the total sales over time for the major continental regions.

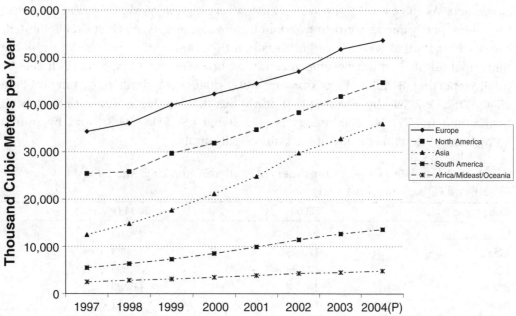

FIGURE WB 1.1 Bottled water consumption by region from 1997 to 2004 (preliminary). While total volume in Europe and North America remains high, consumption is growing very rapidly in Asia.

Source: Data were provided by the Beverage Marketing Corporation to the author in 2005 and are used with permission.

TABLE WB 1.1   Regional Consumption of Bottled Water, 2003 and 2004
(Thousand Cubic Meters per Year)

Region	2003	2004 (P)
Europe	51,768	53,661
North America	41,778	44,715
Asia	32,795	35,977
South America	12,677	13,607
Africa/Mideast/Oceania	4,499	4,823
All others	14,074	15,974
Total	**157,591**	**168,758**

Source: Data were provided by the Beverage Marketing Corporation to the author in 2005 and are used with permission; see also Data Table 11
P, preliminary.

Total use remains especially high in the United States, Mexico, China, and Brazil (Table WB 1.2). Per capita use, however, is still highest in European countries, despite the wide availability of high-quality, reliable tap water, and, in Mexico (Table WB 1.3), where water quality problems are more ubiquitous. Italy reports the highest yearly per capita consumption of bottled water at 184 liters per person, and six of the top ten consumers who drink more than 100 liters per person per year are European.

What drives these sales? No definitive analysis of the reasons has been presented, but three factors play important roles: (1) fear about the quality of tap water, (2) convenience, and (3) preferences and perceptions of taste.

In many parts of the world, tap water is not available or safe to drink. In these regions, the failure of governments to provide basic water services has opened the door to private companies and vendors filling a critical need, albeit at a very high cost to consumers. As Table WB 1.3 shows, total use is very high in Mexico, Brazil, China, and Indonesia, partly due to concern over bad local water quality. As Gleick (2004) noted, the price of bottled water is often literally a thousand times higher than reliable municipal supply. In many developed countries, however, fear of tap water is fueled by public reporting of violations of drinking water quality standards (e.g., Toledo Blade 2006; WISCTV 2006), by advertising that implies that bottled water imparts special health benefits (Water Technology New 2006b; U.S. FDA 2006), and by public ignorance of the actual quality of their municipal supply.

TABLE WB 1.2   Ten Largest Consumers of Bottled Water, 2003 and 2004
(Thousand Cubic Meters per Year)

Country	2003	2004 (P)
United States	24,199	25,893
Mexico	16,495	17,683
China*	10,628	11,894
Brazil	10,758	11,598
Italy	10,350	10,661
Germany	9,950	10,313
France	8,907	8,550
Indonesia	6,945	7,362
Spain	5,098	5,506
India	4,202	5,126

*Includes bottled water use in Taiwan. P, preliminary.

Source: Data were provided by the Beverage Marketing Corporation to the author in 2005 and are used with permission; see also Data Table 10.

TABLE WB 1.3 Fifteen Largest Per Capita Consumers of Bottled Water, 2003 and 2004 (Liters per Person per Year)

Country	2003	2004 (P)
Italy	179	184
Mexico	157	169
United Arab Emirates	145	164
Belgium-Luxembourg	133	148
France	148	142
Spain	127	137
Germany	121	125
Lebanon	96	102
Switzerland	96	100
Cyprus	86	92
United States	85	91
Saudi Arabia	88	88
Czech Republic	84	87
Austria	86	82
Portugal	78	80

P, preliminary.

Source: Data were provided by the Beverage Marketing Corporation to the author in 2005 and are used with permission; see also Data Table 13.

Bottled water is also convenient to use. It is portable, reliable, and widely accessible. Sales at supermarkets, convenience stores, sporting events, and hotels put bottled water in every public location. Companies can even buy generic bottled water with their own labels for sale or distribution to customers.

Finally, some people do not like the taste of their tap water, which is sometimes heavily chlorinated or contains minerals that impart unpleasant tastes. Bottled water, in contrast, is often processed to remove taste. Ironically, many independent blind taste tests have shown that few people can actually identify bottled water by taste or express a preference for bottled water versus tap water when they do not know which they are drinking (Stossel 2005). In 2001 a water company in Yorkshire, England, found that 60 percent of nearly 3,000 people surveyed could not distinguish between local tap water and bottled water. That same year, the *Cincinnati Enquirer* discovered that the city's tap water rated an 8.2, compared with Dannon's 8.3 and Evian's 7.2, on a scale of 1 to 10 (Shermer 2003).

Standards for bottled water remain inconsistent from country to country. In some places, few or no standards are imposed, or monitoring is haphazard, unreliable, or incomplete. Even in the United States, bottled water is regulated and monitored differently from tap water, and, although both are usually safe to drink, both are also periodically found to be contaminated with minerals or other substances in violation of standards. Bottled water violations are not always reported to the public, or are not reported in a timely manner. Table WB 1.4 lists a few examples of bottled water recalls or enforcement reports put out by the U.S. Food and Drug Administration (FDA).[1] As the table indicates, the reports often appear months after the violation is detected and long after the product has been distributed and sold. The violations notices listed in this table appeared at an average of five months after they occurred, but as long as fourteen and fifteen months later, making it impossible for the contaminated product

---

1. A more complete listing of reported violations will be posted at www.worldwater.org. My personal favorite is the 1994 recall of bottles "contaminated with crickets."

TABLE WB 1.4 A Few Examples of Bottled Water Recall and "Field Corrections" Notices from the U.S. Food and Drug Administration

FDA Recall Notice Date	Date of Production	Product	Manufacturer	Reason Given	Volume Recalled	Locations
2/28/90	February 1990	Sparkling water	Perrier Group, Paris, France	Contaminated with benzene	0.5 million cases remain on market as of recall date	Nationwide and U.S. territories
7/18/90	April 1990	Spring water	West Lynn Creamery, Lynn, Massachusetts	Fish smell or taste/burning sensation due to sodium hydroxide contamination	125 cases (12 bottles per case)	Vermont
7/10/91	April 1990	Bottled drinking water	Famous Ramona Water Company, Ramona, California	Contaminated with algae and *Pseudomonas aeruginosa*	Approximately 13,621 cases	Hawaii, California
11/13/91	July–August 1991	Spring water	Idlenot Farm Dairy, Inc., Wilton, New Hampshire	Off-odor and off-taste due to tetrahydrofuran	Approximately 19,500 gallons	New Hampshire, Massachusetts, Vermont
1/29/92	November 1991	Mineral water	Apollinaris Brunnen AG, Federal Republic of Germany	Contaminated with dimethyl disulfide	6,468 cartons (12 bottles per case) were distributed	Nationwide
12/19/94	May 1994	Sparkling water; flavored	Southwest Canners, Inc., Nacogdoches, Texas	Contaminated with crickets	None given	Alabama, Florida, Georgia
10/5/94	July 1994	Bottled water	Southern Beverage Packers, Inc., Harlem, Georgia	Product contained brown precipitates	5 million containers	Southeastern United States
4/26/95	September 1994	Drinking water	All Star Bottling Company, Kansas City, Kansas	Contaminated with mold	1 1/2 to 2 million bottles were distributed; firm estimated none remained on the market when the recall was announced	Montana, Texas, Oklahoma, Missouri, Nebraska, Wyoming, Colorado, North Dakota, Minnesota, Louisiana
5/8/96	March–October 1995	Spring water	North Country Natural Spring Water Company, Kent, New York	Contaminated with mold	226,680 cases (4 1-gallon bottles per case)	Connecticut, Delaware, Massachusetts, Maine, Maryland, North Carolina, New Hampshire, New Jersey, New York, Pennsylvania, Virginia, West Virginia, Washington, DC

Date	Time period	Product type	Company	Reason	Amount	Location
11/27/96	July, August, September 1996	Drinking water, spring water	Marion Pepsi-Cola Bottling Company, Marion, Illinois	Product is adulterated due to the presence of coliforms	403,608 20-ounce bottles and 42,945 1-liter bottles; 67,789 gallon, 3,505 2-1/2-gallon and 980 5-gallon containers	Missouri, Illinois, Arkansas, Kentucky, Tennessee
8/5/98	May 1998	Mineral water, spring water	Calistoga Mineral Water Company, Calistoga, California	Bottled in defective glass bottles whose rims may chip and allow glass to enter the product	172,533 cases or 4,140,792 bottles were distributed	California and other Western states
No public notice[1]	7/19/00	Bottled water	Cott Beverages, Tampa, Florida	Contaminated with unspecified over-the-counter drug	None given	Unavailable
2/7/01	August 2000	Drinking water, purified water	Safeway Bottled Water Division, Tempe, Arizona	Unfit for food because they contain particulate matter	7,560 cases of purified water and 23,100 cases of drinking water. All products are sold in cases of six gallon bottles each.	Arizona and New Mexico
3/13/02	August 2001	Drinking water	Bareman Dairy, Inc., Holland, MI	May contain equipment sanitizer	19,700 gallons	Illinois, Indiana, Michigan
9/1/04	April 2004	Mineral water	Casa Imports, Inc., Utica, NY	Mineral water makes unapproved claim. Label claims to reduce cholesterol.	1 liter (12 bottles); .5 liter (84 bottles); .25 liter (5 bottles)	Maine, New York, Pennsylvania
10/20/04	September 2004	Purified water, distilled water	Spectrum Laboratory Products, Inc., Gardena, CA	Microbial contamination; Burkholderia cepacia	537 containers	Nationwide and Puerto Rico
8/31/05	June 2005	Drinking water	Publix Supermarkets, Inc., Deerfield Beach, Florida	Bottled water has off-odor and off-taste	48,764 1-gallon containers;16,064 2.5-gallon containers	Florida

Notes:

This is a small sampling of FDA-listed recall notices. A more complete listing will be posted at http://www.worldwater.org.

1. This event and additional violations were discovered by the Pacific Institute through a Freedom of Information Act Request (F06-2744) filed February 22, 2006.

Source: FDA Enforcement Report Index. http://www.fda.gov/opacom/Enforce.html.

to be removed from the market. As a result, these "notices" are ineffective at protecting the public from hazardous or mislabeled product.

Bottled water recalls have affected products in every state of the United States. Consumers would benefit from more comprehensive testing, timely detection and reporting, and faster correction of violations, as well as more consistent standards of protection. Serious consideration should be given to making bottled water standards and drinking water standards comparable in all respects.

Some bottled water trade organizations and associations have made an effort to offer standard guidance to their members. The International Bottled Water Association, for example, has created the "IBWA Bottled Water Code of Practice," which applies to all members of the Association in the United States. The code requires that an annual, unannounced independent audit be conducted by a certified, third-party inspection agency. This inspection includes a look at the plant, including areas used for bottled water production as well as nonproduction areas, along with a review of plant records to confirm compliance with FDA and state standards for monitoring/testing, quality, labeling, and Good Manufacturing Practices (Kay 2006). This inspection, however, requires no testing of actual water quality, which is left up to the bottler themselves under the guidelines established by the U.S. FDA.

REFERENCES

Arnold, E. 2006. *Bottled water: Pouring resources down the drain.* Washington, DC: Earth Policy Institute. February 2.

Gleick, P. H. 2004. The myth and reality of bottled water. 17–43. In *The world's water 2004–2005: The biennial report on freshwater.* Gleick, P. H. (editor). Washington, DC: Island Press.

Kay, S. R. 2006. (Personal communication with the author). April 10.

Shermer, M. 2003. Bottled twaddle: Is bottled water tapped out? *Scientific American,* July. http://www.sciam.com/article.cfm?articleID=000007F0-6DBD-1ED9-8E1C809EC588EF21.

Stossel, J. 2005. Is bottled water better than tap? *ABC News—20/20 Commentary,* May 6. http://abcnews.go.com/2020/Health/story?id=728070&page=1.

Toledo Blade. 2006. Water from fields triggers a water alert in defiance. *Toledo Blade,* May 17. http://toledoblade.com/apps/pbcs.dll/article?AID=/20060517/NEWS17/605170362.

United States Food and Drug Administration (U.S. FDA). 2006. Warning Letter SEA 06-21 to Water Oz, Idaho. http://www.fda.gov/foi/warning_letters/g5752d.htm.

Water Technology News. 2006a. Anti-bottled water group confronts Pepsi meeting. *Water Tech Online.* May 3.

Water Technology News. 2006b. Portuguese bottled water makes weight-loss claim. *Water Tech Online.* April 20. http://www.watertechonline.com/news.asp?mode=4&N_ID=61078.

WISCTV. 2006. Officials issue health advisory over tap water. Madison, WI. http://www.channel3000.com/news/9228910/detail.html.

Water on Mars

Peter H. Gleick

Introduction

Previous volumes of *The World's Water* have addressed a wide range of water issues, including a few looking at water literally "out of this world"—the possibility of water in icy comets ("cosmic snowballs") (Gleick 1998), water on the moon, and water in the outer solar system and beyond (Chalecki 2002). In the past two years, significant advances have been made in our understanding of the presence and behavior of water on Mars, as the result of a series of successful exploratory missions by remote satellites, landers, and rovers. As of spring 2006, several satellites were working in orbit, and two rovers, Spirit and Opportunity, were engaged in exploring the surface and sending back vast amounts of data. This In Brief summarizes some of these recent findings.

Mars is the fourth planet of our solar system and an intriguing neighbor. It would be highly inhospitable to humans, yet it may be the least inhospitable of all of the other planets circling our sun. The Martian atmosphere is predominately carbon dioxide, with an atmospheric pressure only one-sixth that of our own. It is cold and dry, with an average surface temperature at the equator of about 220 K, far below the freezing point of water. Despite this, it appears to have had large amounts of liquid water in the past and still retains significant amounts of water in the form of surface and subsurface ice near the polar regions, with water vapor in the atmosphere. The more we learn about the forms, locations, and dynamics of water on Mars, the more scientists can tell us about that planet's complex geophysical history.

The presence of water on Mars is exciting for several other reasons: It can give us insights into the formation of our own planet, and it offers the opportunity to produce hydrogen fuel and breathable oxygen for future manned missions. More dramatic, the presence of significant amounts of liquid water at times in Mars' history suggests the possibility that life in some form existed, or even still exists.

Background

Using telescopic spectroscopy, scientists first detected water vapor in the Martian atmosphere in 1963 (Jakosky and Mellon 2004), at a time when the very first space missions were being attempted. Missions to reach Mars have met with mixed success, with two-thirds of all attempts failing to reach Mars. Table WB 2.1 lists all the Mars missions. The first twenty-one satellite photos were returned in 1965 by Mariner 4,

TABLE WB 2.1   Missions to Mars

Mission	Country	Launched	Type	Purpose/Comments
(Unnamed)	USSR	1960	Flyby	Did not reach Earth orbit
(Unnamed)	USSR	1960	Flyby	Did not reach Earth orbit
(Unnamed)	USSR	1962	Flyby	Achieved Earth orbit only
Mars 1	USSR	1962	Flyby	Radio failed at 65.9 million miles (106 million km)
(Unnamed)	USSR	1962	Flyby	Achieved Earth orbit only
Mariner 3	U.S.	1964	Flyby	Shroud failed to jettison
Mariner 4	U.S.	1964	Flyby	First successful Mars flyby 7/14/65, returned 21 photos
Zond 2	USSR	1964	Flyby	Passed Mars but radio failed, returned no planetary data
Mariner 6	U.S.	1969	Flyby	Mars flyby 7/31/69, returned 75 photos
Mariner 7	U.S.	1969	Flyby	Mars flyby 8/5/69, returned 126 photos
Mariner 8	U.S.	1971	Orbiter	Failed during launch
Kosmos 419	USSR	1971	Lander	Achieved Earth orbit only
Mars 2	USSR	1971	Orbiter and Lander	Arrived 11/27/71, lander destroyed, no useful data
Mars 3	USSR	1971	Orbiter and Lander	Arrived 12/3/71, some data, few photos
Mariner 9	U.S.	1971	Orbiter	In orbit 11/13/71 to 10/27/72, returned 7,329 photos
Mars 4	USSR	1973	Orbiter	Failed to orbit; flew past Mars 2/10/73
Mars 5	USSR	1973	Orbiter	Arrived 2/12/74, lasted only a few days
Mars 6	USSR	1973	Orbiter and Lander	Arrived 3/12/74, little data return
Mars 7	USSR	1973	Orbiter and Lander	Arrived 3/9/74, little data return
Viking 1	U.S.	1975	Orbiter and Lander	Orbit 6/19/76–1980, lander 7/20/76–1982, tens of thousands of photos returned
Viking 2	U.S.	1975	Orbiter and Lander	Orbit 8/7/76–1987, lander 9/3/76–1980, tens of thousands of photos returned
Phobos 1	USSR	1988	Orbiter/ Lander (Mars and Phobos)	Lost 8/88 en route to Mars

Mission	Country	Launched	Type	Purpose/Comments
Phobos 2	USSR	1988	Orbiter/ Lander (Mars and Phobos)	Lost 3/89 near Phobos
Mars Observer	U.S.	1992	Orbiter	Lost just before Mars arrival 8/21/92
Mars Global Surveyor	U.S.	1996	Orbiter	Arrived 9/12/97, still conducting prime mission
Mars 96	Russia	1996	Orbiter and Landers	Launch vehicle failed
Mars Pathfinder	U.S.	1996	Lander/Rover	Landed 7/4/97, last transmission 9/27/97
Nozomi (Planet-B)	Japan	1998	Orbiter	Failed Mars orbit, currently in heliocentric orbit
Mars Climate Orbiter	U.S.	1998	Orbiter	Lost on arrival 9/23/98
Mars Polar Lander/ Deep Space 2	U.S.	1999	Lander/Probes	Lost on arrival 12/3/98
Mars Odyssey	U.S.	2001	Orbiter	Arrived October 2001, still conducting mission; relays rover signals/data
Mars Exploration Rovers	U.S.	2003	Rovers	Landed January 2004, conducting mission
Mars Express	European Union	2003	Orbiter/Lander	Arrived December 2003, lander failed, orbiter conducting mission
Mars Reconnaissance Orbiter	U.S.	2005	Orbiter	Arrived March 2006, conducting mission

Source: NASA 2006c.

which was also the first successful flyby. Early photographs of Mars from subsequent Mariner flyby missions in the 1960s and 1970s first suggested the possibility that water existed at one time on the surface. As more sophisticated instruments reached the planet's orbit, new and increasingly detailed measurements were made.

In the late 1970s, the United States successfully landed Viking on the Martian surface, returning spectacular color pictures and considerable atmospheric and mineralogical information. More recently, new orbiters, such as Mars Global Surveyor (MGS), and two years of phenomenally successful exploration by Mars rovers Spirit and Opportunity, have brought our understanding to new levels. All together, instruments on Earth, in space, and on the Martian surface have now identified and mapped geophysical evidence for past water, the presence of water vapor and ice clouds in the atmosphere, water ice together with frozen carbon dioxide at both poles, and ice mixed with soils near the surface at higher latitudes.

Visual evidence of the past presence of liquid water abounds. Features that look like channels and tributaries can be seen; sediment flows on sloping terrain suggest melting or permafrost movement. Forms of erosion, highly suggestive of the presence of liquids, are evident in photographs and high-resolution laser topography sent back by the MGS and the Mars Orbital Camera (MOC). MOC data include pictures of what appear to be flood gullies near the surface, where water may burst forth before evaporating.[1] The evidence from the MGS was considered so convincing that NASA officially announced the discovery of water on Mars in the summer of 2000.

In December 2003, the European Space Agency's first Mars mission, Mars Express, arrived and promptly provided evidence for ancient catastrophic floods. Large channels appear to be carved by water in the northern plains, suggesting an ancient ocean in the northern hemisphere. Networks of valleys in the southern highlands also appear to have been created by flowing water. Debris surrounding craters suggest "fluidized ejecta" consistent with impacts on wet soils or regions with underground water (ESA 2006).

The tantalizing visual evidence has been further confirmed by instrumental analyses. These include the gamma-ray and neutron spectrometers on the Mars Odyssey, sent into orbit around the planet. These instruments record, among other data, hydrogen signatures as much as 3 feet below the surface. In 2002, Mars Odyssey began returning high-quality data about Mars' surface composition that led NASA to announce that vast quantities of ice appeared to be mixed with some Martian soils (Whitehouse 2002).

The thermal emission imaging system on Mars Odyssey also observed surface water ice in the southern hemisphere near and around the perennial frozen carbon dioxide cap. Observations of the northern hemisphere taken by the high-energy neutron detector on Odyssey, combined with measurements of the thickness of condensed carbon dioxide, are consistent with a layer of water ice under drier soils. Mitrofanov et al. (2003) suggest that these layers may contain as much as 50 to 75 percent water (by weight). Although liquid water is gone from the surface under most conditions, substantial amounts of subsurface ice apparently still remain, and small amounts of exposed water ice have been observed (Titus et al. 2003).

The ESA's Mars Express also carried instruments that quickly identified the presence of both frozen carbon dioxide and frozen water. Although its Beagle lander failed, Express has a range of spectrometers capable of cataloging minerals, atmospheric water vapor, and more, which have been extremely successful and revealing.

Perhaps most exciting was the arrival in January 2004 of Spirit and Opportunity, two robots capable of roving over the surface and performing a wide range of geological and chemical analyses. The public (or at least the author) has been captivated by photographs from these mobile laboratories and cameras, and the rovers were so successful in their studies that in March 2004, NASA scientists announced that they had found rocks that were "once soaked in liquid water."[2]

Martian History of Water

As a result of the accumulation of data, measurements, and observations, scientists can now begin to compile theories of the history of water on Mars through time. Water appears to have played a crucial role in the early years of Mars, just as it did on Earth.

---

1. http://www.msss.com/mars_images/moc/june2000/

2. http://www.cnn.com/2004/TECH/space/03/02/mars.findings/

By looking at the characteristics of ancient surfaces that are still evident, it is possible to speculate about the geological processes that acted to form those surfaces. On Mars, some surfaces can be dated back three to four billion years through mineral analysis and the evidence from craters and asteroid impacts. The billion-year-old portions of the Martian surface show extensive geological features that appear to have been formed or modified by liquid water (Jakosky and Mellon 2004). This indicates that water was either more abundant, or more likely to have existed as a liquid. Among the visual evidence are tributaries and erosion features similar to those seen on Earth.

Squyres et al. (2004) describe the geologic setting in and around Eagle crater in Meridiani Planum, where the rover Opportunity landed. They interpret the rocks of the crater to include a mixture of sediments with a history that is the result of episodic inundation by surface water to shallow depths, followed by evaporation, exposure, and drying. The forms seen in the crater suggest inundation; the mineralogy and geochemistry indicate precipitation of dissolved salts. Moreover, the evidence suggests that these processes were not limited to the immediate vicinity of the landing site but operated over an area that was at least tens of thousands of square kilometers in size.

The rovers confirmed the presence of minerals that had to be the result of immersion in water. This was followed by evidence that suggested that water persisted on the surface for very long periods of time—not just years but eons (Chandler 2004). Among the findings of Opportunity were the salty sedimentary remains of standing water and forms of clay that mark an early warmer and wetter era (Kerr 2005).

In 2005, scientists who worked with the ESA published summary analyses of information from some of the Mars Express instruments mapping the distribution of key minerals on the surface and in the Martian crust. Vast regions on Mars are covered in oxidized materials, but the data suggest that this oxidation did not result from contact with water, and they do not require water to form. This led scientists to conclude that "liquid water is not responsible for Mars being red" (Bibring et al. 2006). The same mineral mapping, however, identified extensive areas of ancient materials that are the result of contact with water, particularly clays formed billions of years ago. As Bibring et al. note, "Surface formation of these clay minerals would indicate a long-lasting wet episode" named the "phyllosian" era over three billion years ago (see also Paige 2005). This was followed by the loss of much of Mars' water and an era in which liquid water was only periodically present on the surface. Current conditions on Mars are very cold and dry, with frozen surface water in limited locations and potentially larger reservoirs in subsurface regions. Liquid water is now probably present only during transient and local events (such as the release of volatiles by impacts or short-term melting of ice), and these events are too limited in scope and duration to leave significant evidence on the surface.

Because of the apparent presence of liquid water in the past, Squyres et al. (2004) explicitly infer that surface conditions at Meridiani may have been habitable for some period of time in Martian history. Several environmental conditions, however, would have been challenging to biological development. The mineral combinations found in the Eagle crater are also found on Earth but are commonly associated with mine drainage and are strongly acidic. The Meridiani geology also suggests high saline waters.

High acidity and saline waters can support life on Earth, but the microscopic and bacterial organisms that accommodate such conditions appear to belong to populations that evolved the ability to do so. Whether such conditions are suitable for the kinds of reactions commonly invoked to explain the origin of life is less certain.

As a result, researchers cannot determine whether life was present or even possible in the waters at Meridiani and other places. Comparable geology on Earth, however, can contain beautifully preserved fossils, so scientists consider this spot an attractive candidate for further study (Squyres et al. 2004).

Future Mars Missions

Water exists on Mars today and since its creation. Although major advances are being made in our understanding of the evolution of the hydrology of Mars, much remains to be discovered and explained. It now appears that any significant amounts of surface water on Mars disappeared long ago, as Mars lost most of its atmosphere and the associated atmospheric pressure and temperature that permitted water to remain liquid. Debate still remains about the temperature regime of ancient Mars and whether a "warm Mars" existed, or existed for long, and about the extent and duration of the presence of liquid water and the formation of associated minerals. New instruments, new data, and new analyses in the coming years may help answer some of these questions.

The Mars Reconnaissance Orbiter arrived in orbit in April 2006 and is already expanding the search for water. Among the instruments onboard are a color imager that will monitor changes in clouds, wind-blown dust, polar caps and other variable features, and a camera that can resolve 6 meters (20 feet) per pixel, allowing surface features "as small as a basketball court to be discerned" (NASA 2006b).[3]

The Mars Scout mission and Phoenix lander are scheduled to be launched in late 2007 and to arrive in May 2008. The Phoenix will land in the high northern latitudes to search for near-surface ice. Phoenix will deploy a robotic arm to dig trenches up to half a meter (1.6 feet) into layers of water ice. These layers could contain organic compounds necessary for life. To analyze soil samples collected by the robotic arm, Phoenix will carry an oven and laboratory to heat and analyze soil so that volatiles can be examined for their chemical composition (NASA 2006a).

Considerable uncertainty still remains about how much liquid water actually existed, where and how long it persisted, and what happened to it over the last three billion years. Also, many questions remain about the current condition and distribution of water. How extensive are the ice deposits? Does groundwater exist? Is there evidence of ancient, or even surviving, forms of life associated with water? As new instruments and observations are put into place, new answers will emerge.

REFERENCES

Bibring, J. P., Langevin, Y., Mustard, J. F., Poulet, F., Arvidson, R., Gendrin, A., Gondet, B., Mangold, N., Pinet, P., Forget, F., Berthé, M., Gendrin, A., Gomez, C., Jouglet, D., Soufflot, A., Vincendon, M., Combes, M., Drossart, P., Encrenaz, T., Fouchet, T., Merchiorri, R., Belluci, G. C., Altieri, F., Formisano, V., Capaccioni, F., Cerroni, P., Coradini, A., Fonti, S., Korablev, O., Kottsov, V., Ignatiev, N., Moroz, V., Titov, D., Zasova, L., Loiseau, D., Mangold, N., Pinet, P., Douté, S., Schmitt, B., Sotin, C., Hauber, E., Hoffmann, H., Jaumann, R., Keller, U., Arvidson, R., Mustard, J. F., Duxbury, T., Forget, F., Neukum, G. 2006. Global mineralogical and aqueous Mars history derived from OMEGA/Mars Express Data. *Science*, 312(5772):400-404, April 21.

---

3. Though discovering a basketball court would cause considerable consternation.

Chalecki, E.L. 2002. "Water and space." In P.H. Gleick (editor), *The World's Water 2002-2003. The Biennial Report on Fresh Water.* Island Press, Washington, D.C. pp. 209-224.

Chandler, D. 2004. Mars rover finds that water persisted. *New Scientist,* July 19. http://www.newscientist.com/article.ns?id=dn6178.

European Space Agency (ESA). 2006. *Was there water on early Mars?* http://www.esa.int.

Gleick, P.H. 1998. "Small comets and the new debate over the origin of water on earth." In P.H. Gleick, *The World's Water 1998-1999. The Biennial Report on Fresh Water.* Island Press, Washington, D.C. pp. 193-199.

Jakosky, B. M., Mellon, M. T. 2004. Water on Mars. *Physics Today,* 71, April. http://www.physicstoday.org/vol-57/iss-4/p71.html.

Kerr, R. A. 2005. An early, muddy Mars just right for life. *Science,* 310(5756):1898-1899.

Mitrofanov, I. G., Zuber, M. T., Litvak, J. L., Boynton, W. V., Smith, D. E., Drake, D., Hamara, D., Kozyrev, A. S., Sanin, A. B., Shinohara, C., Saunders, R. S., Tretyakov, V. 2003. CO_2 snow depth and subsurface water-ice abundance in the northern hemisphere of Mars. *Science,* 300(5628):2081-2084.

National Aeronautics and Space Administration (NASA). 2006a. Missions to Mars: Phoenix. http://mars.jpl.nasa.gov/missions/future/phoenix.html.

National Aeronautics and Space Administration (NASA). 2006b. Mars Reconnaissance Orbiter: Mars cameras debut as NASA craft adjusts orbit. http://marsprogram.jpl.nasa.gov/mro/newsroom/pressreleases/20060413a.html.

National Aeronautics and Space Administration (NASA). 2006c. Missions to Mars. http://marsprogram.jpl.nasa.gov/missions/log/.

Paige, D. A. 2005. Ancient Mars: Wet in many places. *Science,* 307(5715)1575-1576.

Squyres, S. W., Grotzinger, J. P., Arvidson, R. E., Bell, J. F., III, Calvin, W., Christensen, P. R., Clark, B. C., Crisp, J. A., Farrand, W. H., Herkenhoff, K. E., Johnson, J. R., Klingelhöfer, G., Knoll, A. H., McLennan, S. M., McSween, H. Y., Jr., Morris, R. V., Rice, J. W., Jr., Rieder, R., Soderblom, L. A. 2004. In situ evidence for an ancient aqueous environment at Meridiani Planum, Mars. *Science,* 306(5702):1709-1714, December 3.

Titus, T. N., Kieffer, H. H., Christensen, P. R. 2003. Exposed water ice discovered near the South Pole of Mars. *Science,* 229(5609):1048-1051.

Whitehouse, D. 2002. *Ice reservoirs found on Mars.* BBC Online. http://news.bbc.co.uk/1/hi/sci/tech/2009318.stm.

Time to Rethink Large International Water Meetings[1]

Peter H. Gleick

Introduction

International meetings devoted to freshwater problems have become a big business in the last decade, and, although they serve to raise awareness, identify problems, enhance communication, and explore solutions, there are growing concerns about their cost and effectiveness. Unless they are carefully designed, limited in scope, and focused in effort, there is a serious risk that they will distract water experts and divert limited resources available for solving problems. In particular, ministerial meetings associated with many of these conferences are ineffective. They involve considerable preparatory work and lead to posturing, false expectations, and a focus on high-level political statements rather than on effective international and national policy development or small-scale community efforts and solutions. Indeed, reviewing the ministerial statement from the 4th World Water Forum in Mexico City in March 2006 (reprinted following this *In Brief*) reveals absolutely nothing of importance or value in terms of either new commitments on the part of governments or a promise of meeting existing commitments. Although the statement notes (in paragraph 7) that an increase in resources and funding will be needed, no such increase is promised.

Accordingly, Gleick and Lane (2005) recommended that future global water conferences be designed to address only the problems and solutions that really need global consensus. Any "ministerial" water meetings should be organized by the United Nations or national governments, not by separate international water organizations, either to review national progress on accepted targets or to seek formal governmental agreement on global water policies. The resources thus saved could be devoted to smaller sectoral or regional meetings, at which practical progress can be made much more readily and economically.

Background and History

Large water meetings are not new. Salman (2003, 2004) offers a history and background to international conferences. The International Water Association (IWA), the Water

---

1. This Water Brief benefited greatly, and is modified, from joint work done by Peter Gleick and Jon Lane, published as Gleick and Lane (2005).

Environment Federation (WEF), the International Commission on Irrigation and Drainage (ICID), and the International Water Resources Association (IWRA) have held regular water conferences going back to the early 1970s. The annual Stockholm Water Symposium, initiated in 1991, includes science and policy sessions, together with the awarding of the Stockholm Water Prize. In addition to regular association meetings, several special water conferences have been held, either singly or in smaller numbers.

The first major global conference focusing on water was organized by the United Nations in 1977 in Mar del Plata, Argentina. This conference highlighted the growing water crisis and produced the Mar del Plata Action Plan—detailed and comprehensive recommendations for water managers and policy makers that now shine in comparison to the ministerial statements issued at the recent World Water Forums. Among the highlights of the action plan were calls for the International Drinking Water Supply and Sanitation Decade, subsequently initiated in the 1980s, and one of the earliest acknowledgments of the human right to water. The real profusion of large conferences, however, did not begin until the mid-1990s, when two new organizations, the World Water Council (WWC) and the Global Water Partnership (GWP), began to dominate global water discussions. Until this point, most water meetings were organized and coordinated by the diverse United Nations agencies and programs with a role in fresh water, such as the International Hydrological Program (IHP), and by the professional water organizations, such as the International Association of Hydrological Sciences (IAHS) and the IWRA, whose triennial conferences drew much of the water community.

As Salman (2003) notes, the relationships between these new organizations and the other international agencies working on water "seem increasingly to edge on competition . . . despite the apparent efforts of cooperation in some fields." In fact, this tension was evident from the start, when the first World Water Forum was organized by the WWC in Marrakech, Morocco, in 1997, in direct competition with the traditional IWRA triennial congresses. This forum also set the stage for what have become triennial World Water Forums, at The Hague in 2000, in Kyoto in 2003, and in Mexico City in 2006. Professional attendance at these forums has grown from 500 at Marrakech through 5,000 at The Hague to 12,000 to 24,000 at Kyoto. Attendance in Mexico in 2006 was estimated to be just under 20,000 (4th World Water Forum 2006a).

As a measure of the cost (albeit not necessarily the benefit) of these meetings, Gleick and Lane (2005) estimated that the Kyoto Third World Water Forum in 2003 cost more than twice the total water supply and sanitation contribution of the United States Agency for International Development to the entire continent of Africa the same year, not including the cost of people's time and effort to prepare for, travel to, and participate in the meeting.

Outcomes of International Water Meetings

How can we measure whether or not the investments of time, effort, and money in large international water meetings are worthwhile? No formal comprehensive or independent evaluations of these meetings have been conducted, though some informal efforts have been made recently (TWCWM 2005; Varady and Iles-Shih 2005). One recent survey involved around sixty-five responses to a questionnaire, but approximately half of the respondents had not attended any large water meeting, and another quarter had been to only one (TWCWM 2005).

Few official indicators or measures of success of large water conferences are available, and there are many ways to evaluate them. No definitive indicator can be chosen; many outcomes are purely subjective or qualitative—developing new collaborations among water experts or groups, information sharing, public declarations, media attention, and even just meeting old and new friends. Other indicators can be quantitative and formal: adoption of political resolutions, expenditures of money on water-related problems, generation of information and data. Ultimately, the most important objective may be a reduction in the number and severity of water problems themselves. If the goal is increased awareness among the public, policy makers, and politicians, or improved "networking" among water experts, it can strongly be argued that the goal is being accomplished. Less clear is whether or not the ultimate objective of meeting basic human needs for water for the poorest is being achieved. At present, it appears unlikely that the Millennium Development Goals for water will be met.

Addressing Critical Issues

An argument in favor of large international water meetings is the opportunity to advance and debate critical and timely issues. If there are any areas in which international water meetings are important, this is one. A review of the outlines and contents of the World Water Forums, the Stockholm Water Symposia, and the IWRA triennial conferences, for example, shows that major water issues are each addressed in plenary talks or more focused panel discussions. As a tool for public awareness and education, such formats may be valuable. But the design of these meetings is increasingly detrimental to effective discussion and debate, and actual progress on unresolved issues is rare.

Focused, smaller international workshops on specific water challenges are more likely to be successful if the goal is to generate specific principles or actions that can be adopted by either national governments or international organizations working on water. For example, rather than continuing to rhetorically debate water privatization or the human right to water at general international water meetings, separate, focused workshops on privatization standards and principles or implementation of the human right to water as described by UN General Comment 15, are far more likely to produce progress. Such a meeting was coordinated in the fall of 2005 by the German Foreign Office to explore how to implement the human right to water (Auswärtiges Amt 2005).

Ministerial Statements

Ministerial meetings and statements are now routinely expected as part of major global water conferences. Early proponents of comprehensive "World Water Forum" meetings argued that the traditional water meetings held in the 1970s and 1980s were narrowly devoted to academic and scientific issues and lacked integration with policy makers and the public. This led to an effort to first invite water and environment ministers, and, ultimately, to embed ministerial meetings within the forums themselves. This approach has, at its heart, the goal of producing a ministerial declaration or statement to identify new commitments, policies, or agreements at the highest political level. Yet it can be argued that not one of the ministerial declarations produced for the World Water Forums, as a result of this approach, includes specific measurable programs or actions or new commitments.

In fact, despite nearly thirty years of meetings, there has been no more useful or stronger statement generated than that produced at the 1977 Mar del Plata or 1992 Dublin conferences. To make matters worse, the presence of water or environment ministers at these meetings can lead to distortions and disruptions as conflicting interests have attempted to influence wording, content, and outcomes. In the end, the ministerial products have been described by separate observers as "hardly an edifying example of political leadership on water" (Lane 2003), "awash in generalizations and compromising language" (Salman 2004), and "uncritical" and "weak" (Cain 2004).

The ministerial statement that came out of the 4th World Water Forum contains thirteen major paragraphs. Of these, four "reaffirm" and five "take note" or "recognize" past expressions of concern or commitments about water problems. The remaining four "acknowledge," "welcome," and "encourage" participants and existing efforts. The last "thanks" the government of Mexico and the WWC for organizing and hosting the meeting. There are no new contributions, exhortations, calls to action, or promises.

Finally, ministerial statements made at such conferences are not legally binding on countries. Although they may help influence or educate senior policy makers, those outcomes could be achieved at smaller, separately held meetings coordinated by the United Nations, where ministerial functions and activities are well understood and better managed. In particular, this approach would help to carry water-related messages to ministers of finance or foreign ministers, with whom the traditional water sector conferences have failed to communicate. Rather than ask finance ministers to attend water meetings, water professionals should attend finance meetings.

Conclusions and Recommendations

The trend to large and costly international water meetings needs to be reconsidered. The number of such meetings has become excessive, and they have lost focus and effectiveness. Rather than generating positive benefits and real progress toward solving world water problems, additional large-scale meetings are more likely to highlight our failure to make such progress and call attention to the fact that words are no substitute for actions. Global conferences do have a role to play but only if they are well planned with clear purpose, attendance, and objectives. By default, we believe this implies fewer such conferences.

Many of the benefits of international meetings can be obtained through small, detailed technical and policy workshops designed with specific outcomes and objectives. These are more likely to be successful than meetings where the only products are the same exhortations and platitudes we have all heard before. Carefully designed ministerial conferences are also vital, but, when held in conjunction with large international water meetings, they do not result in progress. Ministers of the environment or water or finance should certainly be invited to participate in international water conferences, but it is time to stop requesting ministerial "declarations" that declare nothing of value to our ultimate objective: meeting basic human and environmental water needs for all.

REFERENCES

4th World Water Forum. 2006a. The 4th world water forum concludes: 1,600 local actions were presented by 320 organizations in 205 sessions. Press release, March 22.

Auswärtige Amt. 2005. Stehen wir auf der Leitung?_Wasserpolitik nach dem UN-Gipfel 2005. October 20. http://www.auswaertiges-amt.de/diplo/de/Aussenpolitik/VereinteNationen/VN-Engagements/wasser-konferenz.html.

Gleick, P. H., Lane, J. 2005. Large international water meetings: Time for a reappraisal. *Water International*, 30(3):410–414.

Cain, N. 2004. 3rd world water forum in Kyoto: Disappointment and possibility. 189–197. In *The world's water 2004–2005: The biennial report on freshwater resources*. Gleick, P. H. et al. (editors). Washington, DC: Island Press.

Lane, J. 2003. Invited editorial: The 3rd world water forum. Kyoto. *Water Policy*, 5(4):381–382. March.

Salman, S. M. A. 2003. From Marrakech through The Hague to Kyoto: Has the global debate on water reached a dead end? Part 1. *Water International*, 28(4):491–500.

Salman, S. M. A. 2004. From Marrakech through The Hague to Kyoto: Has the global debate on water reached a dead end? Part 2. *Water International*, 29(1):11–19.

The Third World Centre for Water Management (TWCWM). 2005. Impact of mega-conferences and global water development and management. Conference proceedings. Bangkok, Thailand: TWCWM and Sasakawa Peace Foundation.

Varady, R. G., Iles-Shih, M. 2005. Global water initiatives: *What do the experts think?* Sasakawa Peace Foundation USA, Discussion Paper 7. Impact of mega-conferences and global water development and management. Draft of conference paper. Bangkok, Thailand.

4th WORLD WATER FORUM MINISTERIAL DECLARATION

We, the Ministers assembled in Mexico City on the occasion of the Fourth World Water Forum (4th WWF), "Local Actions for a Global Challenge" on March 21st and 22nd, 2006,

1. Reaffirm the critical importance of water, in particular freshwater, for all aspects on sustainable development, including poverty and hunger eradication, water-related disaster reduction, health, agricultural and rural development, hydropower, food security, gender equality as well as the achievement of environmental sustainability and protection. We underline the need to include water and sanitation as priorities in national processes, in particular national sustainable development and poverty reduction strategies.

2. Reaffirm our commitment to achieve the internationally agreed goals on integrated water resources management (IWRM), access to safe drinking water and basic sanitation, agreed upon in Agenda 21, the Millennium Declaration and the Johannesburg Plan of Implementation (JPOI). We reiterate the continued and urgent need to achieve these goals and to keep track of progress towards their implementation, including the goal to reduce by half, by the year 2015, the proportion of people unable to reach or afford safe drinking water.

3. Reaffirm, in particular, our commitment to the decisions adopted by the 13th session of the United Nations Commission on Sustainable Development (CSD-13), in April 2005, on policy options and practical measures to

expedite implementation in water, sanitation and human settlements. We note with interest the importance of enhancing the sustainability of ecosystems and acknowledge the implementation and importance in some regions of innovative practices such as rain water management and the development of hydropower projects. Further reaffirm the importance of the involvement of relevant stakeholders, particularly women and youth, in the planning and management of water services and, as appropriate, decision-making processes.

4. Take note of the Ministerial Declaration of the Third World Water Forum and recognize the work done within the UN System in support of member States, in order to reach the aforementioned goals. In this regard, we support the coordinating role of UN Water and highlight the need to strengthen its work within its mandate among the relevant UN organizations, funds and programs. We appreciate the inputs from the UN Secretary General's Advisory Board on Water and Sanitation to reinforce ongoing implementation efforts towards reaching water and sanitation targets. We express our continued support to initiatives like the Water for Life Decade and our interest in the United Nations Secretary General's WEHAB Initiative.

5. Recognize the contributions of the 4th WWF and its preparatory regional process to building capacity at international, regional and national levels and promoting the exchange of best practices and lessons learned on international water and sanitation issues.

6. Acknowledge the input of the Forum for the follow up segment on water and sanitation of the 16th Session of the CSD, to be held in 2008 which will play an important role to monitor and follow-up decisions on water and sanitation and their inter-linkages taken at CSD-13, and as an example of coordinated participation and involvement of governments at all levels, civil society, intergovernmental organizations, non-governmental organizations, private sector, scientific 2 / 2 institutions, partnerships, and international financial institutions regarding water issues and other relevant stakeholders.

7. Reaffirm also the decision of the 13th Session of the Commission of Sustainable Development regarding, inter alia: a) that a substantial increase of resources from all sources, including domestic resources, official development assistance and other resources will be required if developing countries are to achieve the internationally agreed development goals and targets, including those contained in the Millennium Declaration and the JPOI, and b) that Governments have the primary role in promoting improved access to safe drinking water, basic sanitation, sustainable and secure tenure, and adequate shelter, through improved governance at all levels and appropriate enabling environments and regulatory frameworks, adopting a pro-poor approach and with the active involvement of all stakeholders.

8. Recognize the importance of domestic and international policies that foster and assist building capacities and cooperation at all levels to

mitigate water-related disasters including prevention, preparedness, risk assessment, community awareness, resilience and response.

9. Recognize the important role that parliamentarians and local authorities are playing in various countries to increase sustainable access to water and sanitation services as well as to support integrated water resource management. An efficient collaboration with and between these actors is a key factor to meet our water related challenges and goals.

10. Note with appreciation the work of stakeholders at the regional preparatory process towards the 4th WWF as well as during the Forum itself, and take note of the documental output of the said regional preparatory stakeholder process, included as annex to this declaration. As appropriate, this output can be used as source of information for our tasks. We also thank participating parliamentarians and local authorities for their valuable views and opinions expressed during our joint working session at the Fourth World Water Forum, and take note of their statements included as annex to this declaration.

11. Welcome the launch at the 4th WWF of the CSD Water Action and Networking Database (CSD WAND), as a means of implementing the decision from CSD-13 to develop "web-based tools to disseminate information on implementation and best practices" on water and sanitation. The CSDWAND will serve as a platform for exchanging information and best practices, lessons learned and relevant international agreements and policy recommendations. We note that the CSD WAND has been built upon information collected during the 2003-2005 CSD Cycle in the Portfolio of Water Actions, as an output of the Ministerial Conference of the Third World Water Forum, and in the database of local actions of the 4th WWF.

12. Encourage all stakeholders, including national and international agencies, and other international and regional fora, such as World Water Week in Stockholm and the water weeks of the regional development banks, to contribute to and exchange information through the WAND.

13. Thank the Government of Mexico and the World Water Council for the organization of the 4th WWF and for their determination to promote better water management through dynamic local actions for a global change.

Mexico City, March 22, 2006.

Source: http://www.worldwaterforum4.org.mx/files/Ministerial_Declaration.pdf.

Environment and Security: Water Conflict Chronology Version, 2006-2007

Peter H. Gleick

The Environment and Security Water Conflict Chronology appears here for the fifth time. It is one of the most popular and regular features of *The World's Water* reports and is available online at http://www.worldwater.org, where regular updates appear. New additions continue to come to me from readers and researchers around the world, and the chronology is used regularly by the media and by academics interested in understanding more about both the history and character of disputes over water resources. A new typology is also presented here, which categorizes water conflicts as military targets, military tools, development disputes, and terrorism.

As we have consistently noted, water resources are rarely the sole source of conflict, and indeed, water is frequently a source of cooperation. Nevertheless, the history of violence over freshwater is long and distressing. Some international security experts ignore or misunderstand the complex and real relationships between water and security by drawing narrow definitions or using semantic twists to exclude water (and often other resources) from the debate over security. Such an approach both misunderstands the connections between water and security and misleads policy makers and the public seeking to reduce tensions and violence.

The Pacific Institute has been evaluating and analyzing these connections since its inception in 1987. A series of papers on these questions have been published ranging from historical reviews to regional case studies to theoretical analyses. We have organized workshops on lessons from regional conflicts in the Middle East, Central Asia, and Latin America, the connections between traditional and nontraditional arms control tools, and the role of science and religion in reducing the risks of water-related violence. In October 2004, the Institute held a workshop in collaboration with the Pontifical Academy of Sciences at the Vatican and Oregon State University, with the support of the Carnegie Corporation of New York, on "Water Conflicts and Spiritual

Transformations."[1] The list of attendees included people not often brought into these discussions: water scientists and managers, a Catholic bishop, leading policy makers and politicians, a Jewish rabbi, experts on Sufi and Hindi philosophy, representatives of the military, a Muslim scholar, and others.

We continue to update, modify, and expand the chronology. In the last print version, we added a series of myths, legends, and history of water conflicts in the Middle East, beginning 5,000 years before the present. The current version focuses a new spotlight on the connections between water and terrorism, as does a complete chapter of this volume of *The World's Water* (see Chapter 1). World events continue to expand the modern list, with examples in Southern Asia, Northern Africa, the Middle East, and elsewhere. Of particular note is a growing trend toward conflicts related to terrorism and to disputes over economic development. In particular, more and more of the entries in the chronology involve subnational players and actors, and fewer are related to transnational conflicts. This supports the thesis identified a decade ago in the first volume of *The World's Water*:

> Traditional political and ideological questions that have long dominated international discourse are now becoming more tightly woven with other variables that loomed less large in the past, including population growth, transnational pollution, resource scarcity and inequitable access to resources and their use.[2]

The updated chronology is presented here, with new entries and a range of corrections and modifications to the older ones. In particular, two old categories have been eliminated. Both "Control of Water Resources" and "Political Tool," which used to include events where water is used by a nation, state, or nonstate actor for a political goal, have been recategorized as military targets or tools, development disputes, or terrorism. These definitions remain imprecise, and single events can fall into more than one category, depending on perception, definition, and history. For example, the concept of the nation-state is a relatively recent one in history, which makes applying this typology to ancient legends, myths, and history challenging. Another challenge is confusion over the term *terrorism*, which presupposes the view that all governments are legitimate and all nonstate actors opposing them are illegitimate. As colloquially stated: one person's terrorist is another's freedom fighter (see Chapter 1).

The current categories or types of conflicts include

- *Military tool* (state actors): Water resources, or water systems themselves, are used by a nation or state as a weapon during a military action.

- *Military target* (state actors): Water resources, or water systems, are targets of military actions by nations or states.

- *Terrorism,* including *cyberterrorism* (nonstate actors): Water resources, or water systems, are the targets or tools of violence or coercion by nonstate actors; a distinction is drawn between environmental terrorism and ecoterrorism (see Chapter 1).

---

1. http://www.vatican.va/roman_curia/pontifical_academies/acdscien/documents/rc_pa_acdscien_doc_20040702_workshop-announcement-2004_en.html.

2. Gleick, P.H. 1998. "Conflict and cooperation over fresh water." In P.H. Gleick, *The World's Water 1998-1999: The Biennial Report on Freshwater Resources.* Island Press, Washington, D.C. pp.105.

- *Development disputes* (state and nonstate actors): Water resources, or water systems, are a major source of contention and dispute in the context of economic and social development.

The importance of water to life means that providing for water needs and demands will never be free of politics. As social and political systems change and evolve, so too will this chronology and the kinds of entries and categories. I look forward to the ongoing debate over water conflicts and to new contributions and comments from readers. Please email any contributions with full citations and supporting information to pgleick@pipeline.com.

W A T E R   B R I E F   4

Water Conflict Chronology[1]
Dr. Peter H. Gleick
Pacific Institute for Studies in Development, Environment, and Security

Date	Parties Involved	Basis of Conflict	Violent Conflict or in the Context of Violence?	Description	Sources
3000 B.C.	Ea, Noah	Religious account	Yes	Ancient Sumerian legend recounts the deeds of the deity Ea, who punished humanity for its sins by inflicting the Earth with a six-day storm. The Sumerian myth parallels the Biblical account of Noah and the deluge, although some details differ.	Hatami and Gleick 1994
2500 B.C.	Lagash, Umma	Military tool	Yes	Lagash-Umma Border Dispute—The dispute over the "Gu'edena" (edge of paradise) region begins. Urlama, King of Lagash from 2450 to 2400 B.C., diverts water from this region to boundary canals, drying up boundary ditches to deprive Umma of water. His son Il cuts off the water supply to Girsu, a city in Umma.	Hatami and Gleick 1994
1790 B.C.	Hammurabi	Development disputes	No	Code of Hammurabi for the State of Sumer—Hammurabi lists several laws pertaining to irrigation that address negligence of irrigation systems and water theft.	Hatami and Gleick 1994
1720–1684 B.C.	Abi-Eshuh, Iluma-Ilum	Military tool	Yes	Abi-Eshuh v. Iluma-Ilum. A grandson of Hammurabi, Abish or Abi-Eshuh, dams the Tigris to prevent the retreat of rebels led by Iluma-Ilum, who declared the independence of Babylon. This failed attempt marks the decline of the Sumerians who had reached their apex under Hammurabi.	Hatami and Gleick 1994
circa 1300 B.C.	Sisra, Barak, God	Religious account, military tool	Yes	This is an Old Testament account of the defeat of Sisera and his "nine hundred chariots of iron" by the unmounted army of Barak on the fabled Plains of Esdraelon. God sends heavy rainfall in the mountains, and the Kishon River overflows the plain and immobilizes or destroys Sisera's technologically superior forces ("...the earth trembled, and the heavens dropped, and the clouds also dropped water," Judges 5:4; "...The river of Kishon swept them away, that ancient river, the river Kishon," Judges 5:21).	New Scofield Reference Bible, KJV; Judges 4:7–15 and Judges 5:4–22.

Date	Parties	Type		Description	Source
1200 B.C.	Moses, Egypt	Military tool, religious account	Yes	Parting of the Red Sea. When Moses and the retreating Jews find themselves trapped between the Pharoah's army and the Red Sea, Moses miraculously parts the waters of the Red Sea, allowing his followers to escape. The waters close behind them and cut off the Egyptians.	Hatami and Gleick 1994
720–705 B.C.	Assyria, Armenia	Military tool	Yes	After a successful campaign against the Halidians of Armenia, Sargon II of Assyria destroys their intricate irrigation network and floods their land.	Hatami and Gleick 1994
705–682 B.C.	Sennacherib, Babylon	Military weapon/target	Yes	In quelling rebellious Assyrians in 695 B.C., Sennacherib razes Babylon and diverts one of the principal irrigation canals so that its waters wash over the ruins.	Hatami and Gleick 1994
6th century B.C.	Assyria	Military target; military tool	Yes	Assyrians poison the wells of their enemies with rye ergot.	Eitzen and Takafuji 1997
Unknown	Sennacherib, Jerusalem	Military tool	Yes	As recounted in Chronicles 32.3, Hezekiah digs into a well outside the walls of Jerusalem and uses a conduit to bring in water. Preparing for a possible siege by Sennacherib, he cuts off water supplies outside of the city walls, and Jerusalem survives the attack.	Hatami and Gleick 1994
681–699 B.C	Assyria, Tyre	Military tool, religious account	Yes	Esarhaddon, an Assyrian, refers to an earlier period when gods, angered by insolent mortals, created destructive floods. According to inscriptions recorded during his reign, Esarhaddon besieges Tyre, cutting off food and water.	Hatami and Gleick 1994
669–626 B.C.	Assyria, Arabia, Elam	Military tool, military target	Yes	Assurbanipal's inscriptions also refer to a siege against Tyre, although scholars attribute it to Esarhaddon. In campaigns against both Arabia and Elam in 645 B.C., Assurbanipal, son of Esarhaddon, dries up wells to deprive Elamite troops. He also guards wells from Arabian fugitives in an earlier Arabian war. On his return from victorious battle against Elam, Assurbanipal floods the city of Sapibel, and ally of Elam. According to inscriptions, he dams the Ulai River with the bodies of dead Elamite soldiers and deprives dead Elamite kings of their food and water offerings.	Hatami and Gleick 1994
612 B.C.	Egypt, Persia, Babylon, Assyria	Military tool	Yes	A coalition of Egyptian, Median (Persian), and Babylonian forces attacks and destroys Ninevah, the capital of Assyria. Nebuchadnezzar's father, Nebopolassar, leads the Babylonians. The converging armies divert the Khosr River to create a flood, which allows them to elevate their siege engines on rafts.	Hatami and Gleick 1994

continues

WATER CONFLICT CHRONOLOGY[1] *continued*

Date	Parties Involved	Basis of Conflict	Violent Conflict or in the Context of Violence?	Description	Sources
605–562 B.C.	Babylon	Military tool	No	Nebuchadnezzar builds immense walls around Babylon, using the Euphrates and canals as defensive moats surrounding the inner castle.	Hatami and Gleick 1994
590–600 B.C.	Cirrha, Delphi	Military tool	Yes	Athenian legislator Solon reportedly had roots of helleborus thrown into a small river or aqueduct leading from the Pleistrus River to Cirrha during a siege of this city. The enemy forces became violently ill and were defeated as a result. Some accounts have Solon building a dam across the Plesitus River cutting off the city's water supply. Such practices were widespread.	Absolute Astronomy 2006
558–528 B.C.	Babylon	Military tool	Yes	On his way from Sardis to defeat Nabonidus at Babylon, Cyrus faces a powerful tributary of the Tigris, probably the Diyalah. According to Herodotus' account, the river drowns his royal white horse and presents a formidable obstacle to his march. Cyrus, angered by the "insolence" of the river, halts his army and orders them to cut 360 canals to divert the river's flow. Other historians argue that Cyrus needed the water to maintain his troops on their southward journey, while another asserts that the construction was an attempt to win the confidence of the locals.	Hatami and Gleick 1994
539 B.C.	Babylon	Military tool	Yes	According to Herodotus, Cyrus invades Babylon by diverting the Euphrates above the city and marching troops along the dry riverbed. This popular account describes a midnight attack that coincided with a Babylonian feast.	Hatami and Gleick 1994
430 B.C.	Athens	Military tool	Yes	During the second year of the Peloponnesian War in 430 B.C. when plague broke out in Athens, the Spartans were accused of poisoning the cisterns of the Piraeus, the source of most of Athens' water.	Strategy Page 2006
355–323 B.C.	Babylon	Military tool	Yes	Returning from the razing of Persepolis, Alexander proceeds to India. After the Indian campaigns, he heads back to Babylon via the Persian Gulf and the Tigris, where he tears down defensive weirs that the Persians had constructed along the river. Arrian describes Alexander's disdain for the Persians' attempt to block navigation, which he saw as "unbecoming to men who are victorious in battle."	Hatami and Gleick 1994

Date	Parties	Type	Realized	Description	References
210–209 B.C.	Rome and Carthage	Military tool	Yes	In 210 B.C., Scipio crossed the Ebro to attack New Carthage. During a short siege, Scipio led a breaching column through a supposedly impregnable lagoon located on the landward side of the city; a strong northerly wind combined with the natural ebb of the tide left the lagoon shallow enough for the Roman infantry to wade through. New Carthage was soon taken.	Fonner 1996; Gowan 2004
537	Goths and Rome	Military tool, military target	Yes	In the 6th century A.D., as the Roman Empire began to decline, the Goths besieged Rome and cut almost all of the aqueducts leading into the city. In 537 A.D. this siege was successful. The only aqueduct that continued to function was that of the Aqua Virgo, which ran entirely underground.	Rome Guide 2004; InfoRoma 2004
1187	Saladin and the Middle East	Military tool	Yes	Saladin was able to defeat the Crusaders at the Horns of Hattin in 1187 by denying them access to water. In some reports, Saladin had sanded up all the wells along the way and had destroyed the villages of the Maronite Christians who would have supplied the Christian army with water.	Lockwood 2006; Priscoli 1998
1503	Florence and Pisa warring states	Military tool	No: plan only	Leonardo da Vinci and Machievelli plan to divert Arno River away from Pisa during conflict between Pisa and Florence.	Honan 1996
1573–1574	Holland and Spain	Military tool	Yes	In 1573 at the beginning of the eighty years war against Spain, the Dutch flooded the land to break the siege of Spanish troops on the town Alkmaar. The same defense was used to protect Lieden in 1574. This strategy became known as the Dutch Water Line and was used frequently for defense in later years.	Dutch Water Line 2002
1642	China; Ming Dynasty	Military tool	Yes	The Huang He's dikes breached for military purposes. In 1642, "toward the end of the Ming dynasty (1368–1644), General Gao Mingheng used the tactic near Kaifeng in an attempt to suppress a peasant uprising."	Hillel 1991
1672	French, Dutch	Military tool	Yes	Louis XIV starts the third of the Dutch Wars in 1672, in which the French overran the Netherlands. In defense, the Dutch opened their dikes and flooded the country, creating a watery barrier that was virtually impenetrable.	Columbia 2000

continues

WATER CONFLICT CHRONOLOGY[1] *continued*

Date	Parties Involved	Basis of Conflict	Violent Conflict or in the Context of Violence?	Description	Sources
1748	United States	Development dispute; terrorism	Yes	Ferry house on Brooklyn shore of East River burns down. New Yorkers accuse Brooklynites of having set the fire as revenge for unfair East River water rights.	Museum of the City of New York (MCNY n.d.)
1777	United States	Military tool	Yes	British and Hessians attacked the water system of New York. ". . . the enemy wantonly destroyed the New York water works" during the War for Independence.	Thatcher 1827
1841	Canada	Development dispute, terrorism	Yes	A reservoir in Ops Township, Upper Canada (now Ontario) was destroyed by neighbors who considered it a hazard to health.	Forkey 1998
1844	United States	Development dispute, terrorism	Yes	A reservoir in Mercer County, Ohio, was destroyed by a mob that considered it a hazard to health.	Scheiber 1969
1850s	United States	Development dispute; terrorism	Yes	Attack on a New Hampshire dam that impounded water for factories downstream by local residents unhappy over its effect on water levels.	Steinberg 1990
1853–1861	United States	Development dispute, terrorism	Yes	Repeated destruction of the banks and reservoirs of the Wabash and Erie Canal in Southern Indiana by mobs regarding it as a health hazard.	Fatout 1972; Fickle 1983
1860–1865	United States	Military tool; military target	Yes	W.T. Sherman's memoirs contain an account of Confederate soldiers poisoning ponds by dumping the carcasses of dead animals into them. Other accounts suggest this tactic was used by both sides.	Eitzen and Takafuji 1997
1870s	China	Development dispute	No	Local construction and government removal (twice) of an unauthorized dam in Hubei, China.	Rowe 1988
1870s to 1881	United States	Development dispute	Yes	Recurrent friction and eventual violent conflict over water rights in the vicinity of Tularosa, New Mexico, involving villagers, ranchers, and farmers.	Rasch 1968
1887	United States	Development dispute, terrorism	Yes	Dynamiting of a canal reservoir in Paulding County, Ohio, by a mob regarding it as a health hazard. State militia called out to restore order.	Walters 1948
1990	Canada	Development dispute, terrorism	Yes	Partly successful attempt to destroy a lock on the Welland Canal in Ontario, Canada, either by Fenians protesting English Policy in Ireland or by agents of Buffalo, New York, grain handlers unhappy at the diversion of trade through the canal.	Styran and Taylor 2001

Date	Parties		Description	Sources
1908–1909	United States	Development dispute / Yes	Violence, including a murder, directed against agents of a land company that claimed title to Reelfoot Lake in northwestern Tennessee who attempted to levy charges for fish taken and threatened to drain the lake for agriculture.	Vanderwood 1969
1863	United States Civil War	Military tool / Yes	General U.S. Grant, during the Civil War campaign against Vicksburg, cut levees in the battle against the Confederates.	Grant 1885; Barry 1997
1898	Egypt; France; Britain	Military and political tools / Military maneuvers	Military conflict nearly ensues between Britain and France in 1898 when a French expedition attempted to gain control of the headwaters of the White Nile. While the parties ultimately negotiated a settlement of the dispute, the incident has been characterized as having "dramatized Egypt's vulnerable dependence on the Nile, and fixed the attitude of Egyptian policy-makers ever since."	Moorhead 1960
1907–1913	Owens Valley, Los Angeles, California	Terrorism, development dispute / Yes	The Los Angeles Valley aqueduct/pipeline suffers repeated bombings in an effort to prevent diversions of water from the Owens Valley to Los Angeles.	Reisner 1986, 1993
1915	German Southwest Africa	Military tool / Yes	Union of South African troops capture Windhoek, capital of German Southwest Africa (May). Retreating German troops poison wells.	Daniel 1995
1935	California, Arizona	Development dispute / Military maneuvers	Arizona calls out the National Guard and militia units to the border with California to protest the construction of Parker Dam and diversions from the Colorado River; dispute ultimately is settled in court.	Reisner 1986, 1993
1938	China and Japan	Military tool, military target / Yes	Chiang Kai-shek orders the destruction of flood-control dikes of the Huayuankou section of the Huang He (Yellow) river to flood areas threatened by the Japanese army. West of Kaifeng dikes are destroyed with dynamite, spilling water across the flat plain. The flood destroyed part of the invading army and its heavy equipment was mired in thick mud, though Wuhan, the headquarters of the Nationalist government, was taken in October. The waters flooded an area variously estimated as between 3,000 and 50,000 square kilometers, and killed Chinese estimated in numbers between "tens of thousands" and "one million."	Hillel 1991; Yang Lang 1989, 1994
1939–1942	Japan, China	Military target, military tool / Yes	Japanese chemical and biological weapons activities reportedly include tests by "Unit 731" against military and civilian targets by lacing water wells and reservoirs with typhoid and other pathogens.	Harris 1994
1940–1945	Multiple parties	Military target / Yes	Hydroelectric dams routinely bombed as strategic targets during World War II.	Gleick 1993

continues

197

WATER CONFLICT CHRONOLOGY[1] *continued*

Date	Parties Involved	Basis of Conflict	Violent Conflict or in the Context of Violence?	Description	Sources
1943	Britain, Germany	Military target	Yes	British Royal Air Force bombed dams on the Möhne, Sorpe, and Eder Rivers, Germany (May 16, 17). Möhne Dam breech killed 1,200, destroying all downstream dams for 50 km. The flood that occurred after breaking the Eder dam reached a peak discharge of 8,500 m^3/s, which is nine times higher than the highest flood observed. Many houses and bridges were destroyed. 68 were killed.	Kirschner 1949; Semann 1950
1944	Germany, Italy, Britain, United States	Military tool	Yes	German forces used waters from the Isoletta Dam (Liri River) in January and February to successfully destroy British assault forces crossing the Garigliano River (downstream of Liri River). The German Army then dammed the Rapido River, flooding a valley occupied by the American Army.	Corps of Engineers 1953
1944	Germany, Italy, Britain, United States	Military tool	Yes	German Army flooded the Pontine Marches by destroying drainage pumps to contain the Anzio beachhead established by the Allied landings in 1944. Over 40 square miles of land were flooded; a 30-mile stretch of landing beaches was rendered unusable for amphibious support forces.	Corps of Engineers 1953
1944	Germany, Allied forces	Military tool	Yes	Germans flooded the Ay River, France (July) creating a lake two meters deep and several kilometers wide, slowing an advance on Saint Lo, a German communications center in Normandy.	Corps of Engineers 1953
1944	Germany, Allied forces	Military tool	Yes	Germans flooded the Ill River Valley during the Battle of the Bulge (winter 1944–45) creating a lake 16 kilometers long, 3–6 kilometers wide, and 1–2 meters deep, greatly delaying the American Army's advance toward the Rhine.	Corps of Engineers 1953
1945	Romania, Germany	Military target	Yes	The only known German tactical use of biological warfare was the pollution of a large reservoir in northwestern Bohemia with sewage in May 1945.	SIPRI 1971
1947– present	Bangladesh, India	Development dispute	No	Partition divides the Ganges River between Bangladesh and India; construction of the Farakka barrage by India, beginning in 1962, increases tension; short-term agreements settle dispute in 1977–82, 1982–84, and 1985–88, and thirty-year treaty is signed in 1996.	Butts 1997; Samson and Charrier 1997
1947–1960s	India, Pakistan	Development dispute	No	Partition leaves Indus basin divided between India and Pakistan; disputes over irrigation water ensue, during which India stems flow of water into irrigation canals in Pakistan; Indus Waters Agreement reached in 1960 after 12 years of World Bank-led negotiations.	Bingham et al. 1994; Wolf 1997

Date	Parties	Type	Yes/No	Description	Sources
1948	Arabs, Israelis	Military tool	Yes	Arab forces cut off West Jerusalem's water supply in first Arab-Israeli war.	Wolf 1995, 1997
1950s	Korea, United States, others	Military target	Yes	Centralized dams on the Yalu River serving North Korea and China are attacked during Korean War.	Gleick 1993
1951	Korea, United Nations	Military tool, military target	Yes	North Korea released flood waves from the Hwachon Dam damaging floating bridges operated by UN troops in the Pukhan Valley. U.S. Navy plans were then sent to destroy spillway crest gates.	Corps of Engineers 1953
1951	Israel, Jordan, Syria	Military tool, development disputes	Yes	Jordan makes public its plans to irrigate the Jordan Valley by tapping the Yarmouk River; Israel responds by commencing drainage of the Huleh swamps located in the demilitarized zone between Israel and Syria; border skirmishes ensue between Israel and Syria.	Wolf 1997; Samson and Charrier 1997
1953	Israel, Jordan, Syria	Development dispute, military target	Yes	Israel begins construction of its National Water Carrier to transfer water from the north of the Sea of Galilee out of the Jordan basin to the Negev Desert for irrigation. Syrian military actions along the border and international disapproval lead Israel to move its intake to the Sea of Galilee.	Naff and Matson 1984; Samson and Charrier 1997
1958	Egypt, Sudan	Military tool, development dispute	Yes	Egypt sends an unsuccessful military expedition into disputed territory amidst pending negotiations over the Nile waters, Sudanese general elections, and an Egyptian vote on Sudan-Egypt unification; Nile Water Treaty signed when pro-Egyptian government elected in Sudan.	Wolf 1997
1960s	North Vietnam, United States	Military target	Yes	Irrigation water supply systems in North Vietnam are bombed during Vietnam War. 661 sections of dikes damaged or destroyed.	IWTC 1967; Gleick 1993; Zemmali 1995
1962	Israel, Syria	Development dispute, military target	Yes	Israel destroys irrigation ditches in the lower Tarfiq in the demilitarized zone. Syria complains.	Naff and Matson 1984
1962–1967	Brazil; Paraguay	Military tool, development dispute	Military maneuvers	Negotiations between Brazil and Paraguay over the development of the Paraná River are interrupted by a unilateral show of military force by Brazil in 1962, which invades the area and claims control over the Guaíra Falls site. Military forces were withdrawn in 1967 following an agreement for a joint commission to examine development in the region.	Murphy and Sabadell 1986
1963–1964	Ethiopia, Somalia	Development dispute, military tool	Yes	Creation of boundaries in 1948 leaves Somali nomads under Ethiopian rule; border skirmishes occur over disputed territory in Ogaden desert, where critical water and oil resources are located; cease-fire is negotiated only after several hundred are killed.	Wolf 1997
1964	Cuba, United States	Military tool	No	On February 6, 1964, the Cuban government ordered the water supply cut off to the U.S. Naval Base at Guantanamo Bay.	Guantanamo Bay Gazette 1964

continues

WATER CONFLICT CHRONOLOGY[1] *continued*

Date	Parties Involved	Basis of Conflict	Violent Conflict or in the Context of Violence?	Description	Sources
1964	Israel, Syria	Military target	Yes	Headwaters of the Dan River on the Jordan River are bombed at Tell El-Qadi in a dispute about sovereignty over the source of the Dan.	Naff and Matson 1984
1965	Zambia, Rhodesia, Great Britain	Military target	No	President Kenneth Kaunda calls on British government to send troops to Kariba Dam to protect it from possible saboteurs from Rhodesian government.	Chenje 2001
1965	Israel, Palestinians	Terrorism	Yes	First attack ever by the Palestinian National Liberation Movement Al-Fatah is on the diversion pumps for the Israeli National Water Carrier. Attack fails.	Naff and Matson 1984; Dolatyar 1995
1965–1966	Israel, Syria	Military tool, development dispute	Yes	Fire is exchanged over "all-Arab" plan to divert the Jordan River headwaters (Hasbani and Banias) and presumably preempt Israeli National Water Carrier; Syria halts construction of its diversion in July 1966.	Wolf 1995, 1997
1966–1972	Vietnam, United States	Military tool	Yes	United States tries cloud-seeding in Indochina to stop flow of materiel along Ho Chi Minh trail.	Plant 1995
1967	Israel, Syria	Military target and tool	Yes	Israel destroys the Arab diversion works on the Jordan River headwaters. During Arab-Israeli War Israel occupies Golan Heights, with Banias tributary to the Jordan; Israel occupies West Bank.	Gleick 1993; Wolf 1995, 1997; Wallenstein and Swain 1997
1969	Israel, Jordan	Military target and tool	Yes	Israel, suspicious that Jordan is overdiverting the Yarmouk, leads two raids to destroy the newly-built East Ghor Canal; secret negotiations, mediated by the United States, lead to an agreement in 1970.	Samson and Charrier 1997
1970	United States	Terrorism	No: threat	The Weathermen, a group opposed to American imperialism and the Vietnam war, allegedly attempted to obtain biological agents to contaminate the water supply systems of U.S. urban centers.	Kupperman and Trent 1979; Eitzen and Takafuji 1997; Purver 1995

200

Date	Parties involved	Basis of conflict	Violent conflict or in the context of violence	Description	Sources
1970s	Argentina, Brazil, Paraguay	Development dispute	No	Brazil and Paraguay announce plans to construct a dam at Itaipu on the Paraná River, causing Argentina concern about downstream environmental repercussions and the efficacy of their own planned dam project downstream. Argentina demands to be consulted during the planning of Itaipu but Brazil refuses. An agreement is reached in 1979 that provides for the construction of both Brazil and Paraguay's dam at Itaipu and Argentina's Yacyreta dam.	Wallenstein and Swain 1997
1972	United States	Terrorism	No: Threat	Two members of the right-wing "Order of the Rising Sun" are arrested in Chicago with 30–40 kg of typhoid cultures that are allegedly to be used to poison the water supply in Chicago, St. Louis, and other cities. It was felt that the plan would have been unlikely to cause serious health problems due to chlorination of the water supplies.	Eitzen and Takafuji 1997
1972	United States	Terrorism	No: threat	Reported threat to contaminate water supply of New York City with nerve gas.	Purver 1995
1972	North Vietnam	Military target	Yes	United States bombs dikes in the Red River delta, rivers, and canals during massive bombing campaign.	Columbia Electronic Encyclopedia 2000
1973	Germany	Terrorism	No: Threat	Threat by a biologist in Germany to contaminate water supplies with bacilli of anthrax and botulinum unless he was paid $8.5 million	Jenkins and Rubin 1978; Kupperman and Trent 1979
1974	Iraq, Syria	Military target, military tool, development dispute	Military maneuvers	Iraq threatens to bomb the al-Thawra dam in Syria and massed troops along the border, alleging that the dam had reduced the flow of Euphrates River water to Iraq.	Gleick 1994
1975	Iraq, Syria	Development dispute, military tool	Military maneuvers	As upstream dams are filled during a low-flow year on the Euphrates, Iraqis claim that flow reaching its territory is "intolerable" and asks the Arab League to intervene. Syrians claim they are receiving less than half the river's normal flow and pull out of an Arab League technical committee formed to mediate the conflict. In May Syria closes its airspace to Iraqi flights and both Syrian and Iraq reportedly transfer troops to their mutual border. Saudi Arabia successfully mediates the conflict.	Gleick 1993, 1994; Wolf 1997
1975	Angola, South Africa	Military goal, military target	Yes	South African troops move into Angola to occupy and defend the Ruacana hydropower complex, including the Gové Dam on the Kunene River. Goal is to take possession of and defend the water resources of southwestern Africa and Namibia.	Meissner 2000

continues

201

WATER CONFLICT CHRONOLOGY[1] *continued*

Date	Parties Involved	Basis of Conflict	Violent Conflict or in the Context of Violence?	Description	Sources
1977	United States	Terrorism	Yes	Contamination of a North Carolina reservoir with unknown materials. According to Clark: "Safety caps and valves were removed, and poison chemicals were sent into the reservoir.... Water had to be brought in."	Clark 1980; Purver 1995
1978–present	Egypt, Ethiopia	Development dispute, political tool	No	Long standing tensions over the Nile, especially the Blue Nile, originating in Ethiopia. Ethiopia's proposed construction of dams on the headwaters of the Blue Nile leads Egypt to repeatedly declare the vital importance of water. "The only matter that could take Egypt to war again is water" (Anwar Sadat 1979). "The next war in our region will be over the waters of the Nile, not politics" (Boutrous Ghali 1988).	Gleick 1991, 1994
1978–1984	Sudan	Development dispute, military target, terrorism	Yes	Demonstrations in Juba, Sudan in 1978 opposing the construction of the Jonglei Canal led to the deaths of two students. Construction of the Jonglei Canal in the Sudan was forcibly suspended in 1984 following a series of attacks on the construction site.	Suliman 1998; Keluel-Jang 1997
1980s	Mozambique, Rhodesia/ Zimbabwe, South Africa	Military target, terrorism	Yes	Regular destruction of power lines from Cahora Bassa Dam during fight for independence in the region. Hydropower dam targeted by RENAMO.	Chenje 2001
1981	Iran, Iraq	Military target and tool	Yes	Iran claims to have bombed a hydroelectric facility in Kurdistan, thereby blacking out large portions of Iraq, during the Iran-Iraq War.	Gleick 1993
1980–1988	Iran, Iraq	Military tool	Yes	Iran diverts water to flood Iraqi defense positions.	Plant 1995
1982	United States	Terrorism	No: threat	Los Angeles police and the FBI arrest a man who was preparing to poison the city's water supply with a biological agent.	Livingstone 1982; Eitzen and Takafuji 1997
1982	Israel, Lebanon, Syria	Military tool	Yes	Israel cuts off the water supply of Beirut during siege.	Wolf 1997
1981–1982	Angola	Military target, military tool	Yes	Water infrastructure, including dams and the major Cunene-Cuvelai pipeline, was targeted during the conflicts in Namibia and Angola in the 1980s.	Turton 2005
1982	Guatemala	Development dispute	Yes	177 civilians killed in Rio Negro over opposition to Chixoy hydroelectric dam.	Levy 2000

Year	Parties	Basis of Conflict	Violent Conflict or in the Context of Violence	Description	Sources
1983	Israel	Terrorism	No	The Israeli government reports that it had uncovered a plot by Israeli Arabs to poison the water in Galilee with "an unidentified powder."	Douglass and Livingstone 1987
1984	United States	Terrorism	Yes	Members of the Rajneeshee religious cult contaminate a city water supply tank in The Dalles, Oregon, using Salmonella. A community outbreak of over 750 cases occurred in a county that normally reports fewer than five cases per year.	Clark and Deininger 2000
1985	United States	Terrorism	No	Law enforcement authorities discovered that a small survivalist group in the Ozark Mountains of Arkansas known as The Covenant, the Sword, and the Arm of the Lord (CSA) had acquired a drum containing 30 gallons of potassium cyanide, with the apparent intent to poison water supplies in New York, Chicago, and Washington, DC. CSA members devised the scheme in the belief that such attacks would make the Messiah return more quickly by punishing unrepentant sinners. The objective appeared to be mass murder in the name of a divine mission rather than to change government policy. The amount of poison possessed by the group is believed to have been insufficient to contaminate the water supply of even one city.	Tucker 2000; NTI 2005
1986	North Korea, South Korea	Military tool	No	North Korea's announcement of its plans to build the Kumgansan hydroelectric dam on a tributary of the Han River upstream of Seoul raises concerns in South Korea that the dam could be used as a tool for ecological destruction or war.	Gleick 1993
1986	Lesotho, South Africa	Military goal, development dispute	Yes	South Africa supports coup in Lesotho over support for ANC and anti-apartheid, and water. New government in Lesotho then quickly signs Lesotho Highlands water agreement.	American University 2000b
1986	Lesotho, South Africa	Development dispute, military goal	Yes	Bloodless coup by Lesotho's defense forces, with support from South Africa, lead to immediate agreement with South Africa for water from the Lesotho Highlands, after 30 previous years of unsuccessful negotiations. There is disagreement over the degree to which water was a motivating factor for either party.	Mohamed 2001
1988	Angola, South Africa, Cuba	Military goal, military target	Yes	Cuban and Angolan forces launch an attack on Calueque Dam via land and then air. Considerable damage inflicted on dam wall; power supply to dam cut. Water pipeline to Owamboland is cut and destroyed.	Meissner 2000
1990	South Africa	Development dispute	No	Pro-apartheid council cuts off water to the Wesselton township of 50,000 blacks following protests over miserable sanitation and living conditions.	Gleick 1993

continues

203

WATER CONFLICT CHRONOLOGY[1] *continued*

Date	Parties Involved	Basis of Conflict	Violent Conflict or in the Context of Violence?	Description	Sources
1990	Iraq, Syria, Turkey	Development dispute, military tool	No	The flow of the Euphrates is interrupted for a month as Turkey finishes construction of the Ataturk Dam, part of the Grand Anatolia Project. Syria and Iraq protest that Turkey now has a weapon of war. In mid-1990 Turkish president Turgut Ozal threatens to restrict water flow to Syria to force it to withdraw support for Kurdish rebels operating in Southern Turkey.	Gleick 1993, 1995
1991– present	Karnataka, Tamil Nadu (India)	Development dispute	Yes	Violence erupts when Karnataka rejects an Interim Order handed down by the Cauvery Waters Tribunal, set up by the Indian Supreme Court. The Tribunal was established in 1990 to settle two decades of dispute between Karnataka and Tamil Nadu over irrigation rights to the Cauvery River.	Gleick 1993; Butts 1997; American University 2000a
1991	Iraq, Kuwait, US	Military target	Yes	During the Gulf War, Iraq destroys much of Kuwait's desalination capacity during retreat.	Gleick 1993
1991	Canada	Terrorism	No: threat	A threat is made via an anonymous letter to contaminate the water supply of the city of Kelowna, British Columbia, with "biological contaminates." The motive was apparently "associated with the Gulf War." The security of the water supply was increased in response and no group was identified as the perpetrator.	Purver 1995
1991	Iraq, Turkey, United Nations	Military tool	Yes	Discussions are held at the United Nations about using the Ataturk Dam in Turkey to cut off flows of the Euphrates to Iraq.	Gleick 1993
1991	Iraq, Kuwait, US	Military target	Yes	Baghdad's modern water supply and sanitation system are intentionally and unintentionally damaged by Allied coalition. "Four of seven major pumping stations were destroyed, as were 31 municipal water and sewerage facilities—20 in Baghdad, resulting in sewage pouring into the Tigris. Water purification plants were incapacitated throughout Iraq" (Arbuthnot 2000). In the first eight months of 1991, after Iraq's water infrastructure was damaged by the Persian Gulf War, the *New England Journal of Medicine* reported that nearly 47,000 more children than normal died in Iraq and the country's infant mortality rate doubled to 92.7 per 1,000 live births.	Gleick 1993; Arbuthnot 2000; Barrett 2003

Date	Parties	Basis of conflict	Violent conflict or in the context of violence	Description	Sources
1992	Czechoslovakia, Hungary	Political tool, development dispute	Military maneuvers	Hungary abrogates a 1977 treaty with Czechoslovakia concerning construction of the Gabcikovo/Nagymaros project based on environmental concerns. Slovakia continues construction unilaterally, completes the dam, and diverts the Danube into a canal inside the Slovakian republic. Massive public protest and movement of military to the border ensue; issue taken to the International Court of Justice.	Gleick 1993
1992	Turkey	Terrorism	Yes	Lethal concentrations of potassium cyanide are reported discovered in the water tanks of a Turkish Air Force compound in Istanbul. The Kurdish Workers' Party (PKK) claimed credit.	Chelyshev 1992
1992	Bosnia, Bosnian Serbs	Military tool	Yes	The Serbian siege of Sarajevo, Bosnia, and Herzegovina, includes a cutoff of all electrical power and the water feeding the city from the surrounding mountains. The lack of power cuts the two main pumping stations inside the city despite pledges from Serbian nationalist leaders to United Nations officials that they would not use their control of Sarajevo's utilities as a weapon. Bosnian Serbs take control of water valves regulating flow from wells that provide more than 80 percent of water to Sarajevo; reduced water flow to city is used to "smoke out" Bosnians.	Burns 1992; Husarska 1995
1993–present	Iraq	Military tool	No	To quell opposition to his government, Saddam Hussein reportedly poisons and drains the water supplies of southern Shiite Muslims, the Ma'dan. The marshes of southern Iraq are intentionally targeted. The European Parliament and UN Human Rights Commission deplore use of water as a weapon in region.	Gleick 1993; American University 2000c; *National Geographic News* 2001
1993	Iran	Terrorism	No	A report suggests that proposals were made at a meeting of fundamentalist groups in Tehran, under the auspices of the Iranian Foreign Ministry, to poison water supplies of major cities in the West "as a possible response to Western offensives against Islamic organizations and states."	Haeri 1993
1993	Yugoslavia	Military target and tool	Yes	Peruca Dam intentionally destroyed during war.	Gleick 1993
1994	Moldavia	Terrorism	No: threat	Reported threat by Moldavian General Nikolay Matveyev to contaminate the water supply of the Russian 14th Army in Tiraspol, Moldova, with mercury.	Purver 1995

continues

WATER CONFLICT CHRONOLOGY[1] *continued*

Date	Parties Involved	Basis of Conflict	Violent Conflict or in the Context of Violence?	Description	Sources
1995	Ecuador, Peru	Military and political tool	Yes	Armed skirmishes arise in part because of disagreement over the control of the headwaters of Cenepa River. Wolf argues that this is primarily a border dispute simply coinciding with location of a water resource.	Samson and Charrier 1997; Wolf 1997
1997	Singapore, Malaysia	Political tool	No	Malaysia supplies about half of Singapore's water and in 1997 threatened to cut off that supply in retribution for criticisms by Singapore of policy in Malaysia.	Zachary 1997
1998	Tajikistan	Terrorism, political tool	No: threat	On November 6, a guerrilla commander threatened to blow up a dam on the Kairakkhum channel if political demands are not met. Col. Makhmud Khudoberdyev made the threat, reported by the ITAR-Tass News Agency.	WRR 1998
1998	Angola	Military and political tool	Yes	In September 1998, fierce fighting between UNITA and Angolan government forces broke out at Gove Dam on the Kunene River for control of the installation.	Meissner 2001
1998/1994	United States	Cyberterrorism	No	The Washington Post reports a 12-year-old computer hacker broke into the SCADA computer system that runs Arizona's Roosevelt Dam, giving him complete control of the dam's massive floodgates. The cities of Mesa, Tempe, and Phoenix, Arizona, are downstream of this dam. No damage was done. This report turns out to be incorrect. A hacker did break into the computers of an Arizona water facility, the Salt River Project in the Phoenix area. But he was 27 years old, not 12, and the incident occurred in 1994, not 1998. And although clearly trespassing in critical areas, investigators concluded that no lives or property were ever threatened.	Gellman 2002; Lemos 2002
1998	Democratic Republic of Congo	Military target, terrorism	Yes	Attacks on Inga Dam during efforts to topple President Kabila. Disruption of electricity supplies from Inga Dam and water supplies to Kinshasa	Chenje 2001; Human Rights Watch 1998
1998–2000	Eritrea and Ethiopia	Military target	Yes	Water pumping plants and pipelines in the border town of Adi Quala were destroyed during the civil war between Eritrea and Ethiopia.	ICRC 2003
1999	Lusaka, Zambia	Terrorism, political tool	Yes	Bomb blast destroyed the main water pipeline, cutting off water for the city of Lusaka, population 3 million.	FTGWR 1999

Date	Parties	Basis of Conflict	Violent Conflict	Description	Sources
1999	Yugoslavia	Military target	Yes	Belgrade reported that NATO planes had targeted a hydroelectric plant during the Kosovo campaign.	Reuters 1999a
1999	Bangladesh	Development dispute, political tool	Yes	50 hurt during strikes called to protest power and water shortages. Protest led by former Prime Minister Begum Khaleda Zia over deterioration of public services and in law and order.	Ahmed 1999
1999	Yugoslavia	Military target	Yes	NATO targets utilities and shuts down water supplies in Belgrade. NATO bombs bridges on Danube, disrupting navigation.	Reuters 1999b
1999	Yugoslavia	Political tool	Yes	Yugoslavia refuses to clear war debris on Danube (downed bridges) unless financial aid for reconstruction is provided; European countries on Danube fear flooding due to winter ice dams will result. Diplomats decry environmental blackmail.	Simons 1999
1999	Kosovo	Political tool	Yes	Serbian engineers shut down water system in Pristina prior to occupation by NATO.	Reuters 1999c
1999	South Africa	Terrorism	Yes	A homemade bomb was discovered at a water reservoir at Wallmansthal near Pretoria. It was thought to have been meant to sabotage water supplies to farmers.	Pretoria Dispatch 1999
1999	Angola	Terrorism, political tool	Yes	100 bodies were found in four drinking water wells in central Angola.	*International Herald Tribune* 1999
1999	Puerto Rico, United States	Political tool	No	Protesters blocked water intake to Roosevelt Roads Navy Base in opposition to U.S. military presence and Navy's use of the Blanco River, following chronic water shortages in neighboring towns.	*New York Times* 1999
1999	China	Development dispute; terrorism	Yes	Around Chinese New Year, farmers from Hebei and Henan provinces fought over limited water resources. Heavy weapons, including mortars and bombs, were used and nearly 100 villagers were injured. Houses and facilities were damaged and the total loss reached $1 million. Parties involved: Huanglongkou Village, Shexian County, Hebei Province and Gucheng Village, Linzhou City, Henan Province	*China Water Resources Daily* 2002
1999	East Timor	Military tool, terrorism	Yes	Militia opposing East Timor independence kill pro-independence supporters and throw bodies in water well.	BBC 1999
1998–1999	Kosovo	Terrorism, political tool	Yes	Contamination of water supplies/wells by Serbs disposing of bodies of Kosovar Albanians in local wells. Other reports of Yugoslav federal forces poisoning wells with carcasses and hazardous materials.	CNN 1999; Hickman 1999
1999 to 2000	Namibia, Botswana, Zambia	Military goal: development dispute	No	Sedudu/Kasikili Island, in the Zambezi/Chobe River. Dispute over border and access to water. Presented to the International Court of Justice	*ICJ* 1999

continues

WATER CONFLICT CHRONOLOGY[1] *continued*

Date	Parties Involved	Basis of Conflict	Violent Conflict or in the Context of Violence?	Description	Sources
2000	Ethiopia	Development dispute	Yes	One man stabbed to death during fight over clean water during famine in Ethiopia.	Sandrasagra 2000
2000	Central Asia: Kyrgyzstan, Kazakhstan, Uzbekistan	Development dispute	No	Kyrgyzstan cuts off water to Kazakhstan until coal is delivered; Uzbekistan cuts off water to Kazakhstan for nonpayment of debt.	Pannier 2000
2000	Belgium	Terrorism	Yes	In July, workers at the Cellatex chemical plant in Northern France dumped 5,000 liters of sulfuric acid into a tributary of the Meuse River when they were denied workers' benefits. A French analyst pointed out that this was the first time "the environment and public health were made hostage in order to exert pressure, an unheard-of situation until now."	Christian Science Monitor 2000
2000	Hazarajat, Afghanistan	Development dispute	Yes	Violent conflicts broke out over water resources in the villages Burna Legan and Taina Legan, and in other parts of the region, as drought depleted local resources.	Cooperation Center for Afghanistan 2000
2000	India: Gujarat	Development dispute	Yes	Water riots reported in some areas of Gujarat to protest against authority's failure to arrange adequate supply of tanker water. Police are reported to have shot into a crowd at Falla village near Jamnagar, resulting in the death of three and injuries to 20 following protests against the diversion of water from the Kankavati dam to Jamnagar town.	FTGWR 2000
2000	Kenya	Development dispute	Yes	A clash between villagers and thirsty monkeys left eight apes dead and ten villagers wounded. The duel started after water tankers brought water to a drought-stricken area and monkeys desperate for water attacked the villagers.	BBC 2000; Okoko 2000
2000	Australia	Cyberterrorism	Yes	In Queensland, Australia, on April 23, 2000, police arrested a man for using a computer and radio transmitter to take control of the Maroochy Shire wastewater system and release sewage into parks, rivers, and property.	Gellman 2002

Year	Parties	Type	Violent conflict	Description	Sources
2000	China	Development dispute	Yes	Civil unrest erupted over use and allocation of water from Baiyangdian Lake, the largest natural lake in northern China. Several people died in riots by villagers in July 2000 in Shandong after officials cut off water supplies. In August 2000, six died when officials in the southern province of Guangdong blew up a water channel to prevent a neighboring county from diverting water.	Pottinger 2000
2001	Israel, Palestine	Terrorism, military target	Yes	Palestinians destroy water supply pipelines to West Bank settlement of Yitzhar and to Kibbutz Kisufim. Agbat Jabar refugee camp near Jericho disconnected from its water supply after Palestinians looted and damaged local water pumps. Palestinians accuse Israel of destroying a water cistern, blocking water tanker deliveries, and attacking materials for a wastewater treatment project.	Israel Line 2001a, 2001b; ENS 2001
2001	Pakistan	Development dispute, terrorism	Yes	Civil unrest over severe water shortages caused by the long-term drought. Protests began in March and April and continued into summer. Riots, four bombs in Karachi (June 13), one death, 12 injuries, 30 arrests. Ethnic conflicts as some groups "accuse the government of favoring the populous Punjab province [over Sindh province] in water distribution."	Nadeem 2001; Soloman 2001
2001	Macedonia	Terrorism, military target	Yes	Water flow to Kumanovo (population 100,000) cut off for 12 days in conflict between ethnic Albanians and Macedonian forces. Valves of Glaznja and Lipkovo Lakes damaged.	AFP 2001; Macedonia Information Agency 2001
2001	China	Development dispute	Yes	In an act to protest destruction of fisheries from uncontrolled water pollution, fishermen in Northern Jiaxing City, Zhejiang province, dammed the canal that carries 90 million tons of industrial wastewater per year for 23 days. The wastewater discharge into the neighboring Shengze Town, Jiangsu province, killed fish and threatened people's health.	China Ministry of Water Resources 2001
2001	Philippines	Terrorism, political tool	No	Philippine authorities shut off water to six remote southern villages after residents complained of a foul smell from their taps, raising fears that Muslim guerrillas had contaminated the supplies. Abu Sayyaf guerrillas, accused of links with Saudi-born militant Osami bin Laden, had threatened to poison the water supply in the mainly Christian town of Isabela on Basilan island if the military did not stop an offensive against them.	World Environment News 2001
2001	Afghanistan	Military target	Yes	U.S. forces bombed the hydroelectric facility at Kajaki Dam in Helmand province of Afghanistan, cutting off electricity for the city of Kandahar. The dam itself was apparently not targeted.	BBC 2001; Parry 2001

continues

WATER CONFLICT CHRONOLOGY[1] *continued*

Date	Parties Involved	Basis of Conflict	Violent Conflict or in the Context of Violence?	Description	Sources
2002	Nepal	Terrorism, political tool	Yes	The Khumbuwan Liberation Front (KLF) blew up a hydroelectric powerhouse of 250 kilowatts in Bhojpur District January 26. The power supply to Bhojpur and adjoining areas was cut off. Estimated repair time was 6 months; repair costs were estimated at 10 million Rs. By June 2002, Maoist rebels had destroyed more than seven microhydro projects as well as an intake of a drinking water project and pipelines supplying water to Khalanga in Western Nepal.	*Kathmandu Post* 2002; FTGWR 2002
2002	Rome, Italy	Terrorism	No: threat	Italian police arrest four Moroccans allegedly planning to contaminate the water supply system in Rome with a cyanide-based chemical, targeting buildings that included the United States embassy. Ties to Al-Qaeda were suggested.	BBC 2002
2002	Kashmir, India	Development dispute	Yes	Two people were killed and 25 others injured in Kashmir when police fired at a group of villagers clashing over water sharing. The incident took place in Garend village in a dispute over sharing water from an irrigation stream.	*The Japan Times* 2002
2002	United States	Terrorism	No: threat	Papers seized during the arrest of a Lebanese national who moved to the US and became an Imam at a Islamist mosque in Seattle included "instructions on poisoning water sources" from a London-based Al-Qaeda recruiter. The FBI issued a bulletin to computer security experts around the country indicating that Al-Qaeda terrorists may have been studying American dams and water-supply systems in preparation for new attacks. "U.S. law enforcement and intelligence agencies have received indications that Al-Qaeda members have sought information on Supervisory Control And Data Acquisition (SCADA) systems available on multiple SCADA-related Web sites," reads the bulletin, according to SecurityFocus. "They specifically sought information on water supply and wastewater management practices in the U.S. and abroad."	McDonnell and Meyer 2002; MSNBC 2002
2002	Colombia	Terrorism	Yes	Colombian rebels in January damaged a gate valve in the dam that supplies most of Bogota's drinking water. Revolutionary Armed Forces of Colombia (FARC), detonated an explosive device planted on a German-made gate valve located inside a tunnel in the Chingaza Dam.	*Waterweek* 2002

Year	Parties	Basis of Conflict	Violent Conflict or in the Context of Violence	Description	Sources
2002	Karnataka, Tamil Nadu, India	Development dispute	Yes	Continuing violence over the allocation of the Cauvery River between Karnataka and Tamil Nadu. Riots, property destruction, more than 30 injuries, arrests through September and October.	*The Hindu* 2002a, 2002b; *The Times of India* 2002a
2002	United States	Terrorism	No: threat	Earth Liberation Front threatens the water supply for the town of Winter Park. Previously, this group claimed responsibility for the destruction of a ski lodge in Vail, Colorado, that threatened lynx habitat.	Crecente 2002; Associated Press 2002
2003	United States	Terrorism	No: threat	Al-Qaeda threatens U.S. water systems via call to Saudi Arabian magazine. Al-Qaeda does not "rule out...the poisoning of drinking water in American and Western cities."	Associated Press 2003a; Waterman 2003; *NewsMax* 2003; *U.S. Water News* 2003
2003	United States	Terrorism	Yes	Four incendiary devices were found in the pumping station of a Michigan water-bottling plant. The Earth Liberation Front (ELF) claimed responsibility, accusing Ice Mountain Water Company of "stealing" water for profit. Ice Mountain is a subsidiary of Nestle Waters.	Associated Press 2003b
2003	Colombia	Terrorism, development dispute	Yes	A bomb blast at the Cali Drinking Water Treatment Plant killed 3 workers May 8. The workers were members of a trade union involved in intense negotiations over privatization of the water system.	PSI 2003
2003	Jordan	Terrorism	No: threat	Jordanian authorities arrested Iraqi agents in connection with a botched plot to poison the water supply that serves American troops in the eastern Jordanian desert near the border with Iraq. The scheme involved poisoning a water tank that supplies American soldiers at a military base in Khao, which lies in an arid region of the eastern frontier near the industrial town of Zarqa.	MJS 2003
2003	Iraq, United States, others	Military Target	Yes	During the U.S.-led invasion of Iraq, water systems were reportedly damaged or destroyed by different parties, and major dams were military objectives of the U.S. forces. Damage directly attributable to the war includes vast segments of the water distribution system and the Baghdad water system, damaged by a missile.	UNICEF 2003; ARC 2003
2003	Iraq	Terrorism	Yes	Sabotage/bombing of main water pipeline in Baghdad. The sabotage of the water pipeline was the first such strike against Baghdad's water system, city water engineers said. It happened around 7 in the morning, when a blue Volkswagen Passat stopped on an overpass near the Nidaa mosque and an explosive was fired at the six-foot-wide water main in the northern part of Baghdad, said Hayder Muhammad, the chief engineer for the city's water treatment plants.	Tierney and Worth 2003

continues

WATER CONFLICT CHRONOLOGY[1] *continued*

Date	Parties Involved	Basis of Conflict	Violent Conflict or in the Context of Violence?	Description	Sources
2003–2004	Sudan	Military tool, military target, terrorism	Yes	The ongoing civil war in the Sudan has included violence against water resources. In 2003, villagers from around Tina said that bombings had destroyed water wells. In Khasan Basao they alleged that water wells were poisoned. In 2004, wells in Darfur were intentionally contaminated as part of a strategy of harassment against displaced populations.	Toronto Daily 2004; Reuters Foundation 2004
2004	Mexico	Development dispute	Yes	Two Mexican farmers argued for years over water rights to a small spring used to irrigate a small corn plot near the town of Pihuamo. In March, these farmers shot each other dead.	The Guardian 2004
2004	Pakistan	Terrorism	Yes	In military action aimed at Islamic terrorists, including Al-Qaeda and the Islamic movement of Uzbekistan, homes, schools, and water wells were damaged and destroyed.	Reuters 2004
2004	India, Kashmir	Terrorism	Yes	Twelve Indian security forces were killed by an IED planted in an underground water pipe during "counter-insurgency operation in Khanabal area in Anantnag district."	TNN 2004
2004	Gaza Strip	Terrorism, development dispute	Yes	The United States halts two water development projects as punishment to the Palestinian Authority for their failure to find those responsible for a deadly attack on a U.S. diplomatic convoy in October 2003.	Associated Press 2004
2004	India	Development dispute	Yes	Four people were killed in October and more than 30 injured in November in ongoing protests by farmers over allocations of water from the Indira Ghandi Irrigation Canal in Sriganganagar district, which borders Pakistan. A curfew was imposed in the towns of Gharsana, Raola and Anoopgarh.	Indo-Asian News Service 2004
2004–2006	Somalia	Development dispute	Yes	At least 250 people killed and many more injured in clashes over water wells and pastoral lands. Villagers call it the "War of the Well" and describe "well warlords, well widows, and well warriors." A three-year drought has led to extensive violence over limited water resources, worsened by the lack of effective government and central planning.	BBC 2004; Wax 2006

2005	Kenya	Development dispute	Yes	Police were sent to the northwestern part of Kenya to control a major violent dispute between Kikuyu and Maasai groups over water. More than 20 people were killed in fighting in January. The tensions arose when Maasai herdsmen accused a local Kikuyu politician of diverting a river to irrigate his farm, depriving downstream livestock. Fighting displaced more than 2,000 villagers and reflects tensions between nomadic and settled communities.	BBC 2005a; Ryu 2005
2006	Yemen	Development dispute	Yes	Local media reported a struggle between Hajja and Amran tribes over a well located between the two governorates in Yemen. According to news reports, armed clashes between the two sides forced many families to leave their homes and migrate. News reports confirmed that authorities arrested 20 people in an attempt to stop the fighting.	Al-Ariqi 2006

Notes:

1. May 2006 version. Conflicts may stem from the drive to possess or control another nation's water resources, thus making water systems and resources a *political or military goal*. Inequitable distribution and use of water resources, sometimes arising from a water development, may lead to *development disputes*, heighten the importance of water as a strategic goal or may lead to a degradation of another's source of water. Conflicts may also arise when water systems are used as instruments of war, either as *targets or tools*. These distinctions are described in detail in Gleick (1993, 1998). In 2001, the Institute began including incidents involving water and *terrorism*. We note, however, the difficulty in defining "terrorism" (as opposed to military target, tool, or goal or other category) and caution users to use care when applying these categories. We use this term when individuals or groups act against governments or official agencies.

Thanks to the many people who have contributed to this over time, including William Meyer who sent nine fascinating items from the 1800s, Patrick Marsh, Hans-Juergen Liebscher, Robert Halliday, Ma Jun, Marcus Moench, and others I've no doubt forgotten.

Sources:

Absolute Astronomy webpage. Reviewed 2006. Incapacitating agent. http://www.absoluteastronomy.com/reference/incapacitating_agent.

Agence France Press (AFP). 2001. Macedonian troops fight for water supply as president moots amnesty. *AFP,* June 8. http://www.balkanpeace.org/hed/archive/june01/hed3454.shtml.

Ahmed, A. 1999. Fifty hurt in Bangladesh strike violence. *Reuters News Service,* Dhaka, April 18.

Al-Ariqi, A. 2006. Water war in Yemen. *Yemen Times,* 14(932). April 24. http://yementimes.com/article.shtml?i=932&p=health&a=1.

American Red Cross (ARC). 2003. Baghdad hospitals reopen but health care system strained. Mason Booth, staff writer, RedCross.org. April 24. http://www.redcross.org/news/in/iraq/030424baghdad.html.

American University (Inventory of Conflict and the Environment ICE). 2000a. Cauvery River dispute. http://www.american.edu/projects/mandala/TED/ice/CAUVERY.HTM.

American University (Inventory of Conflict and the Environment ICE). 2000b. Lesotho "water coup." http://www.american.edu/projects/mandala/TED/ice/LESWATER.HTM.

American University (Inventory of Conflict and the Environment ICE). 2000c. Marsh Arabs and Iraq. http://www.american.edu/projects/mandala/TED/ice/MARSH.HTM.

Arbuthnot, F. 2000. Allies deliberately poisoned Iraq public water supply in Gulf War. *Sunday Herald,* Scotland. September 17.

Associated Press. 2002. Earth Liberation Front members threaten Colorado town's water. *AP,* October 15.

Associated Press. 2003a. Water targeted, magazine reports. *AP,* May 29.

Associated Press. 2003b. Incendiary devices placed at water plant. *AP,* September 25.

Associated Press. 2004. U.S. dumps water projects in Gaza over convoy bomb. *AP,* May 6.

Barrett, G. 2003. Iraq's bad water brings disease, alarms relief workers. *The Olympian,* Olympia Washington, Gannett News Service, June 29. http://www.theolympian.com/home/news/20030629/frontpage/39442.shtml.

Barry, J. M. 1997. *Rising tide: The great Mississippi flood of 1927 and how it changed America*. 67. New York: Simon and Schuster.

Bingham, G., Wolf, A., Wohlegenant, T. 1994. Resolving water disputes: Conflict and cooperation in the United States, the Near East, and Asia. US Agency for International Development (USAID). Bureau for Asia and the Near East. Washington, DC.

BBC. 1999. World: Asia-Pacific Timor atrocities unearthed. September 22. http://news.bbc.co.uk/hi/english/world/asia-pacific/newsid_455000/455030.stm.

BBC. 2000. Kenyan monkeys fight humans for water. BBC News. March 21. http://news.bbc.co.uk/1/hi/world/africa/685381.stm.

BBC. 2001. U.S. bombed Afghan power plant. http://news.bbc.co.uk/1/hi/world/south_asia/1632304.stm.

BBC. 2002. Cyanide attack foiled in Italy. February 20. http://news.bbc.co.uk/hi/english/world/europe/newsid_1831000/1831511.stm.

BBC. 2004. Dozens dead in Somalia clashes. BBC News World Edition online. http://news.bbc.co.uk/2/hi/africa/4073063.stm.

BBC. 2005a. Thousands flee Kenyan water clash. BBC News. January 24. http://news.bbc.co.uk/1/hi/world/africa/4201483.stm.

Burns, J. F. 1992. Tactics of the Sarajevo siege: Cut off the power and water. A1. *New York Times*, September 25.

Butts, K. (editor). 1997. *Environmental change and regional security*. Carlisle, PA: Asia-Pacific Center for Security Studies, Center for Strategic Leadership. U.S. Army War College.

Cable News Network (CNN). 1999. U.S.: Serbs destroying bodies of Kosovo victims. May 5. http://www.cnn.com/WORLD/europe/9905/05/kosovo.bodies.

Chelyshev, A. 1992. Terrorists poison water in Turkish army cantonment. *Telegraph Agency of the Soviet Union (TASS)*, March 29.

Chenje, M. 2001. Hydro-politics and the quest of the Zambezi River basin organization. In *International waters in Southern Africa*. Nakayama, M. (editor). United Nations University, Tokyo, Japan.

China Ministry of Water Resources. 2001. http://shuizheng.chinawater.com.cn/ssjf/20021021/20021016087.htm (the web site of the Policy and Regulatory Department).

China Water Resources Daily. 2002. Villagers fight over water resources. October 24. Citation provided by Ma Jun, personal communication.

Christian Science Monitor. 2000. Ecoterrorism as negotiating tactic. 8. July 21.

Clark, R. C. 1980. *Technological terrorism*. Devin-Adair. Old Greenwich, CT.

Clark, R. M., Deininger, R. A. 2000. Protecting the nation's critical infrastructure: The vulnerability of U.S. water supply systems. *Journal of Contingencies and Crisis Management*, 8(2):73–80.

Columbia Electronic Encyclopedia. 2000. Vietnam: History. http://www.infoplease.com/ce6/world/A0861793.html.

Columbia Encyclopedia. 2000. Netherlands, 6th edition. Columbia Encyclopedia. http://www.bartleby.com/65/ne/Nethrlds.html.

Cooperation Center for Afghanistan. 2000. The social impact of drought in Hazarajat. http://www.ccamata.com/impact.html.

Corps of Engineers. 1953. *Applications of hydrology in military planning and operations and subject classification index for military hydrology data*. Military Hydrology R&D Branch, Engineering Division, Corps of Engineers, Department of the Army, Washington, DC.

Crecente, B. D. 2002. ELF targets water: Group threatens eco-terror attack on Winter Park tanks. *Rocky Mountain News*, October 15. http://www.rockymountainnews.com/drmn/state/article/0,1299,DRMN_21_1479883,00.html.

Daniel, C. (editor) 1995. *Chronicle of the 20th century*. New York: Dorling Kindersley Publishing, Inc.

Dolatyar, M. 1995. Water diplomacy in the Middle East. 256 pp. In *The Middle Eastern environment*. Watson, E. (editor). London: John Adamson Publishing.

Douglass, J. D., Livingstone, N. C. 1987. *America the vulnerable: The threat of chemical and biological warfare*. Lexington, MA: Lexington Books.

Drower, M. S. 1954. Water-supply, irrigation, and agriculture. In *A history of technology*. Singer, C., Holmyard, E. J., Hall, A. R. (editor). New York: Oxford University Press.

Dutch Water Line. 2002. Information on the historical use of water in defense of Holland. http://www.xs4all.nl/~pho/Dutchwaterline/dutchwaterl.htm.

Eitzen, E. M., Takafuji, E. T. 1997. Historical overview of biological warfare. 415–424. In *Textbook of military medicine, medical aspects of chemical and biological warfare*. Published by the Office of The Surgeon General, Department of the Army, Washington, DC.

ENS (Environment News Service). 2001. Environment: A weapon in the Israeli-Palestinian conflict. February 5. http://www.ens-newswire.com/ens/feb2001/2001-02-05-01.asp.

Fatout, P. 1972. *Indiana canals*. 158–162. West Lafayette, IN: Purdue University Studies.

Ferguson, R. B. 2003. The birth of war. *Natural History*, 122(6):28–35.

Fickle, J. E. 1983. The "people" versus "progress": Local opposition to the construction of the Wabash and Erie Canal. *Old Northwest*, 8(4):309–328.

Financial Times Global Water Report. 1999. Zambia: Water cutoff. *FTGWR*, 68:15.

Financial Times Global Water Report. 2000. Drought in India comes as no surprise. *FTGWR*, 94:14.

Financial Times Global Water Report. 2002. Maoists destroy Nepal's infrastructure. *FTGWR*, 146:4–5.

Fonner, D. K. 1996. Scipio Africanus. *Military History Magazine*, March. http://historynet.com/mh/blscipioafricanus/index1.html.

Forkey, N. S. 1998. Damning the dam: Ecology and community in Ops Township, Upper Canada. *Canadian Historical Review*, 79(1):68–99.

Gellman, B. 2002. Cyber-attacks by Al Qaeda feared. A1. *Washington Post*, June 27.

Gleick, P. H. 1991. Environment and security: The clear connections. 17–21. *Bulletin of the Atomic Scientists*, April.

Gleick, P. H. 1993. Water and conflict: Fresh water resources and international security. *International Security*, 18(1):79–112.

Gleick, P. H. 1994. Water, war, and peace in the Middle East. *Environment*, 36(3)6–ff. Washington, DC: Heldref Publications.

Gleick, P. H. 1995. Water and conflict: Critical issues. Presented to the 45th Pugwash Conference on Science and World Affairs. Hiroshima, Japan: July 23–29.

Gleick, P. H. 1998. "Water and conflict." In *The World's Water 1998-1999*. Island Press, Washington.

Gowan, H. 2004. Hannibal Barca and the Punic Wars. http://www.barca.fsnet.co.uk/. Reviewed March 2005.

Grant, U. S. 1885. *Personal memoirs of U. S. Grant*. New York: C. L. Webster. ["On the second of February, [1863] this dam, or levee, was cut....The river being high the rush of water through the cut was so great that in a very short time the entire obstruction was washed away ... As a consequence the country was covered with water."]

Green Cross International. The conflict prevention atlas. http://www.greencrossinternational.net/GreenCrossPrograms/waterres/gcwater/report.html.

Guantanamo Bay Gazette. 1964. The history of Guantanamo Bay: An online edition. http://www.gtmo.net/gazz/hisidx.htm. Chapter XXI: The 1964 water crisis. http://www.gtmo.net/gazz/HISCHP21.HTM.

Guardian. 2004. Water duel kills elderly cousins. *The Guardian Newspapers Limited*, March 11.

Haeri, S. 1993. Iran: Vehement reaction. 8. *Middle East International*, March19.

Harris, S. H. 1994. *Factories of death: Japanese biological warfare 1932–1945 and the American cover-up*. New York: Routledge.

Hatami, H., Gleick, P. 1994. Chronology of conflict over water in the legends, myths, and history of the Ancient Middle East. In *Water, war, and peace in the Middle East. Environment*, 36(3)6–ff. Washington, DC: Heldref Publications.

Hickman, D. C. 1999. A chemical and biological warfare threat: USAF water systems at risk. Counterproliferation Paper No. 3. USAF Counterproliferation Center, Air War College, Maxwell Air Force Base, AL.

Hillel, D. 1991. Lash of the dragon. 28–37. *Natural History*, August.

Hindu, The. 2002a. Ryots on the rampage in Mandya. *The Hindu, India's National Newspaper*, October 31. http://www.hinduonnet.com/thehindu/2002/10/31/stories/2002103106680100.htm.

Hindu, The. 2002b. Farmers go berserk, MLA's house attacked. *The Hindu, India's National Newspaper*, October 30. http://www.hinduonnet.com/thehindu/2002/10/30/stories/2002103004870400.htm.

Honan, W. H. 1996. Scholar sees Leonardo's influence on Machiavelli. 18. *The New York Times*, December 8.

Human Rights Watch. 1998. Human rights watch condemns civilian killings by Congo rebels. http://www.hrw.org/press98/aug/congo827.htm.

Husarska, A. 1995. Running dry in Sarajevo: Water fight. *The New Republic*, July 17 and 24.

InfoRoma. 2004. Roman aqueducts. http://www.inforoma.it/feature.php?lookup=aqueduct. Viewed March 2005.

Indo-Asian News Service. 2004. Curfew imposed in three Rajasthan towns. http://www.hindustantimes.com/news/181_1136315,0009000010008.htm. *Hindustan Times*. December 4, 2004. http://news.newkerala.com/india-news/?action=fullnews&id=46359. India News at newkerala.com.

215

International Committee of the Red Cross. 2003. Eritrea: ICRC repairs war-damaged health centre and water system. December 15. ICRC News No. 03/158. http://www.alertnet.org/thenews/fromthefield/107148342038.htm.

International Court of Justice. 1999. International Court of Justice Press Communiqué 99/53, Kasikili Island/Sedudu Island (Botswana/Namibia). 2. The Hague, Holland, December 13. http://www.icj-cij.org/icjwww/ipresscom/ipress1999/ipresscom9953_ibona_1999121 3.htm.

International Herald Tribune. 1999. 100 bodies found in well. 4. *International Herald Tribune,* August 14–15.

Israel Line. 2001a. Palestinians loot water pumping center, cutting off supply to refugee camp. *Israel Line.* January 5. http://www.israel.org/mfa/go.asp?MFAH0dmp0; http://www.mfa.gov.il/mfa/go.asp?MFAH0izu0.

Israel Line. 2001b. Palestinians vandalize Yitzhar water pipe. *Israel Line,* January 9. http://www.mfa.gov.il/mfa/go.asp?MFAH0iy50.

IWCT (International War Crimes Tribunal). 1967. Some facts on bombing of dikes. http://www.infotrad.clara.co.uk/antiwar/warcrimes/index.html.

Japan Times, The. 2002. Kashmir water clash. 3. *The Japan Times,* May 27.

Jenkins, B. M., Rubin, A. P. 1978. New vulnerabilities and the acquisition of new weapons by nongovernment groups. 221–276. In *Legal aspects of international terrorism.* Evans, A. E., Murphy, J. F. (editors). Lexington, MA: Lexington Books.

Kathmandu Post, The. 2002. KLF destroys micro hydro plant. *The Kathmandu Post,* January 28. http://www.nepalnews.com.np/contents/englishdaily/ktmpost/2002/jan/jan28/index.htm.

Kirschner, O. 1949. Destruction and protection of dams and levees. Military Hydrology, Research and Development Branch, U.S. Corps of Engineers, Department of the Army, Washington District. From Schweizerische Bauzeitung 14 March 1949, Translated by H. E. Schwarz, Washington.

Keluel-Jang, S. A. 1997. Alier and the Jonglei Canal. *Southern Sudan Bulletin,* 2(3). January. http://www.sufo.demon.co.uk/poli007.htm.

Kupperman, R. H., Trent, D. M. 1979. *Terrorism: Threat, reality, response.* Stanford, CA: Hoover Institution Press.

Lemos, R. 2002. Safety: Assessing the infrastructure risk. *CNET/news.com.* http://news.com.com/2009-1001_3-954780.html. August 26.

Levy, K. 2000. Guatemalan dam massacre survivors seek reparations from financiers. 12–13. *World Rivers Review,* International Rivers Network, Berkeley, California. December.

Livingstone, N. C. 1982. *The war against terrorism.* Lexington and Toronto, Canada: Lexington Books.

Lockwood, R. P. Reviewed April 2006. The battle over the Crusades. http://www.catholicleague.org/research/battle_over_the_crusades.htm.

Macedonia Information Agency. 2001. Humanitarian catastrophe averted in Kumanovo and Lipkovo. Republic of Macedonia Agency of Information Archive. June 18. http://www.reliefweb.int/w/rwb.nsf/0/dbd4ef105d93da4ac1256a6f005bc328?OpenDocument.

McDonnell, P. J., Meyer, J. 2002. Links to terrorism probed in Northwest. *Los Angeles Times,* July 13 .

Meissner, R. 2000. Hydropolitical hotspots in Southern Africa: Will there be a water war? The case of the Kunene River. 103-131. In *Water wars: Enduring myth or impending reality?* Solomon, H., Turton, A. (editors). Africa Dialogue Monograph Series No. 2. Accord, Creda Communications, KwaZulu-Natal, South Africa.

Meissner, R. 2001. Interaction and existing constraints in international river basins: The case of the Kunene River Basin. In *International waters in Southern Africa.* Nakayama, M. (editor). Tokyo, Japan: United Nations University.

Milwaukee Journal Sentinel. 2003. Jordan foils Iraqi plot to poison U.S. troops' water, officials say. April 1. http://www.jsonline.com/news/gen/apr03/130338.asp.

Mohamed, A. E. 2001. Joint development and cooperation in international water resources: The case of the Limpopo and Orange River Basins in Southern Africa. In *International waters in Southern Africa.* Nakayama, M. (editor). Tokyo, Japan: United Nations University.

Moorehead, A. 1960. *The white Nile.* England: Penguin Books.

MSNBC. 2002. FBI says al-Qaida after water supply. Numerous wire reports. http://www.ionizers.org/water-terrorism.html.

Murphy, I. L., Sabadell, J. E. 1986. International river basins: A policy model for conflict resolution. *Resources Policy,* 12(1):133–144. United Kingdom: Butterworth and Co. Ltd.

Museum of the City of New York (MCNY). n.d. The greater New York consolidation timeline. http://www.mcny.org/Exhibitions/GNY/timeline.htm.

Nadeem, A. 2001. Bombs in Karachi kill one. *Associated Press.* http://dailynews.yahoo.com/h/ap/20010613/wl/pakistan_strike_3.html.

Naff, T., Matson, R. C. (editors). 1984. Water in the Middle East: Conflict or cooperation? *Westview Press.* Boulder, CO.

National Geographic News. 2001. Ancient fertile crescent almost gone, satellite images show. May 18. http://news.nationalgeographic.com/news/2001/05/0518_crescent.html.

New York Times, The. 1999. Puerto Ricans protest Navy's use of water. 30. *The New York Times,* October 31.

NewsMax. 2003. Al-Qaida threat to U.S. water supply. *NewsMax Wires,* May 29. http://www.newsmax.com/archives/articles/2003/5/28/202658.shtml.

NTI Nuclear Threat Initiative. 2005. A brief history of chemical warfare. http://www.nti.org/h_learnmore/cwtutorial/chapter02_02.html.

Okoko, T. O. 2000. Monkeys, humans fight over drinking water. Panafrican News Agency, March 21.

Pannier, B. 2000. Central Asia: Water becomes a political issue. *Radio Free Europe.* www.rferl.org/nca/features/2000/08/F.RU.000803122739.html.

Parry, R. L. 2001. UN fears "disaster" over strikes near huge dam. *The Independent,* London, November 8.

Plant, G. 1995. Water as a weapon in war. *Water and war, symposium on water in armed conflicts,* Montreux, November 21–23, 1994. Geneva, ICRC.

Pottinger, M. 2000. Major Chinese lake disappearing in water crisis. *Reuters Science News.* http://us.cnn.com/2000/NATURE/12/20/china.lake.reut/.

Pretoria Dispatch Online. 1999. Dam bomb may be "aimed at farmers". July 21. http://www.dispatch.co.za/1999/07/21/southafrica/RESEVOIR.HTM.

Priscoli, J. D. 1998. Water and civilization: Conflict, cooperation and the roots of a new eco realism. A keynote address for the 8th Stockholm World Water Symposium, August 10–13. http://www.genevahumanitarianforum.org/docs/Priscoli.pdf.

PSI. 2003. Urgent action: Bomb blast kills 3 workers at the Cali water treatment plant. *Public Services International.* http://www.world-psi.org. Also http://209.238.219.111/Water.htm.

Purver, R. 1995. Chemical and biological terrorism: The threat according to the open literature. Canadian Security Intelligence Service, Ottawa, Canada. http://www.csis.gc.ca/en/publications/other/c_b_terrorism01.asp.

Rasch, P. J. 1968. The Tularosa ditch war. *New Mexico Historical Review,* 43(3):229–235.

Reisner, M. 1986, 1993. *Cadillac desert: The American West and its disappearing water.* New York: Penguin Books.

Reuters. 1999a. Serbs say NATO hit refugee convoys. April 14. http://dailynews.yahoo.com/headlines/ts/story.html?s=v/nm/19990414/ts/yugoslavia_192.html. http://www.uia.ac.be/u/carpent/kosovo/messages/397.html.

Reuters. 1999b. NATO keeps up strikes but Belgrade quiet. June 5. http://dailynews.yahoo.com/headlines/wl/story.html?s=v/nm/19990605/wl/yugoslavia_strikes_129.html.

Reuters. 1999c. NATO builds evidence of Kosovo atrocities. June 17. http://dailynews.yahoo.com/headlines/ts/story.html?s=v/nm/19990617/ts/yugoslavia_leadall_171.html.

Reuters. 2004. Al Qaeda spy chief killed in Pakistani raid. Reuters Yahoo.

Reuters Foundation. 2004. Darfur: 2.5 million people will require food aid in 2005. Schofield, R. November 22. http://www.medair.org/en_portal/medair_news?news=258.

Rome Guide. 2004. Fontana di Trevi: History. http://web.tiscali.it/romaonlineguide/Pages/eng/rbarocca/sBMy5.htm.

Rowe, W. T. 1988. Water control and the Qing political process: The Fankou Dam controversy, 1876–1883. *Modern China,* 14(4):353–387.

Ryu, A. 2005. Water rights dispute sparks ethnic clashes in Kenya's rift valley. Voice of America. http://www.voanews.com/english/archive/2005-03/2005-03-21-voa28.cfm.

Samson, P., Charrier, B. 1997. International freshwater conflict: Issues and prevention strategies. *Green Cross International.* http://www.greencrossinternational.net/GreenCrossPrograms/waterres/gcwater/report.html.

Sandrasagra. M. J. 2000. Development Ethiopia: Relief agencies warn of major food crisis. *Inter Press Service,* April 11.

Scheiber, H. N. 1969. *Ohio Canal era.* 174–175. Athens, OH: Ohio University Press.

Schofield, C. I. ed. 1967. *New Schofield Reference Bible.* Oxford, United Kingdom: Oxford University Press.

Semann, D. 1950. Die Kriegsbeschädigungen der Edertalspermauer, die Wiederherstellungsarbeiten und die angestellten Untersuchungen über die Standfestigkeit der Mauer. *Die Wasserwirtschaft,* 41. Ig, Nr. 1 u. 2.

Shapiro, C. 2004. A search for flaws deep in the heart of the Surry reactor. *The Virginia-Pilot.* December 6. http://home.hamptonroads.com/stories/print.cfm?story=78992&ran=226100.

Simons, M. 1999. Serbs refuse to clear bomb-littered river. *New York Times,* October 24.

Strategy Page. 2006. http://www.strategypage.com/articles/biotoxin_files/BIOTOXINSINWARFARE.asp. Reviewed April.

Stockholm International Peace Research Institute (SPIRI). 1971. *The rise of CB weapons: The problem of chemical and biological warfare.* New York: Humanities Press.

Soloman, A. 2001. Policeman dies as blasts rock strike-hit Karachi. *Reuters*, June 13. http://dailynews.yahoo.com/h/nm/20010613/ts/pakistan_strike_dc_1.html; http://www.labline.de/indernet/partikel/karachi/bombse.htm.

Steinberg, T. S. 1990. Dam-breaking in the nineteenth-century Merrimack Valley. *Journal of Social History*, 24(1):25–45.

Styran, R. M., Taylor, R. R. 2001. *The great swivel link: Canada's Welland Canal.* The Champlain Society, Toronto, Canada.

Suliman, M. 1998. *Resource access: A major cause of armed conflict in the Sudan. The case of the Nuba Mountains.* Institute for African Alternatives, London, UK. http://srdis.ciesin.org/cases/Sudan-Paper.html.

Thatcher, J. 1827. *A military journal during the American Revolutionary War, from 1775 to 1783.* Second edition, revised and corrected. Boston, MA: Cottons and Barnard. http://www.fortklock.com/journal1777.htm.

Tierney, J., Worth, R. F. 2003. Attacks in Iraq may be signals of new tactics. 1. *The New York Times*, August 18. http://www.nytimes.com/2003/08/18/international/worldspecial/18IRAQ.html?hp.

Times of India. 2002. Cauvery row: Farmers renew stir. October 20. http://timesofindia.indiatimes.com/cms.dll/html/uncomp/articleshow?art_id=26586125.

Times News Network (TNN). 2004. IED was planted in underground pipe. December 5. http://timesofindia.indiatimes.com/articleshow/947432.cms.

Toronto Daily. 2004. Darfur: Too many people killed for no reason. Amnesty International Index: AFR 54/008/2004, February 3.

Tucker, J. B. (editor). 2000. *Toxic terror: Assessing terrorist use of chemical and biological weapons.* Cambridge, MA: MIT Press.

Turton, A. 2005. A critical assessment of the basins at risk in the Southern African hydropolitical complex. Workshop on the management of international rivers and lakes hosted by the Third World Centre for Water Management & Helsinki University of Technology, August 17–19, 2005. Helsinki, Finland. Council for Scientific and Industrial Research (CSIR). African Water Issues Research Unit (AWIRU) CSIR Report Number: ENV-P-CONF 2005-001, Pretoria, South Africa.

UNICEF 2003. Iraq: Cleaning up neglected, damaged water system, clearing away garbage. News note press release, May 27. http://www.unicef.org/media/media_6998.html.

U.S. Water News. 2003. Report suggests al-Qaida could poison U.S. water. *US Water News Online*, June. http://www.uswaternews.com/archives/arcquality/3repsug6.html.

Vanderwood, P. J. 1969. *Night riders of Reelfoot Lake.* Memphis, TN: Memphis State University Press.

Wallenstein, P., Swain, A. 1997. International freshwater resources—Conflict or cooperation? Comprehensive assessment of the freshwater resources of the world: Stockholm: Stockholm Environment Institute.

Walters, E. 1948. *Joseph Benson Foraker: Uncompromising republican.* 44–45. Columbus, OH: Ohio History Press.

Waterman, S. 2003. Al-Qaida threat to U.S. water supply. *United Press International (UPI)*, May 28.

Waterweek. 2002. Water facility attacked in Colombia. *Waterweek*, American Water Works Association. January. http://www.awwa.org/advocacy/news/020602.cfm.

Wax, E. 2006. Dying for water in Somalia's drought: Amid anarchy, warlords hold precious resource. A1. *Washington Post*, April 14. http://www.washingtonpost.com/wp-dyn/content/article/2006/04/13/AR2006041302116.html.

Wolf, A. T. 1995. *Hydropolitics along the Jordan River: Scarce water and its impact on the Arab-Israeli conflict.* Tokyo, Japan: United Nations University Press.

Wolf, A. T. 1997. Water wars' and water reality: Conflict and cooperation along international waterways. NATO Advanced Research Workshop on Environmental Change, Adaptation, and Human Security. Budapest, Hungary. October 9–12.

World Environment News. 2001. Philippine rebels suspected of water "poisoning." http://www.planetark.org/avantgo/dailynewsstory.cfm?newsid=12807.

World Rivers Review (WRR). 1998. Dangerous dams: Tajikistan. 13(6):13, December.

Yang Lang. 1989/1994. High dam: The sword of Damocles. 229–240. In *Yangtze! Yangtze!* Qing, D. (editor). London, UK: Probe International, Earthscan Publications.

Zachary, G. P. 1997. Water pressure: Nations scramble to defuse fights over supplies. A17. *Wall Street Journal*, December 4.

Zemmali, H. 1995. International humanitarian law and protection of water. *Water and war, symposium on water in armed conflicts*, Montreux, November 21–23, 1994. Geneva: ICRC.

The Soft Path in Verse[1]

Gary Wolff

> *I dream*
> *of thirst*
> *of dry rivers*
> *of dead children.*
> *Killed by their parents, who could not see a better way.*
>
> *But soft, moist wind awakens me,*
> *And I see in the dawn, industrial sculptures[2]*
> *   that manage waste and clean clothes and entire cities,*
> *Wrapped in a circle of blue,[3]*
> *   of endless blue water*
> *Moved by the minds and hearts of those who chose to invent a future*
> *worth living for.*

This poem was delivered at the Mexican Museum of Modern Art on March 22, 2006, during the release of *Water,* a book of photography by Antonio Vizcaino and text by Carlos Zolla, Dorion Sagan/Lynne Margulis, Mario Molina/Rafael L. Bras, Jean-Michel Cousteau, Peter Raven, Gerardo Ceballos, Gary Wolff/Peter Gleick (America Natural: Mexico City).

---

1. See P.H. Gleick 2003. "Global Freshwater Resources: Soft-Path Solutions for the 21st Century." *Science* Vol. 302, 28 November, pp. 1524-1528; and G. Wolff and P.H. Gleick. 2002. "The Soft Path for Water." In P.H. Gleick (editor). *The World's Water 2002-2004: The Biennial Report on Freshwater Resources.* Island Press, Washington, D.C. pp. 1-32 for a description of the soft path for water, in prose.

2. Emerging technologies for clothes washing—see Figures WB 5.1 and WB 5.2—inspired the phrase "industrial sculptures." Recently invented and plausible future appliances and plumbing fixtures can create a water use future most people do not imagine, just as the wholesale transformation of patterns of work caused by steam engines and personal computers were unexpected by those without vision. Even the potential of readily available and cost-effective technology is often underestimated. See Gleick, et al., (2003), which demonstrates that 30% of California's urban water use in 2000 was cost-effective to conserve if appliances and plumbing fixtures that were readily available in retail outlets in that year were fully deployed.

3. "Circle of Blue" is a global communications effort to increase awareness of water issues, affiliated with the Pacific Institute. See: http://www.circleofblue.org/

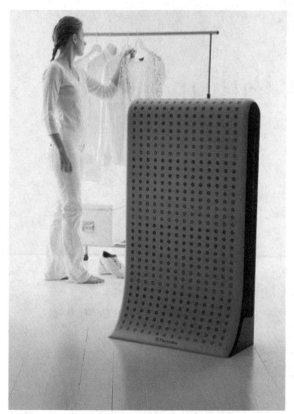

FIGURE WB 5.1   Industrial Sculpture: Waterless Clothes Washers

Two industrial design students from the National University of Singapore (NUS) won an award at the 2005 international Electrolux Design Lab competition. Their product? AirWash: a waterfall-inspired washing unit that requires neither water nor detergents. Ionized air is used to remove dirt and kill bacteria. http://ir.electrolux.com/files/press/electrolux/200511142151en2.pdf

FIGURE WB 5.2

Sanyo Electric Co. has developed a household washing machine that can clean clothes without water, using only ozone and air. When water is used, the water is cleaned and recycled several times within the machine. The electric appliance manufacturer began selling the washer, *Aqua*, on March 11, 2006. http://www.sanyo.co.jp/koho/hypertext4-eng/0602/0202-1e.html

Total Renewable Freshwater Supply by Country

Description

Average annual renewable freshwater resources are listed by country, updating the previous tables in this series from *The World's Water*. These data are typically comprised of both renewable surface water and groundwater supplies, including surface inflows from neighboring countries. The UN FAO refers to this as total natural renewable water resources. Flows to other countries are not subtracted from these numbers. All quantities are in cubic kilometers per year (km^3/yr). These data represent average freshwater resources in a country—actual annual renewable supply will vary from year to year.

Limitations

These detailed country data should be viewed, and used, with caution. The data come from different sources and were estimated over different periods. Many countries do not directly measure or report internal water resources data, so some of these entries were produced using indirect methods. For example, Margat (2001) compiles information from a wide variety of sources and notes that there is a wide variation in the reliability of the data. In the past few years, new assessments have begun to standardize definitions and assumptions.

Not all of the annual renewable water supply is available for use by the countries to which it is credited here; some flows are committed to downstream users. For example, the Sudan is listed as having 154 km^3/yr, but treaty commitments require them to pass significant flows downstream to Egypt. Other countries, such as Turkey, Syria, and France, to name only a few, also pass significant amounts of water to other users. The annual average figures hide large seasonal, interannual, and long-term variations.

Sources

Compiled by P. H. Gleick and H. Cooley, Pacific Institute.

nd = no data

a. Total natural renewable surface and groundwater. Typically includes flows from other countries. (FAO: Natural total renewable water resources)

b. Estimates from Belyaev, Institute of Geography, USSR. 1987.

c. Estimates from FAO. 1995. *Water resources of African countries.* Food and Agriculture Organization, United Nations, Rome, Italy.

d. Estimates from WRI. 1994. See this source for original data source.

e. Estimates from Goscomstat, USSR, 1989 as cited in Gleick 1993, Table A16.

f. Estimates from FAO. 1997. *Water resources of the Near East region: A review.* Food and Agriculture Organization, United Nations, Rome, Italy.

g. Estimates from FAO. 1997. *Irrigation in the countries of the Former Soviet Union in figures.* Food and Agriculture Organization, United Nations, Rome, Italy.

h. UNFAO. 1999. *Irrigation in Asia in figures.* Food and Agriculture Organization, United Nations, Rome, Italy.

i. Nix, H. 1995. *Water/Land/Life: The eternal triangle.* Water Research Foundation of Australia, Canberra, Australia.

j. UNFAO. 2000. *Irrigation in Latin America and the Caribbean.* Food and Agriculture Organization, United Nations, Rome, Italy.

k. AQUASTAT Web site as of February 2003.

l. Margat, J./OSS. 2001. Les ressources en eau des pays de l'OSS. Evaluation, utilisation et gestion. UNESCO/Observatoire du Sahara et du Sahel. (Updating of 1995).

m. Estimates from FAO. 2003. *Review of world water resources by country.* Food and Agriculture Organization, United Nations, Rome, Italy (see specific references in this document for more information).

n. United States Geological Survey Revised: Conterminous US (2071); Alaska (980); Hawaii (18).

o. EUROSTAT, U. Wieland. 2003. *Water resources in the EU and in the candidate countries.* Statistics in Focus, Environment and Energy, European Communities.

p. Margat, J and D. Vallée. 2000. *Blue plan—Mediterranean vision on water, population and the environment for the 21st century.* Sophia Antipolis, France. 62 pp.

q. Geres, D. 1998. *Water resources in Croatia.* International Symposium on Water Management and Hydraulic Engineering. Dubrovnic, Croatia (September 14–19).

r. AQUASTAT Web site as of November 2005.

s. EUROSTAT. 2005.

t. Pearse, P. H., Bertrand, F., MacLaren, J. W. 1985. *Currents of change, final report of inquiry on federal water policy.* Environment Canada. Ottawa, Canada.

DATA TABLE 1   Total Renewable Freshwater Supply, by Country (2006 Update)

Region and Country	Annual Renewable Water Resources[a] (km³/yr)	Year of Estimate	Source of Estimate
AFRICA			
Algeria	14.3	1997	c,f
Angola	184.0	1987	b
Benin	25.8	2001	l
Botswana	14.7	2001	l
Burkina Faso	17.5	2001	l
Burundi	3.6	1987	b
Cameroon	285.5	2003	m
Cape Verde	0.3	1990	c
Central African Republic	144.4	2003	m
Chad	43.0	1987	b
Comoros	1.2	2003	m
Congo	832.0	1987	b
Congo, Democratic Republic (formerly Zaire)	1,283	2001	l
Côte d'Ivoire	81	2001	l
Djibouti	0.3	1997	f
Egypt	86.8	1997	f
Equatorial Guinea	26	2001	l
Eritrea	6.3	2001	l
Ethiopia	110.0	1987	b
Gabon	164.0	1987	b
Gambia	8.0	1982	c
Ghana	53.2	2001	l
Guinea	226.0	1987	b
Guinea-Bissau	31.0	2003	m
Kenya	30.2	1990	c
Lesotho	5.2	1987	b
Liberia	232.0	1987	b
Libya	0.6	1997	c,f
Madagascar	337.0	1984	c
Malawi	17.3	2001	l
Mali	100.0	2001	k
Mauritania	11.4	1997	c,f
Mauritius	2.2	2001	k
Morocco	29.0	2003	m
Mozambique	216.0	1992	c
Namibia	45.5	1991	c
Niger	33.7	2003	m
Nigeria	286.2	2003	m
Reunion	5.0	1988	m
Rwanda	5.2	2003	m
Senegal	39.4	1987	b
Sierra Leone	160.0	1987	b
Somalia	15.7	1997	f
South Africa	50.0	1990	c

continues

DATA TABLE 1   *continued*

Region and Country	Annual Renewable Water Resources[a] (km³/yr)	Year of Estimate	Source of Estimate
AFRICA (*continued*)			
Sudan	154.0	1997	c,f
Swaziland	4.5	1987	b
Tanzania	91	2001	l
Togo	14.7	2001	l
Tunisia	4.6	2003	m
Uganda	66.0	1970	c
Zambia	105.2	2001	l
Zimbabwe	20.0	1987	b
NORTH AND CENTRAL AMERICA			
Antigua and Barbuda	0.1	2000	j
Bahamas	nd	nd	
Barbados	0.1	2003	m
Belize	18.6	2000	j
Canada	3,300.0	1985	t
Costa Rica	112.4	2000	j
Cuba	38.1	2000	j
Dominica	nd	nd	
Dominican Republic	21.0	2000	j
El Salvador	25.2	2001	l
Grenada	nd	nd	
Guatemala	111.3	2000	j
Haiti	14.0	2000	j
Honduras	95.9	2000	j
Jamaica	9.4	2000	j
Mexico	457.2	2000	j
Nicaragua	196.7	2000	j
Panama	148.0	2000	j
St. Kitts and Nevis	0.02	2000	j
Trinidad and Tobago	3.8	2000	j
United States of America	3,069.0	1985	n
SOUTH AMERICA			
Argentina	814.0	2000	j
Bolivia	622.5	2000	j
Brazil	8,233.0	2000	j
Chile	922.0	2000	j
Colombia	2,132.0	2000	j
Ecuador	432.0	2000	j
Guyana	241.0	2000	j
Paraguay	336.0	2000	j
Peru	1,913.0	2000	j
Suriname	122.0	2003	m
Uruguay	139.0	2000	j
Venezuela	1,233.2	2000	j

Region and Country	Annual Renewable Water Resources[a] (km³/yr)	Year of Estimate	Source of Estimate
ASIA			
Afghanistan	65.0	1997	f
Bahrain	0.1	1997	f
Bangladesh	1,210.6	1999	h
Bhutan	95.0	1987	b
Brunei	8.5	1999	h
Cambodia	476.1	1999	h
China	2,829.6	1999	h
India	1,907.8	1999	h
Indonesia	2,838.0	1999	h
Iran	137.5	1997	f
Iraq	96.4	1997	f
Israel	1.7	2001	l,m
Japan	430.0	1999	h
Jordan	0.9	1997	f
Korea DPR	77.1	1999	h
Korea Rep	69.7	1999	h
Kuwait	0.02	1997	f
Laos	333.6	2003	m
Lebanon	4.8	1997	f
Malaysia	580.0	1999	h
Maldives	0.03	1999	h
Mongolia	34.8	1999	h
Myanmar	1,045.6	1999	h
Nepal	210.2	1999	h
Oman	1.0	1997	f
Pakistan	233.8	2003	k
Philippines	479.0	1999	h
Qatar	0.1	1997	f
Saudi Arabia	2.4	1997	f
Singapore	0.6	1975	d
Sri Lanka	50.0	1999	h
Syria	46.1	1997	f
Taiwan	67.0	2000	r
Thailand	409.9	1999	h
Turkey	234.0	2003	k, l, m, o
United Arab Emirates	0.2	1997	f
Vietnam	891.2	1999	h
Yemen	4.1	1997	f
EUROPE			
Albania	41.7	2001	p
Austria	84.0	2005	s
Belgium	20.8	2005	s
Bosnia and Herzegovina	37.5	2003	m
Bulgaria	19.4	2005	s
Croatia	105.5	1998	o, q

continues

DATA TABLE 1 *continued*

Region and Country	Annual Renewable Water Resources[a] (km³/yr)	Year of Estimate	Source of Estimate
EUROPE (*continued*)			
Cyprus	0.4	2005	s
Czech Republic	16.0	2005	s
Denmark	6.1	2003	o
Estonia	21.1	2005	s
Finland	110.0	2005	s
France	189.0	2005	s
Germany	188.0	2005	s
Greece	72.0	2005	s
Hungary	120.0	2005	s
Iceland	170.0	2005	s
Ireland	46.8	2003	o
Italy	175.0	2005	s
Luxembourg	1.6	2005	s
Macedonia	6.4	2001	p
Malta	0.07	2005	s
Netherlands	89.7	2005	s
Norway	381.4	2005	s
Poland	63.1	2005	s
Portugal	73.6	2005	s
Romania	42.3/211.90	2003	o, m
Slovakia	80.3/50.1	2003	o, m
Slovenia	32.1	2005	s
Spain	111.1	2005	s
Sweden	179.0	2005	s
Switzerland	53.3	2005	s
United Kingdom	160.6	2005	s
Serbia-Montenegro*	208.5	2003	m
FORMER SOVIET UNION			
Russia	4,498.0	1997	e,g
Armenia	10.5	1997	g
Azerbaijan	30.3	1997	g
Belarus	58.0	1997	g
Estonia	12.8	1997	g
Georgia	63.3	1997	g
Kazakhstan	109.6	1997	g
Kyrgyzstan	46.5	1997	m
Latvia	49.9	2005	s
Lithuania	24.5	2005	s
Moldova	11.7	1997	g
Tajikistan	99.7	1997	m
Turkmenistan	60.9	1997	m
Ukraine	139.5	1997	g
Uzbekistan	72.2	2003	m

Region and Country	Annual Renewable Water Resources[a] (km³/yr)	Year of Estimate	Source of Estimate
OCEANIA			
Australia	398.0	1995	i
Fiji	28.6	1987	b
New Zealand	397.0	1995	i
Papua New Guinea	801.0	1987	b
Solomon Islands	44.7	1987	b

*referred to as Yugoslavia in previous *The World's Water*

Freshwater Withdrawal by Country and Sector

Description

The use of water varies greatly from country to country and from region to region. Data on water use by regions and by different economic sectors are among the most sought after in the water resources area. Ironically, these data are often the least reliable and most inconsistent of all water-resources information. The following table presents *The World's Water* 2006 update of the data available on total freshwater withdrawals by country in cubic kilometers per year for the year indicated in the table. Per capita withdrawls in cubic meters per person per year were calculated by dividing freshwater withdrawal by 2005 national population estimates. The table also gives the breakdown of that water use by the domestic, agricultural, and industrial sectors, in both percentage of total water use and cubic meters per person per year. The data sources are also explicitly identified.

"Withdrawal" typically refers to water taken from a water source for use. It does not refer to water "consumed" in that use. The domestic sector typically includes household and municipal uses as well as commercial and governmental water use. The industrial sector includes water used for power plant cooling and industrial production. The agricultural sector includes water for irrigation and livestock.

In 2003, the Food and Agriculture Organization of the United Nations published a comprehensive update of its water use estimates in the Aquastat dataset, and it has continued to provide new and updated numbers. As a result, new estimates of water use by country are available, as are new estimates of the sectoral breakdown. An advantage of the Aquastat dataset is that it provides what appears to be a consistent set of information, but users should be very careful to understand which numbers are measured and which are only calculated (see Limitations).

Limitations

Extreme care should be used when applying these data. They come from a wide variety of sources and are collected using a wide variety of approaches, with few formal standards. As a result, this table includes data that are actually measured, estimated, modeled using different assumptions, or derived from other data. The data also come from different years, making direct intercomparisons difficult. For example, some water-use data are over twenty years old. Separate data are now provided for the independent states of the former Soviet Union, but not for the former states of Yugoslavia. Industrial withdrawals for Panama, St. Lucia, St. Vincent, and the Grenadines are included in the domestic category.

The 2003 revision of the FAO Aquastat dataset is the most dramatic change in water use data in recent years, yet these data are marred by inadequate information on sources and assumptions. They should be used with great care, and with appropriate caveats about their quality. As an example, the latest actual measured data on water use in Canada comes from a detailed national assessment published in 1996 by Environment Canada (see later reference "o"). More recent data can be found on the Aquastat database, but these data simply take the 1996 measured value and modify it for the change in population in Canada—thus the 2000 Aquastat value is modeled, not measured. Although we often report the Aquastat values in this table, in the case of Canada, we have chosen to rely on the 1996 measured value, while awaiting a Canadian reassessment (now under way).

Another major limitation of these data is that they do not include the use of rainfall in agriculture. Many countries use a significant fraction of the rain falling on their territory for agricultural production, but this water use is neither accurately measured nor reported in this set. We repeat our regular call for a systematic reassessment of water-use data and for national and international commitments to collect and standardize this information. We note that budgetary constraints continue to delay any major new data initiatives.

Sources

a. Aquastat estimates from www.fao.org. (March 2004).
   World Resources Institute. 1990. *World resources 1990–1991*. New York: Oxford University Press.
c. World Resources Institute. 1994. *World resources 1994–1995*, in collaboration with the United Nations Environment Programme and the United Nations Development Programme. New York: Oxford University Press.
d. Eurostat Yearbook. 1997. *Statistics of the European Union*, EC/C/6/Ser.26GT, Luxembourg.
e. UN FAO. 1999. *Irrigation in Asia in figures*. Food and Agriculture Organization, United Nations, Rome, Italy.
f. Nix, H. 1995. *Water/Land/Life*. Water Research Foundation of Australia, Canberra.
g. UNFAO. 2000. *Irrigation in Latin America and the Caribbean*. Food and Agricultural Organization, United Nations, Rome, Italy.
h. AQUASTAT Web site January 2002. http://www.fao.org.
i. Ministry of Water Resources, China. 2001. *Water resources bulletin of China, 2000*. People's Republic of China, Beijing, China (September).
j. Hutson, S. S., Barber, N. L., Kenny, J. F., Linsey, K. S., Lumia, D. S., Maupin, M. A. 2004. Estimated use of water in the United States in 2000. U.S. Geological Survey, Circular 1268. Reston, VA.
k. Eurostat. 2004. Statistics in focus. http://europa.eu.int/comm/eurostat.
l. New FAO Aquastat estimates. http://www.fao.org (November 2005): See text for details.
m. Eurostat. 2005. Updated July 2005.
n. National land and water resources Audit. 2001.
o. Environment Canada. *Water use in Canada in 1996*. http://www.ec.gc.ca/water/en/manage/use/e_wuse.htm.

Population data. Population Division of the Department of Economic and Social Affairs of the United Nations Secretariat. 2005. World population prospects: The 2004 revision. Highlights. New York: United Nations.

DATA TABLE 2 Freshwater Withdrawal, by Country and Sector (2006 Update)

Region and Country	Year	Total Freshwater Withdrawal (km³/yr)	Per Capita Withdrawal (m³/p/yr)	Domestic Use (%)	Industrial Use (%)	Agricultural Use (%)	Domestic Use m³/p/yr	Industrial Use m³/p/yr	Agricultural Use m³/p/yr	Source	2005 Population (millions)
AFRICA											
Algeria	2000	6.07	185	22	13	65	41	24	120	1	32.85
Angola	2000	0.35	22	23	17	60	5	4	13	1	15.94
Benin	2001	0.13	15	32	23	45	5	4	7	1	8.44
Botswana	2000	0.19	107	41	18	41	44	19	44	1	1.77
Burkina Faso	2000	0.80	60	13	1	86	8	1	52	1	13.23
Burundi	2000	0.29	38	17	6	77	6	2	30	1	7.55
Cameroon	2000	0.99	61	18	8	74	11	5	45	1	16.32
Cape Verde	2000	0.02	39	7	2	91	3	1	36	1	0.51
Central African Republic	2000	0.03	7	80	16	4	6	1	0	1	4.04
Chad	2000	0.23	24	17	0	83	4	0	20	1	9.75
Comoros	1999	0.01	13	48	5	47	6	1	6	1	0.80
Congo, Democratic Republic (formerly Zaire)	2000	0.36	6	53	17	31	3	1	2	1	57.55
Congo, Republic of	2000	0.03	8	59	29	12	4	2	1	1	4.00
Côte d'Ivoire	2000	0.93	51	24	12	65	12	6	33	1	18.15
Djibouti	2000	0.02	25	84	0	16	21	0	4	1	0.79
Egypt	2000	68.30	923	8	6	86	70	55	793	1	74.03
Equatorial Guinea	2000	0.11	220	83	16	1	183	35	2	1	0.50
Eritrea	2000	0.30	68	3	0	97	2	0	66	1	4.40
Ethiopia	2002	5.56	72	6	0	94	4	0	67	1	77.43
Gabon	2000	0.12	87	50	8	42	43	7	37	1	1.38
Gambia	2000	0.03	20	23	12	65	5	2	13	1	1.52
Ghana	2000	0.98	44	24	10	66	11	4	29	1	22.11
Guinea	2000	1.51	161	8	2	90	12	4	144	1	9.40
Guinea-Bissau	2000	0.18	113	13	5	82	15	6	93	1	1.59

Kenya	2000	1.58	46	30	6	64	14	3	29	1	34.26
Lesotho	2000	0.05	28	40	40	20	11	11	6	1	1.80
Liberia	2000	0.11	34	27	18	55	9	6	18	1	3.28
Libya	2000	4.27	730	14	3	83	102	22	606	1	5.85
Madagascar	2000	14.96	804	3	2	96	23	13	769	1	18.61
Malawi	2000	1.01	78	15	5	80	12	4	63	1	12.88
Mali	2000	6.55	484	9	1	90	44	5	436	1	13.52
Mauritania	2000	1.70	554	9	3	88	49	16	489	1	3.07
Mauritius	2000	0.61	488	25	14	60	124	69	294	a	1.25
Morocco	2000	12.60	400	10	3	87	40	12	348	1	31.48
Mozambique	2000	0.63	32	11	2	87	4	1	28	1	19.79
Namibia	2000	0.3	148	24	5	71	35	7	105	1	2.03
Niger	2000	2.18	156	4	0	95	6	0	149	1	13.96
Nigeria	2000	8.01	61	21	10	69	13	6	42	1	131.53
Rwanda	2000	0.15	17	24	8	68	4	1	11	1	9.04
Senegal	2002	2.22	190	4	3	93	8	6	177	1	11.66
Sierra Leone	2000	0.38	69	5	3	92	2	2	63	1	5.53
Somalia	2000	3.29	400	0	0	100	2	0	398	1	8.23
South Africa	2000	12.50	264	31	6	63	82	16	166	1	47.43
Sudan	2000	37.32	1,030	3	1	97	27	7	996	1	36.23
Swaziland	2000	1.04	1,010	2	1	97	20	10	979	1	1.03
Tanzania	2000	5.18	135	10	0	89	14	0	120	1	38.33
Togo	2000	0.17	28	53	2	45	15	1	12	1	6.15
Tunisia	2000	2.64	261	14	4	82	37	10	214	1	10.10
Uganda	2002	0.30	10	43	17	40	4	2	4	1	28.82
Zambia	2000	1.74	149	17	7	76	25	10	113	1	11.67
Zimbabwe	2002	4.21	324	14	7	79	45	23	256	1	13.01
NORTH AND CENTRAL AMERICA											
Antigua and Barbuda	1990	0.005	63	60	20	20	38	13	13	g	0.08
Barbados	2000	0.09	333	33	44	22	111	147	73	1	0.27
Belize	2000	0.15	556	7	73	20	39	406	111	1	0.27

continues

DATA TABLE 2 *continued*

Region and Country	Year	Total Freshwater Withdrawal (km³/yr)	Per Capita Withdrawal (m³/p/yr)	Domestic Use (%)	Industrial Use (%)	Agricultural Use (%)	Domestic Use m³/p/yr	Industrial Use m³/p/yr	Agricultural Use m³/p/yr	Source	2005 Population (millions)
NORTH AND CENTRAL AMERICA *(continued)*											
Canada	1996	44.72	1,386	20	69	12	271	952	163	o	32.27
Costa Rica	2000	2.68	619	29	17	53	182	106	331	1	4.33
Cuba	2000	8.20	728	19	12	69	138	89	500	1	11.27
Dominica	1996	0.02	213							h	0.08
Dominican Republic	2000	3.39	381	32	2	66	122	7	252	1	8.90
El Salvador	2000	1.28	186	25	16	59	46	29	111	1	6.88
Guatemala	2000	2.01	160	6	13	80	10	21	128	1	12.60
Haiti	2000	0.99	116	5	1	94	5	1	110	1	8.53
Honduras	2000	0.86	119	8	12	80	10	14	95	1	7.21
Jamaica	2000	0.41	155	34	17	49	53	26	76	1	2.65
Mexico	2000	78.22	731	17	5	77	127	40	564	1	107.03
Nicaragua	2000	1.30	237	15	2	83	36	5	197	1	5.49
Panama	2000	0.82	254	67	5	28	170	13	72	1	3.23
St. Lucia	1997	0.01	81							g	0.16
St. Vincent and the Grenadines	1995	0.01	83							g	0.12
Trinidad and Tobago	2000	0.31	237	68	26	6	161	62	13	1	1.31
United States of America	2000	477.00	1,600	13	46	41	203	736	660	j	298.21
SOUTH AMERICA											
Argentina	2000	29.19	753	17	9	74	128	71	558	1	38.75
Bolivia	2000	1.44	157	13	7	81	21	11	127	1	9.18
Brazil	2000	59.30	318	20	18	62	65	57	196	1	186.41
Chile	2000	12.55	770	11	25	64	87	194	489	1	16.30

Country	Year										
Colombia	2000	10.71	235	50	4	46	118	9	108	—	45.60
Ecuador	2000	16.98	1,283	12	5	82	160	68	1055	—	13.23
Guyana	2000	1.64	2,187	2	1	98	37	19	2143	—	0.75
Paraguay	2000	0.49	80	20	8	71	16	6	56	—	6.16
Peru	2000	20.13	720	8	10	82	60	73	587	—	27.97
Suriname	2000	0.67	1,489	4	3	93	67	43	1379	—	0.45
Uruguay	2000	3.15	910	2	1	96	22	10	878	—	3.46
Venezuela	2000	8.37	313	6	7	47	19	22	149	—	26.75
ASIA											
Afghanistan	2000	23.26	779	2	0	98	14	0	765	—	29.86
Armenia	2000	2.95	977	30	4	66	293	43	642	—	3.02
Azerbaijan	2000	17.25	2,051	5	28	68	99	567	1385	—	8.41
Bahrain	2000	0.30	411	40	3	57	163	12	233	—	0.73
Bangladesh	2000	79.40	560	3	1	96	18	4	538	—	141.82
Bhutan	2000	0.43	199	5	1	94	10	2	187	—	2.16
Brunei	1994	0.09	243	nd	nd	nd	nd	nd	nd	e	0.37
Cambodia	2000	4.08	290	1	0	98	3	0	284	—	14.07
China*	2000	549.76	415	7	26	68	27	107	281	i	1,323.35
Cyprus	2000	0.21	250	27	1	71	68	4	179	m	0.84
Georgia	2000	3.61	808	20	21	59	161	170	477	—	4.47
India	2000	645.84	585	8	5	86	47	32	506	—	1,103.37
Indonesia	2000	82.78	372	8	1	91	30	3	339	—	222.78
Iran	2000	72.88	1,048	7	2	91	71	24	953	—	69.52
Iraq	2000	42.70	1,482	3	5	92	47	68	1367	—	28.81
Israel	2000	2.05	305	31	7	62	94	20	189	—	6.73
Japan	2000	88.43	690	20	18	62	136	123	431	—	128.09
Jordan	2000	1.01	177	21	4	75	37	8	133	—	5.70
Kazakhstan	2000	35.00	2,360	2	17	82	40	390	1930	—	14.83
Korea Democratic People's Republic	2000	9.02	401	20	25	55	80	101	220	—	22.49
Korea Rep	2000	18.59	389	36	16	48	139	64	186	—	47.82
Kuwait	2000	0.44	164	45	2	52	73	3	86	—	2.69

continues

233

DATA TABLE 2 *continued*

Region and Country	Year	Total Freshwater Withdrawal (km³/yr)	Per Capita Withdrawal (m³/p/yr)	Domestic Use (%)	Industrial Use (%)	Agricultural Use (%)	Domestic Use m³/p/yr	Industrial Use m³/p/yr	Agricultural Use m³/p/yr	Source	2005 Population (millions)
ASIA *(continued)*											
Kyrgyz Republic	2000	10.08	1,916	3	3	94	60	59	1797	1	5.26
Laos	2000	3.00	507	4	6	90	22	29	457	1	5.92
Lebanon	2000	1.38	385	33	1	67	126	2	257	1	3.58
Malaysia	2000	9.02	356	17	21	62	60	75	221	1	25.35
Maldives	1987	0.003	9	98	2	0	9	0	0	e	0.33
Mongolia	2000	0.44	166	20	27	52	34	45	86	1	2.65
Myanmar	2000	33.23	658	1	1	98	8	4	646	1	50.52
Nepal	2000	10.18	375	3	1	96	11	2	362	1	27.13
Oman	2000	1.36	529	7	2	90	38	11	476	1	2.57
Pakistan	2000	169.39	1,072	2	2	96	21	22	1030	1	157.94
Philippines	2000	28.52	343	17	9	74	57	32	254	1	83.05
Qatar	2000	0.29	358	24	3	72	86	10	257	1	0.81
Saudi Arabia	2000	17.32	705	10	1	89	69	8	628	1	24.57
Singapore	1975	0.19	44	45	51	4	20	22	2	c	4.33
Sri Lanka	2000	12.61	608	2	2	95	14	15	579	1	20.74
Syria	2000	19.95	1,048	3	2	95	34	19	994	1	19.04
Tajikistan	2000	11.96	1,837	4	5	92	68	86	1683	1	6.51
Thailand	2000	82.75	1,288	2	2	95	32	32	1225	1	64.23
Turkey	2001	39.78	544	15	11	74	80	59	404	m	73.19
Turkmenistan	2000	24.65	5,104	2	1	98	86	39	4978	1	4.83
United Arab Emirates	2000	2.30	511	23	9	68	118	44	349	1	4.50
Uzbekistan	2000	58.34	2,194	5	2	93	104	45	2045	1	26.59
Vietnam	2000	71.39	847	8	24	68	66	205	577	1	84.24
Yemen	2000	6.63	316	4	1	95	13	2	301	1	20.98

EUROPE

Albania	2000	1.71	546	27	11	62	146	61	339	l	3.13
Austria	1999	3.67	448	35	64	1	157	286	4	m	8.19
Belarus	2000	2.79	286	23	47	30	67	134	86	l	9.76
Belgium	1998	7.44	714	13	85	1	95	610	9	k	10.42
Bosnia and Herzegovina											3.91
Bulgaria	2003	6.92	895	3	78	19	27	700	168	m	7.73
Croatia											4.55
Czech Republic	2002	1.91	187	41	57	2	76	107	4	m	10.22
Denmark	2002	0.67	123	32	26	42	40	32	52	m	5.43
Estonia	2002	1.41	1,060	56	39	5	591	418	52	m	1.33
Finland	1999	2.33	444	14	84	3	61	371	12	m	5.25
France	2000	33.16	548	16	74	10	86	408	54	m	60.50
Germany	2001	38.01	460	12	68	20	57	312	91	m	82.69
Greece	1997	8.70	782	16	3	81	128	25	630	m	11.12
Hungary	2001	21.03	2,082	9	59	32	192	1222	668	m	10.10
Iceland	2003	0.17	567	34	66	0	193	373	1	m	0.30
Ireland	1994	1.18	284	23	77	0	64	220	0	m	4.15
Italy	1998	41.98	723	18	37	45	131	265	326	m	58.09
Latvia	2003	0.25	108	55	33	12	59	35	13	m	2.31
Lithuania	2003	3.33	971	78	15	7	758	149	64	m	3.43
Luxembourg	1999	0.06	121	42	45	13	51	55	16	m	0.47
Macedonia	2000	2.27	1,118							m	2.03
Malta	2000	0.02	50	74	1	25	37	0	13	m	0.40
Moldova	2000	2.31	549	10	58	33	55	316	181	l	4.21
Netherlands	2001	8.86	544	6	60	34	33	326	184	m	16.30
Norway	1996	2.40	519	23	67	10	118	347	54	k	4.62
Poland	2002	11.73	304	13	79	8	40	240	25	m	38.53
Portugal	1998	11.09	1,056	10	12	78	101	128	827	m	10.50
Romania	2003	6.50	299	9	34	57	26	103	171	m	21.71
Russian Federation	2000	76.68	535	19	63	18	100	340	95	m	143.20
Serbia-Montenegro										l	10.50

continues

235

Region and Country	Year	Total Freshwater Withdrawal (km³/yr)	Per Capita Withdrawal (m³/p/yr)	Domestic Use (%)	Industrial Use (%)	Agricultural Use (%)	Domestic Use m³/p/yr	Industrial Use m³/p/yr	Agricultural Use m³/p/yr	Source	2005 Population (millions)
EUROPE (*continued*)											
Slovakia	2003	1.04	193							m	5.40
Slovenia	2002	0.90	457							m	1.97
Spain	2002	37.22	864	13	19	68	116	160	588	m	43.06
Sweden	2002	2.68	296	37	54	9	109	161	26	m	9.04
Switzerland	2002	2.52	348	24	74	2	84	257	7	m	7.25
Ukraine	2000	37.53	807	12	35	52	98	286	424	l	46.48
United Kingdom	1994	11.75	197	22	75	3	43	148	6	d	59.67
OCEANIA											
Australia	2000	24.06	1,193	15	10	75	176	120	898	n	20.16
Fiji	2000	0.07	82	14	14	71	12	12	58	l	0.85
New Zealand	2000	2.11	524	48	9	42	251	50	220	l	4.03
Papua New Guinea	1987	0.10	17	56	43	1	9	7	0	c	5.89
Solomon Islands	1987			40	20	40	0	0	0	c	0.48

Notes:

Figures may not add to totals due to independent rounding.

2005 Population numbers: medium UN Variant.

*Includes Hong Kong and Macao.

Access to Safe Drinking Water by Country, 1970 to 2002

Description

Safe drinking water is one of the most basic human requirements, and one of the Millennium Development Goals (MDGs) by 2015 is to reduce by half the proportion of people unable to reach or afford safe drinking water. As a result, estimates of access to safe drinking water are a cornerstone of most international assessments of progress, or lack thereof, toward solving global and regional water problems.

Data are given here for the percent of urban, rural, and total populations, by country, with access to safe drinking water for 1970, 1975, 1980, 1985, 1990, 1994, 2000, and 2002—the most recent year for which data are available. The World Health Organization (WHO) collected the data presented here over various periods. Most of the data presented were drawn from responses by national governments to WHO questionnaires. Participants in data collection include the Joint Monitoring Programme (JMP) of WHO, the United Nations Children's Fund, and the Water Supply and Sanitation Collaborative Council, which has continued sector monitoring and aims to support and strengthen the monitoring efforts of individual countries.

The most recent data (WHO 2000, 2002) reflect a significant change in definition. Data are now reported for populations without access to "improved" water supply. The forty largest countries in the developing world account for 90 percent of population in these regions. As a result, WHO spent extra effort to collect comprehensive data for these countries. According to WHO, the following technologies were included in the assessment as representing "improved" water supply:

Household connection
Public standpipe
Borehole
Protected dug well
Protected spring
Rainwater collection

In comparison, "unimproved" drinking water sources refers to

Unprotected well
Unprotected spring
Rivers or ponds

Vendor-provided water
Bottled water
Tanker truck water

Limitations

A review of water and sanitation coverage data from the 1980s and 1990s shows that the definition of safe, or improved, water supply and sanitation facilities differs from one country to another and for a given country over time. Indeed, some of the data from individual countries often showed rapid and implausible changes in the level of coverage from one assessment to the next. This indicates that some of the data are also unreliable, irrespective of the definition used. Countries used their own definitions of "rural" and "urban."

For the 1996 data, two-thirds of the countries reporting indicated how they defined "access." At the time, the definition most commonly centered on walking distance or time from household to water source, such as a public standpipe, which varied from 50 to 2,000 meters and 5 to 30 minutes. Definitions sometimes included considerations of quantity, with the acceptable limit ranging from 15 to 50 liters per capita per day. The WHO considers safe drinking water to be treated surface water or untreated water from protected springs, boreholes, and wells.

The 2000 and 2002 WHO Assessments attempts to shift from gathering information from water providers strictly to include consumer-based information. The current approach uses household surveys in an effort to assess the actual use of facilities. "Reasonable access" was broadly defined as the availability of at least 20 liters per person per day from a source within one kilometer of the user's dwelling. A drawback of this approach is that household surveys are not conducted regularly in many countries. Thus, direct comparisons between countries, and across time within the same country, are difficult. Direct comparisons are additionally complicated by the fact that these data hide disparities between regions and socioeconomic classes.

Access to water, as reported by WHO, does not imply that the level of service or quality of water is "adequate" or "safe." The assessment questionnaire did not include any methodology for discounting coverage figures to allow for intermittence or poor quality of the water supplies. However, the instructions stated that piped systems should not be considered "functioning" unless they were operating at over 50 percent capacity on a daily basis; and that hand pumps should not be considered functioning unless they were operating for at least 70 percent of the time with a lag between breakdown and repair not exceeding two weeks. These aspects were taken into consideration when estimating coverage for countries for which national surveys had not been conducted. More details of the methods used, and their limitations, can be found at http://www.who.int/docstore/water_sanitation_health/ Globassessment/GlobalTOC.htm.

Sources

United Nations Environment Programme (UNEP). 1989. *Environmental data report*, GEMS Monitoring and Assessment Research Centre, Basil Blackwell, Oxford.

United Nations Environment Programme (UNEP). 1993–94. *Environmental data report*, GEMS Monitoring and Assessment Research Centre in cooperation with the World Resources Institute and the UK Department of the Environment, Basil Blackwell, Oxford.

World Health Organization (WHO). 1996. *Water supply and sanitation sector monitoring report: 1996 (Sector status as of 1994),* in collaboration with the Water Supply and Sanitation Collaborative Council and the United Nations Children's Fund, UNICEF, New York.

World Health Organization (WHO). 2000. *Global water supply and sanitation assessment 2000 report.* http://www.who.int/docstore/water_sanitation_health/Globassessment/GlobalTOC.htm.

World Resources Institute (WRI). 1988. World Health Organization data, cited by the World Resources Institute, *World resources 1988–89,* World Resources Institute and the International Institute for Environment and Development in collaboration with the United Nations Environment Programme, Basic Books, New York.

DATA TABLE 3   Access to Safe Drinking Water, by Country, 1970 to 2002

Fraction of Population with Access to Safe Drinking Water

Region and Country	URBAN								RURAL								TOTAL							
	1970	1975	1980	1985	1990	1994	2000	2002	1970	1975	1980	1985	1990	1994	2000	2002	1970	1975	1980	1985	1990	1994	2000	2002
AFRICA	66	68	69	77						61		55						77		68				
Algeria	84	100		85			88	92						15	94	80						32	94	87
Angola			85	87	73	69	34	70	20	20		15	20		40	40	29		26	33	35	32	38	50
Benin	83	100	26	80	73	41	74	79	26	39	15	34	43	53	55	60	29	34	18	50	54	50	63	68
Botswana	71	95		84	100		100	100	10			46	88			90		45		53	91	78		95
Burkina Faso	35	50	27	43			84	82		23	31	69	70	49		44	12	25	31	67		52		51
Burundi	77		90	98	92	92	96	90	21		20	21	43			78			23	25	45			79
Cameroon	77			43	42		82	84	21			24	45	34	42	41	32			32	44	51	62	63
Cape Verde			100	83		70	64	86				50		34	89	73			25	52		51	74	80
Central African Republic				13	19	18	80	93			21		26	18	43	61					23	18	60	75
Chad	47	43				48	31	40	24	23				17	26	32	27	26				24	27	34
Comoros			42				98	90	6	9	7				95	90			20				96	94
Congo	63	81					71	72							17	17	27	38					51	46
Congo, Democratic Rep.	33	38		52	68	37	89	83	4	12		21	24	23	26	29	11	19		32	36	27	45	46
Côte d'Ivoire	98				57	59	90	98	29				80	81	65	74	44				71	72	77	84
Djibouti			50	50		77	100	82			20	20		100	100	67			43	45		90	100	80
Egypt	94		88		95	82	96	100	93		64		86	50	94	97	93		84		90	64	95	98
Equatorial Guinea			47		65	88	45	45					18	100	42	42					32	95	43	44
Eritrea							63	72							42	54							46	57
Ethiopia	61	58		69			77	81		1		9			13	11	6	8		16			24	22
Gabon							73	95							55	47							70	87
Gambia	97		85	97	100		80	95	12			50	48		53	77	12			59	60	76	62	82
Ghana	86	86	72	93	63	70	87	93	35	14	2	39	21	62	49	68	35	35	45	56	21	56	64	79
Guinea	68	69	69	41	100	61	72	78			2	12	37	57	36	38		14	15	18	53	62	48	51
Guinea-Bissau			18	17	53	38	29	79			8	22			55	49			10	21		53	49	59
Kenya	100	100	85			67	87	89	2	4	15			49	31	46	15	17	26			53	49	62
Lesotho	100	65	37	65		14	98	88	1	14	11	30		64	88	74	3	17	15	36		52	91	76
Liberia	100	100	100	100		58		72	6	14		23		8		52	15			53		30		62
Libya	100	100	100				72	72	42	82	90				68	68	58	87	96				72	72

Region / Country																					
Madagascar	67	76	80	81	83	85	75	1	14	7	17	10	31	34	11	25	21	31	29	47	45
Malawi		77	97	52	95	96			37	50	44	44	62			41	41	56	45	57	67
Mali	29	37	37	46	36	74	76		10	38	61	35		16	37	65	37	16	45	
Mauritania	98		80	73	84	34	63	85		69	40	45	84	60		99	76	37	56	
Mauritius	100	100	100	100	95	100	100	98	100	100	100	100		99	100	98	100	100	
Morocco	92		100	100	98	100	99	28	25	18	14	58	56	51	59	52	52	82	80
Mozambique			38		17	86	76		9		40	43	24		15	32	60	42
Namibia	37	36		90	87	100	98			37	42	67	72	57	53	77	80
Niger		41	35	46	46	70	80	19	26	49	45	55	56	20	27	47	57	53	59	46
Nigeria	37		100		63	81	72			20	22	26	39	49	33	38	39	57	60
Reunion																					
Rwanda	81	84	79	84		60	92	66	68	48	67	40	69	67	55	50	68	50	41	73
Sao Tome and Principe						89			45	73	45		53	79
Senegal	87	56	77	79	82	92	90		25	38	28	65	54	43	53	42	50	78	72
Seychelles		77				100		95	95			75	95	39	87
Sierra Leone	75		68	58	58	23	75	1	2	20	21	31	46	12	14	24	39	34	28	57
Somalia	17						32	14	22	22	38	27	15	34	29
South Africa					92	92	98				80	73	70	86	87
Sudan	61	96	100		66	86	78	13	43	31	45	69	64	19	50	50	75	69
Swaziland		83	100	90	41		87	29	7	7	44	42	31	37	31	43	52
Tanzania	61	88			80	80	92	9	36	42	44	42	62	13	39	53	54	54	73
Togo	100	49	70	100	74	85	80	5	10	41	31	58	38	17	16	54	63	51
Tunisia	92	93	100	100	100	94	17	31	31	89	89	60	49	60	70	99	82
Uganda	88	100	37	60	47	72	87	17	29	18	32	46	52	22	35	20	34	50	56
Zambia	70	86	76		64	88	90	22	16	41	27	48	36	37	42	58	43	64	55
Zimbabwe			95		100	100	32	80		77	74			84	85	83
NORTH & CENTRAL AMERICA & CARIBBEAN																					
Anguilla						95															91
Antigua and Barbuda						100	89														100
Aruba							100														
Bahamas	100	100	100	98	98	99	98	12	13	75	75	86	86	65	65	90	50	96	97
Barbados	95	100	99	100	100	100	100	100	100	100	99	100	100	100	99	100	100	100	100
Belize			95	96	83		82		53	82	69	64	68	74	89	76	91
British Virgin Islands			100		100		98	100	100	98									98

continues

DATA TABLE 3 continued

Fraction of Population with Access to Safe Drinking Water

Region and Country	URBAN								RURAL								TOTAL							
	1970	1975	1980	1985	1990	1994	2000	2002	1970	1975	1980	1985	1990	1994	2000	2002	1970	1975	1980	1985	1990	1994	2000	2002
NORTH & CENTRAL AMERICA & CARIBBEAN (continued)																								
Canada							100	100							99	99							100	100
Cayman Islands			100	98																				
Costa Rica	98	100	100	100		85	98	100	59	56	82	83		99	98	92				91		92	98	97
Cuba	82	96			100	96	99	95	15				91	85	82	78					98	93	95	91
Dominican Republic	72	88	85	85	82	74	83	98	14	27	34	33	45	67	70	85	37	55	60	62	67	71	79	93
Dominica							100	100							100	90							100	97
El Salvador	71	89	67	68	87	78	88	91	20	28	40	40	15	37	61	68	40	53	50	51	47	55	74	82
Grenada	100	100					97	97	47	77					93	93							94	95
Guadeloupe							94	98							94	93							94	98
Guatemala	88	85	90	72	92		97	99	12	14	18	14	43		88	92	38	39	46	37	62		92	95
Haiti		46	51	59	56	37	46	91		3	8	30	35	23	45	59		12	19	38	41	28	46	71
Honduras	99	99	93	56	85	81	97	99	10	13	40	45	48	53	82	82	34	41	59	49	64	65	90	90
Jamaica	100	100	55	99			81	98	48	79	46	93			59	87	62	86	51	96			71	93
Martinique																								
Mexico	71	70	90	99	94	91	94	97	29	49	40	47		62	63	72	54	62	73	83	69	83	86	91
Montserrat								100								100								100
Netherlands Antilles																								
Nicaragua	58	100	67	76		81	95	93	16	14	6	11		27	59	65	35	56	39	48		61	79	81
Panama	100	100	100	100			88	99	41	54	62	64			86	79	69	77	81	82		83	87	91
Puerto Rico																								
St Kitts								99								99								99
St Lucia								98								98								98
St Vincent																								
Trinidad/Tobago	100	79	100	100	100			92	95	100	93	95	88			88	96	93	97	98	96		86	91
Turks/Caicos Islands				87				100				68				100				77				100
United States of America							100	100							100	100							100	100
United States Virgin Islands																								

	1	2	3	4	5	6	7	8	9	10	11	12	13	14	15	16	17	18	19	20	21	22	23	24
SOUTH AMERICA																								
Argentina	69	76	61	63		78	85	97	12	26	17	17	30	22	30	68		56	66	54	56	55	79	85
Bolivia	92	81	69	75	76	85	93	95	2	6	10	13	55	31	61	58	33	34	36	43	72		79	89
Brazil	78	87	83	85	85	94	95	96	28		51	56	54	37		59	55	70	72	77	87	85	87	95
Chile	67	78	100	98	94		99	100	13	28	17	29	66	48	82	71	56		84	87	85	94	95	
Colombia	88	86	93	100	87	88	98	99	28	33	73	76	73	55	44	77	63	64	86	86	76	91	92	86
Ecuador	76	67	79	81	63	82	81	92	7	8	20	31	51			51	34	36	50	57	55	71		86
Falkland Islands (Malvinas)																								
French Guiana	100	100	100	100	100	90	88	88	63	75	60	65	71	45	71	71	75	84	72	76	81	61	84	84
Guyana	22	25	39	53	61		98	83	5	5	9	8	91	9	83	62	11	13	21	28	34	81	94	83
Paraguay	58	72	68	73	68	74	95	100	8	15	18	17	58	24	62	66	35	47	50	55	55	60	79	83
Peru			100	71	100		87	87	59	87	79	94	51		66	73	92	98	88	83	89		95	81
Suriname	100	100	96	95		80	94	98	38		2	27	96	75	93	93	75	89	81	85	89		98	92
Uruguay	100	96	95	93			98	98			53	65	58		58	70	86	89	86	81		79	84	83
Venezuela	92	93	93	93			88	85			65	38												
ASIA																								
Afghanistan	18	40	28	38	40	39	19	19	1	5	8	17	19	5	11	11	3	9	8	17	23	12	13	13
Armenia								99					0		80	80								92
Azerbaijan					100	100	95	95				100	89		59	59	100	100						77
Bahrain	100	100	100	100			100	100	94	100	40	49	30	89	72	72	99	100	39	46	81	97	97	
Bangladesh	13	22	24	26	39	75	99	82	47	61	5	19	54	97	97	60	45	56	7		32	64	62	75
Bhutan				50	60		86	86			95												62	62
Brunei Darus			100	100																				
Cambodia				87	60	93	53	58					68	68	25	29				73	90		30	34
China	100	94	100	100	100		94	92	92	96	100	100	100	89	66	66	95	95	100	100	100		75	77
Cyprus				100			100	100							51	51							100	100
East Timor							73	73																52
Gaza Strip															61	61								
Georgia		100						90			95	95	96											76
Hong Kong																	100			73			
India	60	80	77	76	86	85	92	96	6	18	31	50	69	79	86	82	17	31	42	56	73	81	88	86
Indonesia	10	41	35	43	35	78	91	89	1	4	19	36	33	54	65	69	3	11	23	38	34	62	76	78
Iran	68	76	82	100	100	89	99	98	11	30	50	54	75	77	89	83	35	51	66	86	89	83	95	93
Iraq	83	100		93	93		96	97	7	11	54	54	41	48	48	50	51	66			78	44	85	81

continues

243

DATA TABLE 3 *continued*

Fraction of Population with Access to Safe Drinking Water

Region and Country	URBAN								RURAL								TOTAL							
	1970	1975	1980	1985	1990	1994	2000	2002	1970	1975	1980	1985	1990	1994	2000	2002	1970	1975	1980	1985	1990	1994	2000	2002
ASIA (*continued*)																								
Israel								100								100								100
Japan								100								100								100
Jordan	98		100	100	100		100	91	59		65	88	97		84	91	77		86	96	99	89	96	91
Kazakhstan							98	96							82	72							91	86
Korea DPR							100	100							100	100							100	100
Korea Rep	84	95	86	90	100		97	97	38	33	61	48	76		71	71	58	66	75	75	93		92	92
Kuwait	60	100	86	97							100						51	89	87					
Kyrgyzstan							98	98							66	66							77	76
Laos	97	100	28		47	40	59	66	39	32	20		25	39	38	38	48	41	21		29	39		43
Lebanon						100	100	100						100	100	100						100	100	100
Macau																								
Malaysia	100	100	90	96	96			96	1	6	49	76	66	66	94	94	29	34	63	84	79		94	95
Maldives			11	58	77	98	100	99			3	12	68	86	100	78			2	21		89	100	84
Mongolia					100		77	87					58		30	30					82		60	62
Myanmar (Burma)	35	31	38	36	79	36	88	95	13	14	15	24	72	39	60	74	18	17	21	27	74	38	68	80
Nepal	53	85	83	70	66	66	85	93		5	7	25	34	41	80	82	2	8	11	28	38	44	81	84
Oman		100		90			41	81		48		49			30	72		52		53			39	79
Pakistan	77	75	72	83	82	77	96	95	4	5	20	27	42	52	84	87	21	25	35	44	55	60	88	90
Philippines	67	82	49	49	93	93	92	90	20	31	43	54	72	77	80	77	36	50	45	52	81	85	87	85
Qatar	100	100	76		100			100	75	83	43					100	95	97	71					100
Saudi Arabia	100	97	92	100	100		100	97	37	56	87	88			64	97	49	64	90	94			95	
Singapore			100	100	100		100	100			100					100			100	100	100		100	100
Sri Lanka	46	36	65	82	80	43	91	99	14	13	18	29	55	47	80	72	21	19	28	40	60	46	83	78
Syria	98		98			92	94	94	50		54			78	64	64	71		74			85	80	79
Tajikistan								93								47								58
Thailand	60	69	65	56			89	95	10	16	63	66	85		77	80	17	25	63	64			80	85
Turkey			95				82	96			62				84	87			76				83	93
Turkmenistan								93								54								71
United Arab Emirates			95								81								92					
Uzbekistan							96	97							78	84							85	89

Country																			
Vietnam	45	70			93	2	32	39	33	32	50	67	4		20	16	31	45	36	36	56	73
Yemen A R	88	100			74	43	18	25			64	68	57				52	40			69	69
Yemen Dem		85			85		25				64										69	
OCEANIA																						
American Samoa				100	100						100	100									100	100
Australia				100	100			88	100		100	100									100	100
Cook Islands	100	99		100	98						100	88						92	80		100	95
Fiji	78	94	96	100	43	15	66		69		51		37				77			100	47	
French Polynesia			100	100	100	56			18	100	100	100			69						100	100
Guam				100	100					100	100	100									100	100
Kiribati		93	91	82	77		25		63		25	53									47	64
Marshall Islands			100		80				45			95										85
Micronesia				100	95				38	100		94								100		94
Nauru																						
New Caledonia																						
New Zealand			0	100	100			100	100	100	100	100								100	100	100
Niue			100	100	100				0		0	97										98
Northern Mariana Islands					98																	
Palau	44		100	100	79	72	19	10	97	20	20	94	70	20		16		26	32	28	79	84
Papua New Guinea			94	88	88		23	15	20	17	32	32	17	43					32		42	39
Pitcairn		100		100						100												
Samoa	86	97	100	95	91		94		77	94	100	88	17						62		99	88
Solomon Islands	96	82	100	94	94		45	58	58		65	65									71	70
Tokelau				100	100			100		100		89								100		
Tonga	100	86	92	100	100	53	70	99	98	100	100	100	63	83				99	96	100	100	100
Tuvalu	100		100	100	94			100		95	100	92				17				98	100	93
Vanuatu	65	95	63		85		53	54			94	52						64			88	60
Wallis and Futuna Islands																						
Western Samoa	97		100				94	67										69				
EUROPE																						
Albania					99						95	95										97
Hungary					100							98										99

continues

DATA TABLE 3 *continued*

Fraction of Population with Access to Safe Drinking Water

Region and Country	URBAN								RURAL								TOTAL							
	1970	1975	1980	1985	1990	1994	2000	2002	1970	1975	1980	1985	1990	1994	2000	2002	1970	1975	1980	1985	1990	1994	2000	2002
EUROPE (*continued*)																								
Netherlands								100								100								100
Republic of Moldova								97								88								92
Romania								91								16								57
Russian Federation								99								88								96
Serbia and Montenegro								99								86								93
Ukraine								100								94								98
Sources:	UNEP 1989; WRI 1988	UNEP 1989; WRI 1988	UNEP 1989; WRI 1988	UNEP 1989; WRI 1988	UNEP 1993	WHO 1996	WHO 2000	WHO/UNICEF 2004	UNEP 1989; WRI 1988	UNEP 1989; WRI 1988	UNEP 1989; WRI 1988	UNEP 1989; WRI 1988	UNEP 1993	WHO 1996	WHO 2000	WHO/UNICEF 2004	UNEP 1989; WRI 1988	UNEP 1989; WRI 1988	UNEP 1989; WRI 1988	UNEP 1989; WRI 1988	Calculated from UNEP 1993	WHO 1996	WHO 2000	WHO/UNICEF 2004

Note: The UN considers all European countries, except those shown, to have 100 percent water supply and sanitation coverage.

Access to Sanitation by Country, 1970 to 2002

Description

Adequate sanitation is also a fundamental requirement for basic human well-being, and improving access is one of the MDGs. Data are given here for the percent of urban, rural, and total populations, by country, with access to sanitation services for 1970, 1975, 1980, 1985, 1990, 1994, 2000, and 2002—the most recent year for which data are available. The World Health Organization (WHO) collected these data over various periods. Most of the data presented were drawn from responses by national governments to WHO questionnaires. Participants in data collection include the Joint Monitoring Programme, the United Nations Children's Fund, and the Water Supply and Sanitation Collaborative Council, which has continued sector monitoring and aims to support and strengthen the monitoring efforts of individual countries. Countries used their own definitions of "rural" and "urban."

For the 2000 and 2002 WHO assessments, new definitions were provided of "improved" sanitation with allowance for acceptable local technologies. The forty largest countries in the developing world account for 90 percent of population. As a result, WHO spent extra effort to collect comprehensive data for these countries. The excreta disposal system was considered adequate if it was private or shared (but not public) and if it hygienically separated human excreta from human contact. The following technologies were included in the 2000 assessment as representing improved sanitation:

Connection to a public sewer
Connection to septic system
Pour-flush latrine
Simple pit latrine
Ventilated improved pit latrine

In comparison, unimproved sanitation facilities refer to

Public or shared latrine
Open pit latrine
Bucket latrine

Limitations

As is the case with drinking water data, definitions for access to sanitation vary from country to country, and from year to year within the same country. Countries generally regard sanitation facilities that break the fecal-oral transmission route as adequate. In urban areas, adequate sanitation may be provided by connections to public sewers or by household systems such as pit privies, flush latrines, septic tanks, and communal toilets. In rural areas, pit privies, pour-flush latrines, septic tanks, and communal toilets are considered adequate. Direct comparisons between countries and across time within the same country are difficult and are additionally complicated by the fact that these data hide disparities between regions and socioeconomic classes.

The 2000 and 2002 WHO Assessments attempt to shift from gathering information from water providers only to include consumer-based information. The current approach uses household surveys in an effort to assess the actual use of facilities. Access to sanitation services, as reported by WHO, does not imply that the level of service is "adequate" or "safe." The assessment questionnaire did not include any methodology for discounting coverage figures to allow for intermittence or poor quality of the service provided. More details of the methods used, and their limitations, can be found at http://www.who.int/docstore/water_sanitation_health/Globassess-ment/GlobalTOC.htm.

Sources

United Nations Environment Programme (UNEP). 1989. *Environmental data report*, GEMS Monitoring and Assessment Research Centre, Basil Blackwell, Oxford.

United Nations Environment Programme (UNEP). 1993–94. *Environmental data report*, GEMS Monitoring and Assessment Research Centre in cooperation with the World Resources Institute and the UK Department of the Environment, Basil Blackwell, Oxford.

World Health Organization (WHO). 1996. *Water supply and sanitation sector monitoring report: 1996 (Sector status as of 1994)*, in collaboration with the Water Supply and Sanitation Collaborative Council and the United Nations Children's Fund, UNICEF, New York.

World Health Organization (WHO). 2000. *Global water supply and sanitation assessment 2000 report*. http://www.who.int/docstore/water_sanitation_health/Globassessment/GlobalTOC.htm.

World Resources Institute (WRI). 1988. World Health Organization data, cited by the World Resources Institute, *World resources 1988–89*, World Resources Institute and the International Institute for Environment and Development in collaboration with the United Nations Environment Programme, Basic Books, New York.

DATA TABLE 4 Access to Sanitation, by Country, 1970 to 2002

Fraction of Population with Access to Sanitation

Region and Country	URBAN								RURAL								TOTAL							
	1970	1975	1980	1985	1990	1994	2000	2002	1970	1975	1980	1985	1990	1994	2000	2002	1970	1975	1980	1985	1990	1994	2000	2002
AFRICA	47	75	57	75			90	99				40			47	82	6	50						
Algeria	13	100		80			70	56			15	16	20	8	30	16	9	67	20	57			73	92
Angola			40	29	25	34	70	56			4	20	35	6	6	16				19	21	16	44	30
Benin	83		48	58	60	54	46	58	1							12	14		16	33	45	20	23	32
Botswana				93	100		88	25				28	85							40	89			41
Burkina Faso	49	47	38	44		42	88	45			5	6	16	11	16	5	4	4	7	9		18	29	12
Burundi	96		40	84	64	60	79	47			35	56	16	50	85	35			35	58	18	51		36
Cameroon				100			99	63				1				33				43			92	48
Cape Verde			34	32	45	40	95	61			10	9		10	32	19			11	10		24	71	42
Central African Republic	64	100			45		43	47	96	100			46		23	12	72	100			46	46	31	27
Chad	7	9				73	81	30		1				7	13	0	1	1				21	29	8
Comoros							98	38							98	15							98	23
Congo	8	10					14	14	6	9						2	6	9						9
Congo, Democratic Republic	5	65			46	23	53	43	5	6		9	11	4	6	23	5	22			21	9	20	29
Côte d'Ivoire	23				81	59		61					100	51	50	23	5				92	54		40
Djibouti			43	78	80	77	99	55			20	17	26	100		27			39	64		90	91	50
Egypt					80	20	98	84			10			5	91	56					50	11	94	68
Equatorial Guinea					54	61	60	60					24	48	46	46					33	54	53	53
Eritrea							66	34							1	3							13	9
Ethiopia	67	56		96			58	19	8	8	19	96			6	4	14	14					15	6
Gabon							25	37							4	30							21	36
Gambia					100	83	41	72					27	23	35	46					44	37	37	53
Ghana	92	95	47	51	63	53	62	74	40	40	17	16	60	36	64	46	55	56	26	30	61	42	63	58
Guinea	70		54				94	25	2		1		0		41	6	13		11			70	58	13
Guinea-Bissau			21	29		32	88	57			13	18		17	34	23			15	21		20	47	34
Kenya	85	98	89			69	96	56	45	48	19	14		81	81	43	50	55	30			77	86	48
Lesotho	44	51	13	22		1	93	61	10	12	14	14		7	96	32	11	13	14	15		6	92	37
Liberia	100	100		6		38		49	9			2		2		7	19					18		26
Libya	100	100	100				97	97	54	69	72				30	96	67	79	88				97	97
Madagascar	88	9		55		50	70	49		9				3	70	27		9				15	42	33

continues

249

DATA TABLE 4   *continued*

Fraction of Population with Access to Sanitation

Region and Country	URBAN								RURAL								TOTAL							
	1970	1975	1980	1985	1990	1994	2000	2002	1970	1975	1980	1985	1990	1994	2000	2002	1970	1975	1980	1985	1990	1994	2000	2002
Malawi			100			70	96	66			81			51	98	42			83			53	77	46
Mali	63		79	90	81	58	93	59				3	10	21	100	38				19	27	31	69	45
Mauritania	100		5	8			44	64							19	9							33	42
Mauritius	51	63	100	100	100	100	100	100	99	100	90	86	100	100	99	99	77	82	94	92	100	100	99	99
Morocco	75			62	100	69	100	83	4			16		18	42	31	29			20		40	75	61
Mozambique				53		70	69	51				12	11		26	14							43	27
Namibia					24		96	66					11		17	14					15		41	30
Niger	10	30	36		71	71	79	43		1			4	4	5	4	1	3	7		17	15	20	12
Nigeria					80	61	85	48				5	11	21	45	30					35	36	63	38
Reunion																								
Rwanda	83	87	60	77	88		12	56	52	56	50	55	17		8	38	53	57	51	56	21		8	41
Sao Tome and Principe								32				15				20				15				24
Senegal			100	87	57	83	94	70			2		38	40	48	34			36		46	58	70	52
Seychelles																100								
Sierra Leone			31	60	55	17	23	53			6	10	31	8	31	30			12	24	39	11	28	39
Somalia		77		44				47		35		5				14		47		18				25
South Africa				73		79	99	86						12	73	44						46	86	67
Sudan	100	100	73			79	87	50	4	10				4	48	24	16	22					62	34
Swaziland		99		100		36	98	78		25		25		37		44		36		45		36		52
Tanzania		88		93			98	54		14		58			86	41		17	66				90	46
Togo	4	36	24	31		57	69	71	1	12	10	9		13	17	15	1	15	13	14		26	34	34
Tunisia	100	82	100	84		100	96	90	34	95		16		85		62	62	94		55		96		80
Uganda	84	82		32	32	75	96	53	76			30	60	55	72	39	76			30	57	57	75	41
Zambia	12	87		76		40	99	68	18	16		34		10	64	32	16	42		55		23	78	45
Zimbabwe					95		99	69				15	22		51	51					43		68	57
NORTH & CENTRAL AMERICA & CARIBBEAN																								
Anguilla																								
Antigua and Barbuda								98								94								95

Country	Block 1	Block 2	Block 3
Aruba	100 100	13 13	66 65 88 100 63 93 100
Bahamas	100 100	100 100	100 100 69 66 100 100 99
Barbados	62 87		42 47
Belize	76 59 23 100 100	75 45 87 22 100	50 57 71 100
British Virgin Islands	100		100
Canada	100 100	99 99	100 100
Cayman Islands	94 96 99 98 85 89	94 89 99 95 97	96 52 95 96 66 92
Costa Rica	94 99	84 93 84	95 93 91 66 95 92
Cuba	100 100 71	16	92
Dominican Republic	74 25 41 95 75 76 67	4 10 83 64 43	58 42 15 23 78 71 57
Dominica	86		83 83
El Salvador	71 48 82 85 78 78	26 43 59 78 40	37 39 35 58 68 83 63
Grenada	96 96		97 97
Guadeloupe	61 61		61 60
Guatemala	45 41 72 98 72	20 12 12 76 52	30 24 85 61
Haiti	53 42 44 42 50 52	10 13 17 16 23	19 21 25 28 34
Honduras	49 24 89 94 81 89	26 34 42 53 57 52	35 30 63 77 68
Jamaica	100 12 92 98 92 90	2 90 91 66 68	94 94 7 91 84 80
Martinique			
Mexico	77 77 85 87 81 90	12 13 13 26 32 39	55 58 66 73 77
Montserrat	96 96		96
Netherlands Antilles			
Nicaragua	34 35 34 96 78	8 24 16 27 68 51	77 27 31 84 66
Panama	78 83 99 87 89	69 76 61 59 94 51	81 71 86 99 72
Puerto Rico	96 96		96
St Kitts	89 89		89
St Lucia	96		89
St Vincent	100		
Trinidad/Tobago	100 100	92	97 88 100
Turks/Caicos Islands	51 83 96 100 98 100	96 97 88 95 92 94	81 92 93 98 100 96
United States of America	100 100 100	100 100	100 100
United States Virgin Islands			

continues

251

DATA TABLE 4 *continued*

Fraction of Population with Access to Sanitation

Region and Country	URBAN								RURAL								TOTAL							
	1970	1975	1980	1985	1990	1994	2000	2002	1970	1975	1980	1985	1990	1994	2000	2002	1970	1975	1980	1985	1990	1994	2000	2002
SOUTH AMERICA																								
Argentina	87	100	80	75			89		79	83	35	35	35		48		85	97		69			85	
Bolivia	25		37	33	38	58	82	58	4	9	4	10	14	16	38	23	12		18	21	26	41	66	45
Brazil	85			86	84	55	85	83	24		1	1	32	3	40	35	58			63	71	44	77	75
Chile	33	36	100	100		82	98	96	10	11	10	4			93	64	29	32	83	84			97	92
Colombia	75	73	93	96	84	76	97	96	8	13	4	13	18	33	51	54	47	48	61	65	64	63	85	86
Ecuador			73	98	56	87	70	80		7	17	29	38	34	37	59			43		48	64	59	72
Falkland Islands (Malvinas)																								
French Guiana							85								57								79	
Guyana	95	99	73	100	97		97	86	92	94	80	79	81		81	60	93	96	78	86	86		87	70
Paraguay	16	28	95	89	31		95	94	6		80	83	60		95	58	6	10	86	85	46		95	78
Peru	52		57	67	76	62	90	72	16		0	12	20	10	40	33	36		36	49	59	44	76	62
Suriname			100	78			100	99			79	48			34	76			88	62			83	93
Uruguay	97	97	59	59			96	95	13	17	6	59			89	85	82	83	51	59			95	94
Venezuela			60	57		64	86	71	45	17	12	5	72	30	69	48	45		52	50		58	74	68
ASIA																								
Afghanistan	69	63		5	13	38	25	16	16	15				1	8	5	21	21				8	12	8
Armenia								96								61								84
Azerbaijan								73								36								55
Bahrain				100	100			100				100								100				
Bangladesh	87	40	21	24	40	77	82	75	6	5	1	3	4	30	44	39	6	5	3	5	10	35	53	48
Bhutan					80	66	65	65					3	18	70	70					7	41	69	70
Brunei Darus																								
Cambodia							58	53			76				10	8							18	16
China							68	69					81	7	24	29					86	21	38	44
Cyprus	100	94	100	100	96		100	100	92	95		100	100		100	100	95	95		100	98		100	100
East Timor																30								33
Gaza Strip								65																

252

Country	1	2	3	4	5	6	7	8	9	10	11	12	13	14	15	16	17	18	19	20	21	22	23	24
Georgia									69	14	14	50					96			90				85
Hong Kong	83			88	9	7	20	18	18	52	40	3	2	1	2	1	58	73	70	44	31	27	87	50
India	30	31	29	14	37	23	15	12	38	74	37	30	38	21	5	4	71	87	73	79	33	29	60	100
Indonesia	52	66	51	44					78	31		35					86	86		100	100	96	100	82
Iran	84	81	67	72	74	69	78	70	48					43	59	48	95	93	89	96		100	75	
Iraq	80	79	36				47	47					11		1									
Israel																	100							59
Japan	100			100					100			100	100				100	100		100				
Jordan	93	99	95		70	70			85	98				34			94	100			92	94		
Kazakhstan	72	99				100			52	98							87	99			100	100		
Korea DPR	59	99				100			60	100				100			58	99			100			
Korea Rep		63		52	100		64	25		4		12	100		50	50		76		67			80	
Kuwait															100		75	100						
Kyrgyzstan	60	100	24	12					51	100				100			61	84						
Laos	24	46			5	5	3		14	34	13	8	2	4	2				30		13	10		
Lebanon	98	99	100						87	92	100						100	100	70					
Macau																	100		100					
Malaysia	58	56	44	94	75	70	60	59	98	41		94	60	55	43		75	100		94	100	100		
Maldives		30			22	13	3			58	26		2	1	1		96		95	95	60	21	21	
Mongolia	59	46	41	78	24	20	33			2	40	4					68	46		100	38	38	38	
Myanmar (Burma)	73	27	20	22	31	1		35	63	39		47	1	1	5		97	65	42	50	17	16	14	45
Nepal	27	92	76	6	19	13	1	1	20	20	16	13	25				92	75	51	34	88	42	21	14
Oman	89	61	30		67	75	12	12	61	61	35	3	6		44		81	98	53	53		81	76	100
Pakistan	54	83	70	25	82	70	6	3	35	42	19	12					100	94		79	51	81	91	12
Philippines	73	100	52	70	99	80	56	57	61	71	58	63	56	67		40	100	92			83	80		90
Qatar	100	100	56	50			100	83	100			85			100		100			100	100			100
Saudi Arabia	100	83			44	67	47	21	89	100	35		33	50	35	16	98	100					68	67
Singapore	91	90			52	50	59		56	80		45	39	63	55	11	97	100		99	99		58	
Sri Lanka	77	96				45	40	64						28		61	71	91	33	68	65	80	56	76
Syria	53	91				80				96			46	41	36		97	98	77		78			
Tajikistan	99	96							50															
Thailand	83							17	100						8	8	94	97						65
Turkey	62									70		86	22				77			64	56	64		
Turkmenistan	100																			56		74		
United Arab Emirates																	100	98		93				

continues

Fraction of Population with Access to Sanitation

Region and Country	URBAN								RURAL								TOTAL							
	1970	1975	1980	1985	1990	1994	2000	2002	1970	1975	1980	1985	1990	1994	2000	2002	1970	1975	1980	1985	1990	1994	2000	2002
ASIA (continued)																								
Uzbekistan	100						100	73							100	48							100	57
Vietnam					23	43	87	84		2	55		10	15	70	26	26				13	21	73	41
Yemen A R			60	83			99	76							31	14							45	30
Yemen Dem			70				99				15				31				35				45	
OCEANIA																								
American Samoa																								
Australia							100	100							100	100							100	100
Cook Islands	100	100	100	100	100		100	100			76	99	100		100	100				99			100	100
Fiji			85		91	100	75	99	87	93	60		65	85	12	98	91	96	70		75	92	43	98
French Polynesia					98		99	99					95		97	97							98	98
Guam								99								98								99
Kiribati					91	100	54	59					49		44	22						100	48	39
Marshall Islands					100			93					45			59								82
Micronesia					99	100		61					46	100		14						100		28
Nauru																								
New Caledonia																								
New Zealand																								
Niue					0	100	100	100							100	100						100	100	100
Northern Mariana Islands					100			94					71		92	96								94
Palau					95		100	96					100		100	52							100	83
Papua New Guinea	100	100	96	99	57	82	92	67	5	5	3	35		11	80	41	14		15	44		22	82	45
Pitcairn																								
Samoa	100	100	86		100	100	95	100	80	99	83		92	17	100	100	84	99					99	100
Solomon Islands			80		73	100	98	98			21		2		18	18					13		34	31
Tokelau						100					41			100		74						100		

254

Country	WHO/UNICEF 2004	WHO 2000	WHO 1996	Calculated from UNEP 1993	UNEP 1989; WRI 1988	UNEP 1989; WRI 1988	UNEP 1989; WRI 1988	UNEP 1989; WRI 1988	WHO/UNICEF 2004	WHO 2000	WHO 1996	UNEP 1993	UNEP 1989; WRI 1988	UNEP 1989; WRI 1988	UNEP 1989; WRI 1988	UNEP 1989; WRI 1988	WHO/UNICEF 2004	WHO 2000	WHO 1996	UNEP 1993	UNEP 1989; WRI 1988	UNEP 1989; WRI 1988	UNEP 1989; WRI 1988	UNEP 1989; WRI 1988
Tonga	97						100	100	96				40	94	100	100	98				99	97	100	100
Tuvalu	88	100	100						83	100	100		73	80			92	100	100		81	100		
Vanuatu	50	100	87	82	52	19			42	100	85	78	25	68			78	100	90	88	86	95		
Wallis and Futuna Islands					40																			
Western Samoa					84								83	83							88	86		
EUROPE																								
Albania	89								81								99							
Hungary	95								85								100							
Republic of Moldova	68								52								86							
Romania	51								10								86							
Russian Federation	87								70								93							
Serbia-Montenegro	87								77								97							
Ukraine	99								97								100							

Sources:

Note: The UN considers all European countries, except those shown, to have 100 percent water supply and sanitation coverage.

Access to Water Supply and Sanitation by Region, 1990 and 2002

Description

Table 5 shows the total population and the population without access to improved water supply or sanitation services ("unserved") in urban and rural areas for 1990 and 2002. The regions shown correspond with those of the Millennium Development Goals (MDGs) and are not comparable to those in previous editions of *The World's Water*. Overall, global water supply coverage for the year 2002 is estimated to be 83 percent, and global sanitation coverage is estimated to be 58 percent. Since 1990, the fraction of the population with access to an improved water supply increased from 77 percent to 83 percent in 2002, and the unserved population declined by over 150 million. During this same period, the fraction of the population with access to improved sanitation systems increased from 49 percent to 58 percent, and the unserved population declined by 70 million. These estimates are based on data from 93 and 97 percent of the global population in 1990 and 2002, respectively.

These data form the basis for all major international policy statements on lack of access to water. Because 1990 was used as the base year for establishing the 2015 MDGs, the 2002 data represents the halfway point and shows progress toward achieving these goals. The data indicate that significant work remains, particularly in the sanitation sector. And although some regions are on target, greater effort must be placed on improving water supply and sanitation coverage in sub-Saharan Africa and parts of Asia. The World Health Organization and UNICEF intend to report coverage data more frequently to allow better tracking of progress toward the MDGs. Go to the original source (later section) for a list of which countries belong to which regions shown here.

Limitations

These data give a good picture of the current lack of access to improved water and sanitation services, but comparison from different assessments should be done with extreme care, or not at all, because of changing definitions.

Country-reported data may reflect national definitions of "improved," unlike survey data, which were standardized as much as possible. For example, in many African countries the population "without access" to improved sanitation means people with no access to any sanitary facility. In Latin America and the Caribbean, however, it is more likely that those "without access" have a sanitary facility, but the facility is deemed unsatisfactory by local or national authorities. Low coverage figures for Latin America and the Caribbean, in part, may be a reflection of the comparatively narrow definitions used within that region.

Changes in the source of data also complicate comparisons over time. Before 2000, for example, data collected by WHO were provider-based and collected from service providers, such as utilities, ministries, and water agencies. The data shown here, however, is user-based and was collected from household surveys and censuses. User-based data is more likely to include improvements installed by households or local communities and gives a more complete picture of water supply and sanitation coverage.

Source

World Health Organization/UNICEF. 2004. *Meeting the MDG drinking water and sanitation target: A mid-term assessment of progress.* http://wssinfo.org/pdf/JMP_04_tables.pdf.

DATA TABLE 5 Access to Water Supply and Sanitation by Region, 1990 and 2002

Region	1990 Population (millions)				2002 Population (millions)			
	Total Population	Population Served	Population Unserved	Percent Served	Total Population	Population Served	Population Unserved	Percent Served
GLOBAL								
Urban water supply	2,263	2,150	113	95	2,988	2,839	149	95
Rural water supply	3,000	1,890	1,110	63	3,237	2,331	906	72
Total water supply	5,263	4,053	1,211	77	6,225	5,167	1,058	83
Urban sanitation	2,263	1,788	475	79	2,988	2,420	568	81
Rural sanitation	3,000	750	2,250	25	3,237	1,198	2,039	37
Total sanitation	5,263	2,579	2,684	49	6,225	3,610	2,614	58
DEVELOPED REGIONS								
Urban water supply	672	672	0	100	745	745	0	100
Rural water supply	262	259	3	99	248	233	15	94
Total water supply	934	934	0	100	993	973	20	98
Urban sanitation	672	672	0	100	745	745	0	100
Rural sanitation	262	259	3	99	248	228	20	92
Total sanitation	934	934	0	100	993	973	20	98
EURASIA								
Urban water supply	183	178	5	97	180	178	2	99
Rural water supply	99	82	17	83	101	83	18	82
Total water supply	282	259	23	92	281	261	20	93
Urban sanitation	183	170	13	93	180	165	14	92
Rural sanitation	99	67	32	68	101	66	35	65
Total sanitation	282	237	45	84	281	233	48	83

NORTHERN AFRICA								
Urban water supply	58	55	3	95	77	74	3	96
Rural water supply	60	49	11	82	71	59	11	84
Total water supply	118	104	14	88	147	133	15	90
Urban sanitation	58	49	9	84	77	68	8	89
Rural sanitation	60	28	32	47	71	40	30	57
Total sanitation	118	77	41	65	147	108	40	73
SUB-SAHARAN AFRICA								
Urban water supply	141	116	25	82	240	197	43	82
Rural water supply	363	131	232	36	445	200	245	45
Total water supply	504	247	257	49	685	397	288	58
Urban sanitation	141	76	65	54	240	132	108	55
Rural sanitation	363	87	276	24	445	116	329	26
Total sanitation	504	161	343	32	685	247	438	36
LATIN AMERICA & THE CARIBBEAN								
Urban water supply	313	292	22	93	407	387	20	95
Rural water supply	128	74	54	58	129	89	40	69
Total water supply	442	366	75	83	536	477	59	89
Urban sanitation	313	257	56	82	407	342	65	84
Rural sanitation	128	45	83	35	129	57	72	44
Total sanitation	442	305	137	69	536	402	134	75
EASTERN ASIA								
Urban water supply	368	364	4	99	550	511	38	93
Rural water supply	858	515	343	60	825	561	264	68
Total water supply	1,226	883	343	72	1,375	1,072	302	78

continues

259

DATA TABLE 5 *continued*

Region	1990 Population (millions)				2002 Population (millions)			
	Total Population	Population Served	Population Unserved	Percent Served	Total Population	Population Served	Population Unserved	Percent Served
EASTERN ASIA (*continued*)								
Urban sanitation	368	235	132	64	550	379	170	69
Rural sanitation	858	60	798	7	825	247	577	30
Total sanitation	1,226	294	932	24	1,375	619	756	45
SOUTH ASIA								
Urban water supply	317	285	32	90	444	417	27	94
Rural water supply	857	549	309	64	1,036	829	207	80
Total water supply	1,175	834	341	71	1,480	1,243	237	84
Urban sanitation	317	171	146	54	444	293	151	66
Rural sanitation	857	60	797	7	1,036	249	788	24
Total sanitation	1,175	235	940	20	1,480	548	933	37
SOUTH-EASTERN ASIA								
Urban water supply	141	128	13	91	220	200	20	91
Rural water supply	299	194	105	65	316	221	95	70
Total water supply	440	321	119	73	536	423	112	79
Urban sanitation	141	94	46	67	220	173	46	79
Rural sanitation	299	117	182	39	316	155	161	49
Total sanitation	440	211	229	48	536	327	209	61
WESTERN ASIA								
Urban water supply	85	80	5	94	121	115	6	95
Rural water supply	52	34	18	65	63	46	16	74
Total water supply	136	113	23	83	184	162	22	88

Urban sanitation	85	81	3	96	121	115	6	95
Rural sanitation	52	27	25	52	63	31	32	49
Total sanitation	136	108	29	79	184	145	39	79
OCEANIA								
Urban water supply	1	1	0	92	2	2	0	91
Rural water supply	5	2	3	39	6	3	4	40
Total water supply	6	3	3	51	8	4	4	52
Urban sanitation	1	1	0	83	2	2	0	84
Rural sanitation	5	2	2	50	6	3	3	46
Total sanitation	6	4	3	58	8	5	4	55

Due to rounding, coverage figures might not total 100%, even if the population unserved is shown as 0.
Definitions have changed over time. See the description of this table.

Annual Average ODA for Water, by Country, 1990 to 2004 (Total and Per Capita)

Description

The annual "ODA for Water" is listed here, by countries receiving such assistance, averaged from 1990 to 2004 (in 2003 constant dollars). Shown are the total amount received on an annual average basis (in million 2003 U.S. dollars), the country population in 2004 (in thousands), and the per capita ODA (in U.S. dollars per person per year).

Overseas Development Assistance, or Official Development Assistance (ODA), is the term given to funding that flows to countries or to multilateral institutions for the purpose of providing aid to countries. This funding is provided by official agencies and governments for the purpose of promoting economic development and welfare, and is "concessional in character and conveys a grant element of at least 25 percent." ODA can take several forms, including technical assistance, investment projects, debt forgiveness or rescheduling, equity investments, and other assistance.

The term "ODA for Water" in Tables 6, 7, and 8 encompasses official development assistance for a broad range of water-related projects, including water supply and sanitation, but excluding amounts committed for large water-related infrastructures. In fact, a relatively small (albeit hard to separate) amount of ODA is devoted to projects that will provide basic water supply and sanitation to populations not currently served, and hence targets of the Millennium Development Goals for water. One of the concerns of many analysts is that much ODA is directed to serving wealthier populations or improving services to populations that are already, at least, partly served by existing systems. The OECD Development Assistance Committee (DAC) identifies seven headings in the "water supply and sanitation" category:

Water resources policy and administrative management
Water resources protection
Water supply and sanitation: large systems
Water supply and sanitation: small systems
River development

Waste management/disposal
Education and training in water supply and sanitation

Figure DT 6.1 shows the breakdown of expenditures of ODA for water into these seven categories. By far, the largest expenditure is for large-scale water supply and sanitation projects.

The data in this table exclude ODA for:

Hydroelectric power plants
Agricultural water resources
Water transportation
Flood prevention/control

Limitations

ODA does not constitute all the funding that flows to developing countries, such as other public sector or private sector flows, hence these numbers do not reflect all funding for water projects. ODA levels vary substantially from year to year, depending on politics, funding priorities, and economic health of donor countries. The averages shown in this table hide these variations.

Sources

Clermont, F. 2006. Official Development Assistance for water from 1990 to 2004. Figures and trends. World Water Council. Marseilles, France.
OECD. 2002. *Creditor reporting system reporting directives.* July 30. Paris, France.
OECD Database. 2006. http://www.oecd.org/dac/stats/idsonline.

Data Table 6 Annual Average ODA for Water, by Country, 1990 to 2004
(Total and Per Capita)

Country	Population in 2004 (thousands)	ODA for Water Annual Average 1990-2004 (millions of US$)	Average ODA for Water ($/capita/yr)
Afghanistan	29,929	4.78	0.16
Albania	3,170	17.7	5.58
Algeria	31,833	10.55	0.33
Angola	13,523	8.47	0.63
Anguilla	14	0.25	17.86
Argentina	38,227	9.89	0.26
Armenia	3,056	7.67	2.51
Azerbaijan	8,233	10.68	1.30
Bangladesh	138,067	64.78	0.47
Barbados	272	0.05	0.18
Belize	274	0.21	0.77
Benin	6,721	20.76	3.09
Bhutan	874	2.51	2.87
Bolivia	8,815	36.74	4.17
Bosnia-Herzegovina	3,832	13.12	3.42

continues

DATA TABLE 6   *continued*

Country	Population in 2004 (thousands)	ODA for Water Annual Average 1990-2004 (millions of US$)	Average ODA for Water ($/capita/yr)
Botswana	1,723	7.22	4.19
Brazil	176,597	38.64	0.22
Burkina Faso	12,110	41.46	3.42
Burundi	7,206	7.34	1.02
Cambodia	13,404	14.52	1.08
Cameroon	16,088	8.39	0.52
Cape Verde	470	5.36	11.40
Central African Republic	3,881	3.66	0.94
Chad	8,582	14.77	1.72
Chile	15,774	3.75	0.24
China	1,288,400	251.06	0.19
Colombia	44,584	6.24	0.14
Comoros	601	1.42	2.36
Congo Dem. Republic (Zaire)	53,154	6.48	0.12
Congo Rep.	3,758	0.52	0.14
Cook Islands	22	0.36	16.36
Costa Rica	4,005	2.68	0.67
Cote d'Ivoire	16,836	9.93	0.59
Croatia	4,445	3.11	0.70
Cuba	11,326	1.13	0.10
Djibouti	706	5.56	7.88
Dominica	72	1	19.44
Dominican Republic	8,739	6.53	0.75
Ecuador	13,008	22.45	1.73
Egypt	67,560	167.99	2.49
El Salvador	6,534	9.1	1.39
Equatorial Guinea	494	1.34	2.71
Eritrea	4,390	3.7	0.84
Ethiopia	68,614	28.85	0.42
Fiji	835	1.42	1.70
Gabon	1,345	5.9	4.39
Gambia	1,421	3.61	2.54
Georgia	4,568	0.58	0.13
Ghana	20,670	64.72	3.13
Grenada	105	0.72	6.86
Guatemala	12,308	12.76	1.04
Guinea	7,909	19.1	2.41
Guinea-Bissau	1,490	2.46	1.65
Guyana	769	6.18	8.04
Haiti	8,440	5.66	0.67
Honduras	6,969	32.69	4.69
India	1,064,399	257.11	0.24
Indonesia	214,675	103.79	0.48
Iran	66,393	0.22	0.00
Iraq	24,700	61.27	2.48
Jamaica	2,643	9.7	3.67

Country	Population in 2004 (thousands)	ODA for Water Annual Average 1990-2004 (millions of US$)	Average ODA for Water ($/capita/yr)
Jordan	5,308	65.95	12.42
Kazakstan	14,879	18.32	1.23
Kenya	31,916	33.67	1.05
Kiribati	97	1.09	11.24
Korea Dem. Rep.	22,613	0.15	0.01
Kyrgyz Rep.	5,052	2.78	0.55
Laos	5,660	10.05	1.78
Lebanon	4,498	21.89	4.87
Lesotho	1,793	7.1	3.96
Liberia	3,374	0.15	0.04
Macedonia	2,062	13.53	6.56
Madagascar	16,894	9.47	0.56
Malawi	10,963	16.22	1.48
Malaysia	24,775	36.83	1.49
Maldives	294	0.59	2.01
Mali	11,652	24.08	2.07
Marshall Islands	57	0.63	11.05
Mauritanie	2,848	10.65	3.74
Mauritius	1,223	11.98	9.80
Mayotte	166	0.71	4.28
Mexico	103,796	47.55	0.46
Micronesia	125	0.72	5.76
Moldova	4,238	2.57	0.61
Mongolia	2,480	3.97	1.60
Montserrat	10	0.29	29.00
Morocco	30,113	82.71	2.75
Mozambique	18,792	44.86	2.39
Myanmar (Burma)	49,363	0.79	0.02
Namibia	2,015	11.17	5.54
Nepal	24,660	41.79	1.69
Nicaragua	5,480	23.65	4.32
Niger	11,763	19.05	1.62
Nigeria	136,461	35.92	0.26
Niue	3	0.04	13.33
Oman	2,660	0.01	0.00
Pakistan	148,439	60.68	0.41
Palau	20	0.001	0.05
Palestinian adm. Areas	3,367	71.47	21.23
Panama	2,985	0.68	0.23
Papua New Guinea	5,502	4.34	0.79
Paraguay	5,644	4.89	0.87
Peru	27,148	62.45	2.30
Philippines	81,503	69.39	0.85
Rwanda	8,395	11.09	1.32
Sao Tome and Principe	158	0.69	4.37
Saudi Arabia	23,215	0.01	0.00

TABLE 6 *Continues*

continues

DATA TABLE 6   *continued*

Country	Population in 2004 (thousands)	ODA for Water Annual Average 1990-2004 (millions of US$)	Average ODA for Water ($/capita/yr)
Senegal	10,240	43.44	4.24
Serbia and Montenegro	8,152	7.87	0.97
Seychelles	85	0.21	2.47
Sierra Leone	5,337	5.47	1.02
Solomon Islands	457	1.77	3.87
Somalia	9,626	5.14	0.53
South Africa	45,829	17.85	0.39
Sri Lanka	19,232	57.27	2.98
St. Helena	8	0.47	58.75
St. Kitts-Nevis	47	0.005	0.11
St. Lucia	161	2.55	15.84
St.Vincent & Grenadines	110	1.08	9.82
Sudan	33,546	4.56	0.14
Suriname	439	4.94	11.25
Swaziland	1,106	2.33	2.11
Syria	17,385	6.74	0.39
Tajikistan	6,360	1.87	0.29
Tanzania	35,889	50.11	1.40
Thailand	62,015	47.47	0.77
Timor-Leste	877	2.48	2.83
Togo	4,862	5.05	1.04
Tokelau	2	0.01	5.00
Tonga	102	1.63	15.98
Trinidad and Tobago	1,324	0.56	0.42
Tunisia	9,896	53.95	5.45
Turkey	70,712	101.68	1.44
Turkmenistan	4,864	0.14	0.03
Turks and Caicos	21	0.38	18.10
Tuvalu	12	0.15	12.50
Uganda	25,280	37.03	1.46
Uruguay	3,400	0.3	0.09
Uzbekistan	25,590	4.83	0.19
Vanuatu	211	0.14	0.66
Venezuela	25,674	4.64	0.18
Viet Nam	81,315	150.34	1.85
Western Samoa	178	2.26	12.70
Yemen	19,174	41.8	2.18
Zambia	10,403	30.57	2.94
Zimbabwe	13,102	16.77	1.28

Sources: Clermont, F. 2006. Official Development Assistance for water from 1990 to 2004. Figures and trends. World Water Council. Marseilles, France.

OECD. 2002. *Creditor reporting system reporting directives.* July 30. Paris, France.

OECD database. 2006. http://www.oecd.org/dac/stats/idsonline.

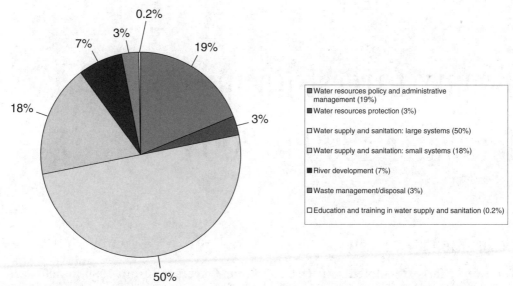

FIGURE DT 6.1 Breakdown of ODA for water by project type, 1990-2004.

Twenty Largest Recipients of ODA for Water, 1990 to 2004

Description

The twenty largest recipients of "ODA for Water," averaged from 1990 to 2004, are shown in this table. India and China received on average more than $250 million per year (in US 2003 dollars), whereas Indonesia, Egypt, Turkey, and Vietnam also received at least $100 million annually.

Overseas Development Assistance, or Official Development Assistance (ODA), is the term given to funding that flows to countries or to multilateral institutions for the purpose of providing aid to countries. This funding is provided by official agencies and governments for the purpose of promoting economic development and welfare, and is "concessional in character and conveys a grant element of at least 25 percent." ODA can take several forms, including technical assistance, investment projects, debt forgiveness or rescheduling, equity investments, and other assistance.

The term "ODA for Water" in Tables 6, 7, and 8 encompasses official development assistance for a broad range of water-related projects, including water supply and sanitation, but excluding amounts committed for large water-related infrastructures. In fact, a relatively small (albeit hard to separate) amount of ODA is devoted to projects that will provide basic water supply and sanitation to populations not currently served, and hence targets of the Millennium Development Goals for water. One of the concerns of many analysts is that much ODA is directed to serving wealthier populations, or improving services to populations that are already at least partly served by existing systems. The OECD Development Assistance Committee (DAC) identifies seven headings in the "water supply and sanitation" category:

Water resources policy and administrative management
Water resources protection
Water supply and sanitation: large systems
Water supply and sanitation: small systems
River development
Waste management/disposal
Education and training in water supply and sanitation

The data in this table exclude ODA for:

Hydroelectric power plants
Agricultural water resources
Water transportation
Flood prevention/control

Limitations

As might be expected, this table is dominated by the largest countries in terms of population, China and India, though both also have very serious water challenges. Other countries with serious water problems, however, do not appear on this list, especially countries in Africa, where water-related ODA contributions have historically been far below the levels needed to address problems there.

ODA does not constitute all the funding that flows to developing countries, such as other public sector or private sector flows, hence these numbers do not reflect all funding for water projects. ODA levels vary substantially from year to year, depending on politics, funding priorities, and economic health of donor countries.

Sources

Clermont, F. 2006. Official Development Assistance for water from 1990 to 2004. Figures and trends. World Water Council. Marseilles, France.
OECD. 2002. *Creditor reporting system reporting directives.* July 30. Paris, France.
OECD Database. 2006. http://www.oecd.org/dac/stats/idsonline.

DATA TABLE 7   Twenty Largest Recipients of ODA for Water, 1990-2004

Country	Population in 2004 (thousands)	ODA for Water Annual Average 1990-2004 (millions of US$)	Average ODA for Water ($/capita/yr)
India	1,064,399	257.11	0.24
China	1,288,400	251.06	0.19
Egypt	67,560	167.99	2.49
Vietnam	81,315	150.34	1.85
Indonesia	214,675	103.79	0.48
Turkey	70,712	101.68	1.44
Morocco	30,113	82.71	2.75
Palestinian administration areas	3,367	71.47	21.23
Philippines	81,503	69.39	0.85
Jordan	5,308	65.95	12.42
Bangladesh	138,067	64.78	0.47
Ghana	20,670	64.72	3.13
Peru	27,148	62.45	2.30
Iraq	24,700	61.27	2.48
Pakistan	148,439	60.68	0.41
Sri Lanka	19,232	57.27	2.98
Tunisia	9,896	53.95	5.45
Tanzania	35,889	50.11	1.40
Mexico	103,796	47.55	0.46
Thailand	62,015	47.47	0.77

Twenty Largest Per Capita Recipients of ODA for Water, 1990 to 2004

Description

The amount of "ODA for Water" given, per person, averaged between 1990 and 2004, is shown here in 2003 US dollars per person per year for the twenty largest national recipients. Most of these countries have small populations, boosting their per capita values, but this list includes both the Palestinian administered territories and Jordan, two regions with large and growing populations and severe water constraints.

Overseas Development Assistance, or Official Development Assistance (ODA), is the term given to funding that flows to countries or to multilateral institutions for the purpose of providing aid to countries. This funding is provided by official agencies and governments for the purpose of promoting economic development and welfare, and is "concessional in character and conveys a grant element of at least 25 percent." ODA can take several forms, including technical assistance, investment projects, debt forgiveness or rescheduling, equity investments, and other assistance.

The term "ODA for Water" in Tables 6, 7, and 8 encompasses official development assistance for a broad range of water-related projects, including water supply and sanitation, but excluding amounts committed for large water-related infrastructures. In fact, a relatively small (albeit hard to separate) amount of ODA is devoted to projects that will provide basic water supply and sanitation to populations not currently served, and hence targets of the Millennium Development Goals for water. One of the concerns of many analysts is that much ODA is directed to serving wealthier populations or improving services to populations that are already, at least, partly served by existing systems. The OECD Development Assistance Committee (DAC) identifies seven headings in the "water supply and sanitation" category:

Water resources policy and administrative management
Water resources protection
Water supply and sanitation: large systems
Water supply and sanitation: small systems
River development

Waste management/disposal
Education and training in water supply and sanitation

The data in this table exclude ODA for:
Hydroelectric power plants

Agricultural water resources
Water transportation
Flood prevention/control

Limitations

As noted previously, the per capita values in this table are distorted by several low-population countries. However, almost no countries from Africa appear here, a further indication of the paucity of funding available for that continent, compared to the magnitude of the water problems there. Note that the per capita value is simply an average of the annual ODA from 1990 to 2004, divided by the population in 2004, somewhat distorting the final figure. A time series can be found at the OECD database online, cited later.

ODA does not constitute all the funding that flows to developing countries, such as other public sector or private sector flows, hence these numbers do not reflect all funding for water projects. ODA levels vary substantially from year to year, depending on politics, funding priorities, and economic health of donor countries.

Sources

Clermont, F. 2006. Official Development Assistance for water from 1990 to 2004. Figures and trends. World Water Council. Marseilles, France.
OECD. 2002. *Creditor reporting system reporting directives.* July 30. Paris, France.
OECD Database. 2006. http://www.oecd.org/dac/stats/idsonline.

DATA TABLE 8 Twenty Largest Per Capita Recipients of ODA for Water, 1990-2004

Country	Population in 2004 (thousands)	ODA for Water Annual Average 1990-2004 (millions of US$)	Average ODA for Water ($/capita/yr)
St. Helena*	8	0.47	58.75
Montserrat	10	0.29	29.00
Palestinian administration areas	3,367	71.47	21.23
Dominica	72	1.00	19.44
Turks and Caicos	21	0.38	18.10
Anguilla	14	0.25	17.86
Cook Islands	22	0.36	16.36
Tonga	102	1.63	15.98
St. Lucia	161	2.55	15.84
Niue	3	0.04	13.33
Western Samoa	178	2.26	12.70
Tuvalu	12	0.15	12.50

continues

DATA TABLE 8   *continued*

Country	Population in 2004 (thousands)	ODA for Water Annual Average 1990-2004 (millions of US$)	Average ODA for Water ($/capita/yr)
Jordan	5,308	65.95	12.42
Cape Verde	470	5.36	11.40
Suriname	439	4.94	11.25
Kiribati	97	1.09	11.24
Marshall Islands	57	0.63	11.05
St.Vincent & Grenadines	110	1.08	9.82
Mauritius	1,223	11.98	9.80
Guyana	769	6.18	8.04

Investment in Water and Sewerage Projects with Private Participation, by Region, in Middle- and Low-Income Countries, 1990 to 2004

Description

Table 9 contains data on total investment (in US$ million nominal dollars) in infrastructure projects with private participation from 1990 to 2004 in middle- and low-income countries. For this table, only projects that are designed to provide potable water production and distribution and sewerage collection and treatment are included.

Data are grouped by region and listed in the year the project contract was finalized (defined by the World Bank as "financial closure"). Total investment between 1990 and 2004 was over $41 billion. By 2004, however, contracts representing 37 percent of this investment were either canceled or in distress. These data indicate that total investment in all countries is substantially less than during the peak years of the late 1990s. In addition, investment in middle- and low-income countries has been concentrated in two regions, East Asia and the Pacific and Latin America and the Caribbean. Investment in the rest of the world, especially Africa, has been minor.

The database includes only projects that have reached financial closure. The definition of financial closure varies among types of private participation but typically require a final, legally binding commitment for funding, management, or asset acquisition.

The World Bank PPI database covers infrastructure projects that meet three criteria:

- Projects that are owned or managed by private companies in low- and middle-income countries
- Projects that directly or indirectly serve the public
- Projects that reached financial closure after 1983

Figure DT 9.1 shows the breakdown by region of the sum of these investments over this period. Figure DT 9.2 shows total investment by year.

Limitations

Investments are added to the database upon financial closure. Projects that are cancelled are not removed from the database, thus these data overestimate actual investments. These data do not track private investment alone; rather they track total investment in projects with private participation. It is not possible using this database to separate out public and private investment totals. Some of the projects covered here are 100 percent private; others may be less than 50 percent private. Details on specific projects can be found in the database.

Source

World Bank. 2006. Private Participation in Infrastructure (PPI) database. World Bank, Washington, DC. http://www.ppi.worldbank.org.

DATA TABLE 9   Investment in Water and Sewerage Projects with Private Participation, by Region, in Middle- and Low-Income Countries, 1990 to 2004

Investment in US$ Millions (nominal)

Year of Investment	Sub-Saharan Africa	East Asia and Pacific	Europe and Central Asia	Latin America and the Caribbean	Middle East and North Africa	South Asia	Total Public and Private Investment
1990	0	0	0	0	0	0	0
1991	1	0	0	75	0	0	76
1992	0	1,559	0	0	6	0	1,565
1993	0	2,332	0	4,084	0	0	6,416
1994	0	133	16	263	0	0	412
1995	0	224	31	1,293	0	0	1,548
1996	6	623	942	148	10	0	1,729
1997	0	6,354	168	1,899	0	0	8,421
1998	13	631	108	1,471	0	0	2,223
1999	197	1,156	31	5,156	0	0	6,540
2000	0	94	1,432	3,026	0	216	4,768
2001	8	577	517	1,289	0	7	2,398
2002	0	1,218	68	643	32	0	1,961
2003	5	444	499	237	189	0	1,374
2004	0	531	10	1,385	0	0	1,926
Total	**230**	**15,876**	**3,822**	**20,969**	**237**	**223**	**41,357**

Source: World Bank PPI database. http://www.ppi.worldbank.org.

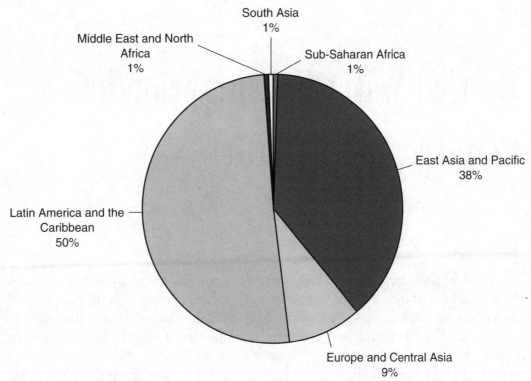

FIGURE DT 9.1 Regional investments in water and sewerage with private participation, 1990-2004.

Annual Investments in Middle and Low-Income Countries in Water and Sewerage Projects with Private Participation, 1990-2004

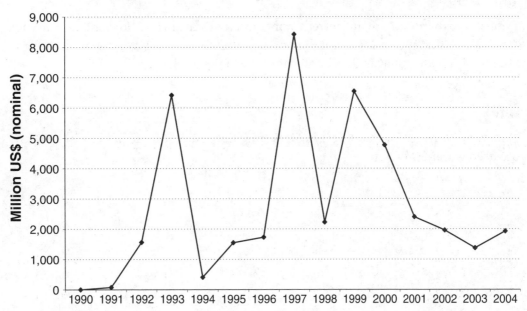

FIGURE DT 9.2 Annual investments in middle- and low-income countries in water and sewerage projects with private participation, 1990-2004.

Bottled Water Consumption by Country, 1997 to 2004

Description

Total bottled water consumption is reported by region and country for the years from 1997 to 2004. Data for 2004 are preliminary estimates. Units are in thousand cubic meters per year. The greatest consumption occurs in the United States and Mexico. In 1997, China was the ninth largest consumer of bottled water. By 2004, consumption in China had risen to third greatest, on a par with Brazil. Total global consumption exceeds 154 million cubic meters. Data for Fiji show almost no bottled water consumption, though, ironically, Fiji is the source of a premium bottled water sold in the United States. The Beverage Marketing Corporation (BMC) assigns countries to a continental region, shown in the first column.

Limitations

Data earlier than 1997 are not currently or consistently available. No distinction among types of bottled water is provided, and such definitions may vary from country to country. Data are not available for all countries of the world.

Source

Data were provided by the Beverage Marketing Corporation to the author in 2005 and are used with permission.

DATA TABLE 10   Bottled Water Consumption, by Country, 1997 to 2004

Thousand Cubic Meters

Region	Countries	1997	1998	1999	2000	2001	2002	2003	2004 (P)
N. America	United States	14,362	15,635	17,348	18,563	20,535	22,893	24,199	25,893
N. America	Mexico	10,484	10,883	11,579	12,424	13,244	14,767	16,495	17,683
Asia	China*	2,750	3,540	4,610	5,993	7,605	9,887	10,628	11,894
S. America	Brazil	3,932	4,742	5,658	6,817	8,166	9,628	10,758	11,598
Europe	Italy	7,558	7,722	8,925	9,221	9,480	9,690	10,350	10,661
Europe	Germany	8,207	8,216	8,313	8,401	8,552	8,680	9,950	10,313
Europe	France	6,053	6,565	6,947	7,462	7,820	8,430	8,907	8,550
Asia	Indonesia	2,262	2,736	3,436	4,300	5,122	6,146	6,945	7,362
Europe	Spain	3,543	3,716	4,077	4,208	4,344	4,513	5,098	5,506
Asia	India	1,047	1,364	1,682	2,150	2,668	3,361	4,202	5,126
Asia	Thailand	3,567	3,842	4,064	4,286	4,539	4,837	4,934	4,962
Europe	Turkey	932	1,185	1,369	1,667	1,871	2,007	2,332	2,460
Mideast	Saudi Arabia	1,298	1,490	1,610	1,770	1,938	2,116	2,140	2,270
Europe	United Kingdom	724	813	1,200	1,415	1,590	1,780	2,070	2,205
Asia	Korea, Republic of	892	1,009	1,110	1,191	1,274	1,359	1,631	1,957
Europe	Russian Federation	524	611	791	968	1,162	1,406	1,691	1,944
Europe	Poland	829	944	1,106	1,279	1,461	1,723	1,852	1,873
Asia	Japan	647	790	923	1,149	1,231	1,461	1,539	1,566
Europe	Belgium-Luxembourg	1,213	1,234	1,299	1,262	1,265	1,330	1,429	1,532
Asia	Philippines	728	837	999	1,119	1,213	1,292	1,363	1,414
N. America	Canada	541	650	755	848	939	1,027	1,063	1,116
Europe	Romania	406	448	583	640	800	839	890	993
Europe	Czech Republic	555	598	639	701	763	821	862	891
Europe	Portugal	647	646	706	719	736	761	789	846
Europe	Switzerland	623	653	652	654	657	668	705	744
Europe	Austria	569	610	605	610	632	645	704	672
S. America	Argentina	569	575	594	599	600	603	648	669
Europe	Hungary	202	245	300	398	467	515	618	662

continues

277

DATA TABLE 10 *continued*

Thousand Cubic Meters

Region	Countries	1997	1998	1999	2000	2001	2002	2003	2004 (P)
Asia	Pakistan	69	108	158	242	360	548	580	637
Europe	Ukraine	242	274	316	362	421	479	547	612
Oceania	Australia	304	355	390	443	488	566	585	595
S. America	Colombia	563	579	560	549	548	557	568	577
Europe	Greece	393	410	436	450	463	483	501	522
Europe	Serbia	—	—	297	313	335	392	439	469
Mideast	United Arab Emirates	230	245	256	270	285	326	359	413
Asia	Hong Kong SAR	191	222	245	271	298	331	364	401
Europe	Bulgaria	54	67	142	180	228	286	324	384
Mideast	Lebanon	181	215	240	275	310	346	358	383
Mideast	Israel	100	112	133	170	225	283	340	378
S. America	Venezuela, Boliv Rep of	202	221	230	248	263	290	317	346
Europe	Netherlands	249	241	273	286	296	317	333	344
Europe	Croatia	136	158	177	200	224	247	275	309
Mideast	Egypt	134	146	168	188	209	235	259	290
Asia	Malaysia	138	158	180	199	218	237	257	288
Asia	Viet Nam	115	140	159	180	200	219	239	248
Europe	Slovakia	159	162	169	170	173	178	183	190
Europe	Sweden	127	127	143	151	161	175	183	189
Mideast	Kuwait	68	78	96	112	128	144	158	172
S. America	Peru	76	81	89	104	118	132	147	162
Europe	Slovenia	68	81	93	109	124	138	150	162
S. America	Chile	67	81	102	116	117	118	123	130
Europe	Ireland	51	62	66	86	99	107	119	129
Asia	Singapore	57	64	70	76	82	88	95	103
Africa	South Africa	31	41	58	69	81	96	96	101
Europe	Norway	67	76	79	79	88	91	88	101
Europe	Denmark	72	72	82	82	82	84	95	97

Region	Country								
Europe	Finland	45	51	58	62	66	68	72	74
Europe	Latvia	2	3	31	39	47	56	72	74
S. America	Paraguay	46	50	54	57	61	65	69	73
Europe	Cyprus	48	49	51	55	58	62	67	71
Mideast	Qatar	33	38	43	47	52	56	61	66
Mideast	Jordan	28	31	36	40	44	48	53	57
Europe	Lithuania	15	18	20	24	29	35	40	46
Mideast	Bahrain	27	28	31	34	37	40	42	45
Mideast	Oman	16	18	20	23	26	29	33	37
Europe	Estonia	15	18	21	24	27	30	34	37
S. America	Uruguay	17	19	20	22	23	25	26	27
S. America	Nicaragua	13	15	16	17	19	20	22	23
N. America	Cuba	11	12	13	15	17	19	21	23
Asia	Brunei Darussalam	10	11	12	14	15	16	18	19
Oceania	Pacific Islands**	10	11	12	13	14	15	16	18
	Subtotal	80,141	87,244	97,722	107,280	117,831	131,265	143,517	152,784
	All Others	508	595	737	891	1,033	1,234	1,407	1,597
	TOTAL	80,649	87,839	98,459	108,171	118,864	132,499	144,925	154,381

P, Preliminary

* Includes Taiwan.

** Includes the Caroline Islands (Micronesia excluding Palau), the Marshall Islands, and the Northern Marianas (excluding Guam).

Global Bottled Water Consumption, by Region, 1997 to 2004

Description

Consumption of bottled water by continent is shown from 1997 through 2004. Data for 2004 are preliminary estimates. Units are in thousand cubic meters per year. In 2004, total consumption of bottled water was estimated at 154 million cubic meters, or 154 billion liters—around 25 liters for everyone on the planet—up nearly 25 percent from just two years ago. Regions include Europe, Asia, North America, South America, and Africa/Oceania/Middle East. Bottled water consumption is greatest in Europe, though the rate of increase in North America and Asia is higher. The countries that are assigned to each region are shown in Table 10.

Limitations

Earlier data are not currently available, limiting the significance of observed trends. No distinction is made among types of bottled water, and such definitions may vary from country to country. Data are not available for all countries of the world.

Source

Data were provided by the Beverage Marketing Corporation to the author in 2005 and are used with permission.

DATA TABLE 11 Total Bottled Water Consumption by Region, 1997 to 2004

Thousand Cubic Meters

Regions	1997	1998	1999	2000	2001	2002	2003	2004 (P)
Europe	34,328	36,074	39,965	42,276	44,520	47,037	51,768	53,661
North America	25,398	25,822	29,695	31,850	34,734	38,349	41,778	44,715
Asia	12,472	14,820	17,647	21,170	24,824	29,783	32,795	35,977
South America	5,484	6,362	7,323	8,528	9,915	11,437	12,677	13,607
Africa/Middle East/ Oceania	2,459	2,808	3,092	3,456	3,837	4,302	4,499	4,823
All Others	508	1,953	737	891	1,033	1,592	1,407	1,597
TOTAL	80,649	87,838	98,459	108,171	118,864	132,499	144,925	154,381

P, Preliminary

Per Capita Bottled Water Consumption by Region, 1997 to 2004

Description

Per capita consumption of bottled water, by continental region, is shown from 1997 through 2004. Units are liters per person per year. Global average per capita consumption is also shown. Global per capita consumption in 1996 was 12.6 liters per person per year, growing to over 24 liters per person per year in 2004. As the table shows, there are vast disparities in the use of bottled water by region. Europeans drink an average of over 97 liters per person per year, whereas average use in Africa, the Middle East, and Oceania is just over 4 liters per person per year. Data for 2004 are preliminary estimates. Regions include Europe, Asia, North America, South America, and Africa/Oceania/Middle East. The countries that are assigned to each region are shown in Table 10.

Limitations

Earlier data are not currently available, limiting the significance of observed trends. No distinction is made among types of bottled water, and such definitions may vary from country to country. Data are not available for all countries of the world.

Source

Data were provided by the Beverage Marketing Corporation to the author in 2005 and are used with permission.

DATA TABLE 12   Per Capita Bottled Water Consumption by Region, 1997 to 2004

Liters per capita

Regions	1997	1998	1999	2000	2001	2002	2003	2004 (P)
North America	59.4	59.7	68	72.2	77.9	85.2	91.8	97.5
Europe	47.8	50.3	55.6	58.8	61.9	65.5	72.1	74.7
South America	14.8	16.9	19.2	22.1	25.3	28.5	31.2	33.2
Asia	3.7	4.3	5.1	6	7	8.3	9	9.7
Africa/Middle East/Oceania	2.6	2.9	3.1	3.4	3.7	4.1	4.2	4.4
TOTAL	**13.8**	**14.8**	**16.4**	**17.8**	**19.3**	**21.3**	**23**	**24.2**

P, Preliminary

Per Capita Bottled Water Consumption, by Country, 1999 to 2004

Description

Per capita consumption of bottled water, by country, is shown from 1999 through 2004. Units are liters per person per year. Italy, Mexico, and the United Arab Emirates lead the table with over 160 liters per person per year. European countries dominate the top of the list; poorer countries congregate at the bottom, providing some indication of both the limited availability and the high cost of bottled water in developing countries. Data for 2004 are preliminary estimates.

Limitations

Earlier data are not currently available, limiting the significance of observed trends. No distinction is made among types of bottled water, and such definitions may vary from country to country. Data are not available for all countries of the world.

Source

Data were provided by the Beverage Marketing Corporation to the author in 2005 and are used with permission.

DATA TABLE 13   Per Capita Bottled Water Consumption by Country, 1999 to 2004

Liters per person

Countries	1999	2000	2001	2002	2003	2004 (P)
Italy	155	160	164	167	179	184
Mexico	117	124	130	143	157	169
United Arab Emirates	110	114	119	133	145	164
Belgium-Luxembourg	122	118	118	124	133	148
France	118	126	131	141	148	142
Spain	102	105	109	112	127	137
Germany	101	102	103	105	121	125
Lebanon	68	77	85	94	96	102
Switzerland	90	90	90	92	96	100
Cyprus	67	72	76	81	86	92
United States	64	67	74	82	85	91
Saudi Arabia	76	80	85	90	88	88
Czech Republic	62	68	74	80	84	87
Austria	75	75	78	79	86	82
Portugal	70	72	73	76	78	80
Slovenia	48.4	56.3	64.4	71.4	77.7	80.3
Qatar	59.5	63.3	67.1	70.8	74.3	78
Thailand	67	70	73.4	76	76.8	76.5
Kuwait	50.2	57	62.9	68.2	72.6	76.1
Pacific Islands*	48.3	50.8	54.1	57.3	66.6	72.1
Croatia	41.6	46.7	51.6	56.3	62.3	68.6
Bahrain	50.5	53.7	56.9	60.2	63.3	66.6
Hungary	29.5	39.2	46.2	51.1	61.5	66
Brazil	33.1	39.4	46.8	53.5	59.1	63
Israel	23.1	29.1	37.9	47	55.6	60.9
Hong Kong SAR	35.1	38.1	41.4	45.3	49.2	58.4
Brunei Darussalam	36.9	40.3	43.5	46.6	49.7	53.2
Bulgaria	17.9	23.1	29.6	37.5	43	51.1
Greece	41.2	42.5	43.6	45.4	47	49
Poland	28.6	33.1	37.8	44.6	47.9	48.5
Serbia	29.6	31.4	33.5	39.1	43.7	46.6
Romania	26	28.6	35.8	37.6	40	44.4
Korea, Republic of	23.6	25.1	26.6	28.3	33.8	40.3
United Kingdom	20.2	23.8	26.6	29.7	34.4	36.6
Turkey	21.1	25.4	28.1	29.8	34.2	35.7
Slovakia	31.2	31.5	32	32.9	33.8	35
Canada	24.4	27.1	29.7	32.2	33	34.3
Ireland	17.6	22.6	25.6	27.6	30.3	32.4
Latvia	12.7	16.1	19.7	23.7	30.7	31.9
Indonesia	15.5	19.1	22.4	26.6	29.6	30.9
Australia	20.5	23.1	25.2	29	29.6	29.9
Estonia	14.5	16.7	18.8	21.4	24	27.2
Singapore	17.4	18.2	19	19.8	20.7	23.6
Norway	17.7	17.6	19.5	20.1	19.4	22.1
Netherlands	17.3	18	18.5	19.7	20.6	21.1
Sweden	16.2	17	18.1	19.7	20.6	21
Denmark	15.4	15.4	15.4	15.7	17.7	17.9

continues

DATA TABLE 13   *continued*

Liters per person

Countries	1999	2000	2001	2002	2003	2004 (P)
Argentina	16.3	16.2	16.1	15.7	16.7	17.1
Philippines	12.6	13.8	14.6	15.6	16.1	16.4
Finland	11.3	12	12.7	13.2	13.9	14.1
Venezuela, Boliv Rep of	9.9	10.5	11	11.9	12.9	13.8
Colombia	14.4	13.8	13.6	13.6	13.6	13.6
Russian Federation	5.4	6.6	8	9.7	11.7	13.5
Ukraine	6.4	7.4	8.6	9.9	11.4	12.8
Lithuania	5.6	6.6	7.9	9.6	11.2	12.8
Oman	8.3	9.2	10	10.9	11.8	12.6
Japan	7.3	9.1	9.7	11.5	12.1	12.3
Malaysia	8.4	9.1	9.8	10.4	11.1	12.2
Paraguay	9.8	10.2	10.6	11	11.4	11.8
Jordan	7.4	8.1	8.6	9.1	9.6	10.1
China**	3.6	4.7	5.9	7.6	8.1	9
Chile	6.8	7.6	7.6	7.6	7.8	8.2
Uruguay	6.2	6.6	6.9	7.2	7.5	7.9
Peru	3.4	3.8	4.3	4.7	5.2	5.9
India	1.7	2.1	2.6	3.3	4	4.8
Nicaragua	3.3	3.6	3.9	4.1	4.3	4.4
Pakistan	1.1	1.7	2.5	3.7	3.8	4
Egypt	2.5	2.8	3	3.2	3.5	3.8
Viet Nam	2.1	2.3	2.5	2.7	2.9	3
Cuba	1.2	1.3	1.5	1.6	2.6	2.8
South Africa	1.3	1.6	1.9	2.2	2.2	2.4
Subtotal	**20**	**21.8**	**23.7**	**26.1**	**28.3**	**29.8**
All Others	0.7	0.8	0.9	1	1.1	1.3
TOTAL	**16.4**	**17.8**	**19.3**	**21.3**	**23**	**24.2**

P, Preliminary

* Includes the Caroline Islands (Micronesia excluding Palau), the Marshall Islands, and the Northern Marianas (excluding Guam).

** Includes Taiwan.

Global Cholera Cases and Deaths Reported to the World Health Organization, 1970 to 2004

Description

Annual cases and deaths from cholera are shown here for all continental regions, along with the number of countries reporting. Cholera is one of many waterborne diseases that are a consequence of the lack of sanitation services and access to clean drinking water. Some data on cholera cases and deaths are available as early as 1922 from reports to the League of Nations, but consistent reports are available from World Health Organization (WHO) since 1950 only. The data here are given annually from 1970 to 2004, with a sum of total cases for Asia from 1950 to 1969.

Figure DT 14.1 and Figure DT 14.2 show the annual number of deaths and total cases from 1970 to 2004. These data clearly reflect the rapid and severe spread of cholera in Latin America during the seventh pandemic, starting with the appearance of cholera in Peru in 1991 and spreading rapidly. That sharp spike in cholera was the first manifestation of the seventh pandemic in the Americas (defined as North, Central, and South America and the Caribbean) and was the first time that the Americas has seen more than a handful of indigenous cases in over a century. Figure DT 14.3 shows the total number of countries reporting to the WHO. As shown, the number has fluctuated over time and has dropped significantly in recent years, which may have an important role in the quality of the overall totals (see Limitations).

Limitations

These data are only the cases identified and reported to the WHO and do not include undiagnosed or unreported cases. The number of countries reporting cholera cases and deaths increased in Asia in the 1960s and 1970s, declining in the 1980s and then peaking in the 1990s. The number of countries reporting in the past five years is far below the peak, suggesting that totals shown in the table and figures could seriously underrepresent total cholera cases and deaths.

Source

The WHO maintains surveillance and reporting programs for cholera and provides online data in several places. These data were collected, sorted, and summed from http://www.who.int/ emc-documents/surveillance/docs/whocdscsrisr2001.html/cholera/cholera.htm.

All 1999 data: http://www.who.int/emc/diseases/cholera/choltbl1999.html.

All 2000 data: http://www.who.int/wer/pdf/2001/wer7631.pdf.

All 2001 to 2004 data: http://www.who.int/topics/cholera/wer/en/index.html.

DATA TABLE 14   Global Cholera Cases and Deaths Reported to the World Health Organization, 1970 to 2004

	1950-1969	1970	1971	1972	1973	1974	1975	1976	1977	1978	1979	1980	1981	1982	1983	1984	1985	1986
Africa																		
Total no. of cases		11,086	72,654	5,137	6,337	6,074	6,650	3,180	9,502	24,643	21,586	18,742	19,415	46,924	37,383	17,504	31,884	35,585
Total no. of deaths		747	11,427	386	636	582	504	194	462	1,591	1,869	1,185	1,581	2,988	1,903	1,711	3,837	3,490
No. of countries reporting		16	23	18	16	17	14	14	13	20	19	16	18	16	15	19	21	18
The Americas																		
Total no. of cases																		
Total no. of deaths																		
No. of countries reporting																		
Asia																		
Total no. of cases	1,355,747	52,429	103,308	75,064	102,148	89,077	85,316	65,469	52,460	52,511	35,732	24,015	32,343	17,991	27,877	11,809	13,389	17,131
Total no. of deaths	6,787	6,787	14,701	10,271	9,422	7,019	6,567	3,754	1,694	1,763	1,602	769	860	833	765	119	276	477
No. of countries reporting	23	22	15	16	15	14	18	19	21	21	22	17	18	14	11	9	11	11
Europe																		
Total no. of cases		726	97	4	303	2,484	1,089	16	8	5	289	16	46	22	12	11	6	52
Total no. of deaths		1	4	0	23	48	8	2	0	0	8	0	0	0	0	0	1	0
No. of countries reporting		4	6	2	5	6	6	8	6	2	5	5	9	4	5	4	3	5
Oceania																		
Total no. of cases				43					1,310	533	64	3	6	2,217	326	20	7	3
Total no. of deaths				1					21	0	0	0	0	17	1	0	0	0
No. of countries reporting		0		2	0	1	0	0	3	3	3	2	2	3	3	2	2	1
Global Totals																		
Total no. of cases	1,355,747	64,241	176,059	80,248	108,788	97,641	93,055	68,665	63,280	77,692	57,671	42,776	51,810	67,154	65,598	29,344	45,286	52,771
Total no. of deaths	6,787	7,535	26,132	10,658	10,081	7,650	7,079	3,950	2,177	3,354	3,479	1,954	2,441	3,838	2,669	1,830	4,114	3,967
No. of countries reporting	23	42	44	38	36	38	38	41	43	46	49	40	47	37	34	34	37	35

continues

DATA TABLE 14 *continued*

Africa	1987	1988	1989	1990	1991	1992	1993	1994	1995	1996	1997	1998	1999	2000	2001	2002	2003	2004
Total no. of cases	31,324	23,583	35,951	38,683	155,358	91,081	76,713	161,983	71,081	108,535	118,349	211,748	206,746	118,932	173,359	137,866	108,067	95,560
Total no. of deaths	2,658	1,500	1,445	2,288	13,998	5,291	2,532	8,128	3,024	6,216	5,853	9,856	8,728	4,610	2,590	4,551	1,884	2,331
No. of countries reporting	17	13	16	11	22	20	16	28	26	28	25	29	28	27	27	27	29	31

The Americas	1987	1988	1989	1990	1991	1992	1993	1994	1995	1996	1997	1998	1999	2000	2001	2002	2003	2004
Total no. of cases					391,220	354,089	209,192	113,684	85,809	24,643	17,760	57,106	8,126	3,101	535	23	33	36
Total no. of deaths					4,002	2,401	2,438	1,107	845	351	225	558	103	40	0	0	0	0
No. of countries reporting					16	21	21	17	15	18	16	16	12	11	8	4	4	5

Asia	1987	1988	1989	1990	1991	1992	1993	1994	1995	1996	1997	1998	1999	2000	2001	2002	2003	2004
Total no. of cases	17,558	20,871	18,007	31,003	49,791	16,299	90,862	106,100	53,159	10,142	11,293	24,212	39,417	11,246	10,340	4,409	3,463	5,764
Total no. of deaths	238	378	224	628	1,286	372	1,809	1,393	1,158	122	196	172	344	232	138	13	10	14
No. of countries reporting	10	12	13	13	16	18	25	26	18	13	18	16	14	12	13	13	8	12

Europe	1987	1988	1989	1990	1991	1992	1993	1994	1995	1996	1997	1998	1999	2000	2001	2002	2003	2004
Total no. of cases	14	14	11	349	320	18	73	2,630	937	25	18	47	16	35	58	7	12	21
Total no. of deaths	0	0	0	2	9	0	0	0	0	0	1	0	0	0	0	0	0	0
No. of countries reporting	5	4	6	9	6	7	15	20	17	6	6	10	5	2	5	6	4	7

Oceania	1987	1988	1989	1990	1991	1992	1993	1994	1995	1996	1997	1998	1999	2000	2001	2002	2003	2004
Total no. of cases	2	1	0	67		296	5	6	7	4	5	8	5	3,757	19	6	0	2
Total no. of deaths	0	0	0	1		8	0	0	0	0	0	0	0	26	0	0	0	0
No. of countries reporting	2	1	0	5	0	2	1	3	2	3	2	3	2	4	3	2	0	1

Global Totals	1987	1988	1989	1990	1991	1992	1993	1994	1995	1996	1997	1998	1999	2000	2001	2002	2003	2004
Total no. of cases	48,898	44,469	53,969	70,102	596,689	461,783	376,845	384,403	210,993	143,349	147,425	293,121	254,310	137,071	184,311	142,311	111,575	101,383
Total no. of deaths	2,896	1,878	1,669	2,919	19,295	8,072	6,779	10,628	5,027	6,689	6,275	10,586	9,175	4,908	2,728	4,564	1,894	2,345
No. of countries reporting	34	30	35	38	60	68	78	94	78	68	67	74	61	56	56	52	45	56

http://www.who.int/emc-documents/surveillance/docs/whocdscsrisr2001.html/cholera/cholera.htm
http://www.who.int/emc/diseases/cholera/choltbl1999.html for all 1999 data
http://www.who.int/wer/pdf/2001/wer7631.pdf for all 2000 data
http://www.who.int/topics/cholera/wer/en/index.html for 2001-2004 data

Global Cholera Deaths Reported 1970 to 2004

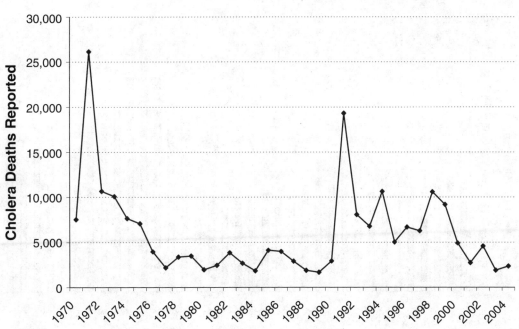

FIGURE DT 14.1 Reported global cholera deaths, 1970-2004.

Global Cholera Cases Reported 1970 to 2004

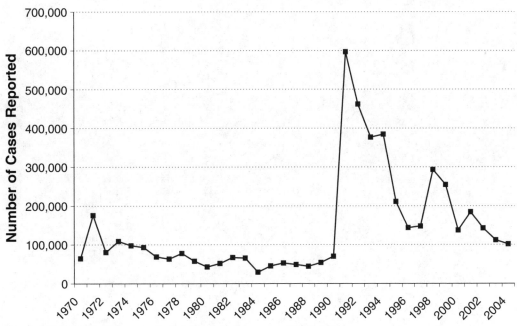

FIGURE DT 14.2 Reported global cholera cases, 1970-2004.

Number of Reporting Countries

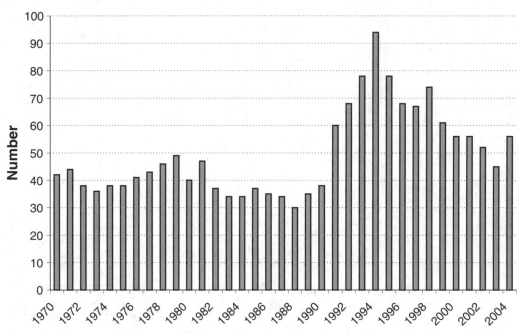

FIGURE DT 14.3 Number of reporting countries, 1970-2004.

Reported Cases of Dracunculiasis by Country, 1972 to 2005

Description

Dracunculiasis, or guinea worm disease, is the result of contact with water contaminated with a parasite, and hence directly related to drinking unclean water. A global campaign is under way to eradicate guinea worm. Although the original goal of elimination by the year 2000 has failed, substantial progress continues to be made. We have reported on this disease in *The World's Water 1998–1999* and in *The World's Water 2002–2003*. This update adds the last five years of data and summarizes the successes made in reducing the number of cases and endemic countries. Complete elimination could occur within five to seven years.

Guinea worm cases are reported here by country. Figure DT 15.1 shows reported cases from 1972 to 2005. Comprehensive monitoring began in the late 1980s. By 2005, the number of cases had declined to just 10,674 in twelve countries, down from a peak of nearly a million in the late 1980s in twenty countries. Moreover, nine of the twelve countries that still report the disease have fewer than 200 cases each. The vast majority of cases are in the Sudan and Ghana, with several hundred still reported annually in Mali. The Sudan continues to be racked by a genocidal civil war, limiting the ability of health workers to implement a comprehensive guinea worm detection and prevention program. The total number of cases reported from 1990 to 1996 may not equal the totals obtained by adding the country numbers. In these cases, the World Health Organization (WHO) adjusted totals to reflect estimates for countries not reporting at the time, or reporting by a different system than that used by the WHO.

Limitations

Dracunculiasis cases occur primarily in remote, rural areas. Reporting has improved tremendously in the past decade through implementation of comprehensive monitoring programs, but some cases may still be missed or underreported. Data for early years should be viewed with care, because many national reporting programs were not established before the late 1980s or early 1990s. For some countries, the totals reflect imported cases only. Cameroon, Chad, and Central African Republic have stopped reporting, or report zero cases, but they have yet to be certified as free of the disease—a process that takes several years.

Sources

World Health Organization. Dracunculiasis surveillance. *Weekly Epidemiological Record,* 57(9):65–67, March 5, 1982; 60(9):61–63, March 1, 1985; 61(5)31–36, January 31, 1986; 66(31):225–230, August 2, 1991; 68(18):125–126, April 30, 1993; 69(17):121–128, April 29, 1994; 70(18):125–126, May 5, 1995; 71(19):141–147, May 10, 1996; 72(19):133–139, May 9, 1997.

World Health Organization (WHO). 2001. Dracunculiasis global surveillance summary, 2000. *Weekly Epidemiological Record,* 76(18):133–140, May 4, 2001. http://childinfo.org/eddb/gw/countdata/

United States Department of Health and Human Services (DHHS). 2006. Public Health Service, Centers for Disease Control and Prevention (CDC).

Guinea Worm Wrap-Up #121. March 25, 2002. http://www.cdc.gov/ncidod/dpd/parasites/dracunculiasis/wrapup/word121.pdf.

Guinea Worm Wrap-Up #132. April 25, 2003. http://www.cdc.gov/ncidod/dpd/parasites/dracunculiasis/wrapup/word132.pdf.

Guinea Worm Wrap-Up #142. April 12, 2004. http://www.cdc.gov/ncidod/dpd/parasites/dracunculiasis/wrapup/word142.pdf.

Guinea Worm Wrap-Up #152. April 15, 2005. http://www.cdc.gov/ncidod/dpd/parasites/dracunculiasis/wrapup/word152.pdf.

Guinea Worm Wrap-Up #161. April 10, 2006. http://www.cdc.gov/ncidod/dpd/parasites/dracunculiasis/wrapup/161.pdf.

DATA TABLE 15 Reported Dracunculiasis Cases, by Country, 1972 to 2005

Region and Country	1972	1973	1974	1975	1976	1977	1978	1979	1980	1981	1982	1983	1984	1985	1986	1987	1988
AFRICA																	
Benin	1,480		820													400	33,962
Burkina Faso	5,822	4,404	4,008	6,277	1,557		2,885	2,694	2,565			4,362	1,739	458	2,558	1,957	1,266
Cameroon				251									0	168	86		752
Central African Republic																	
Chad							172						1,472	9	314		
Cote d'Ivoire	4,891	4,654	6,283	4,971	4,656	5,207	6,993		6,712	7,978		2,259	2,573	1,889	1,177	1,272	1,370
Ethiopia													2,882	1,467	3,385	2,302	1,487
Ghana	693	1,606	1,226	4,052	1,421	1,617	1,676		2,703	853	3,413	3,040	4,244	4,501	4,717	18,398	71,767
Kenya																	
Mali	668	786	737	542	760	1,084			816	777	401	428	5,008	4,072	5,640	435	564
Mauritania						127			651	663	903	1,612	1,241	1,291		227	608
Niger					2,600	3,000	5,560		1,906	2,113	1,530			1,373		699	
Nigeria	98			1,007					1,693				8,777	5,234	2,821	216,484	653,492
Senegal		334	208	65	137				161					62	128	132	138
Sudan															822		542
Togo				3,261	1,648		2,617	2,673	1,748	951	2,592		1,839	1,456	1,325	399	178
Uganda													6,230	4,070			1,960
ASIA AND THE MIDDLE EAST																	
India						7,052	6,827	2,846	2,729	5,406	42,926	44,818	39,792	30,950	23,070	17,031	12,023
Pakistan							250		14,155							2,400	1,110
Yemen		25															
Number of countries reporting	6	6	6	8	7	6	8	3	11	7	6	6	12	14	12	13	15
Number of Cases	13,652	11,809	13,282	20,426	12,779	18,087	26,980	8,213	35,839	18,741	51,765	56,519	75,797	57,000	46,043	262,136	781,219

continues

DATA TABLE 15 *continued*

Region and Country	1989	1990	1991	1992	1993	1994	1995	1996	1997	1998	1999	2000	2001	2002	2003	2004	2005
AFRICA																	
Benin	7,172	37,414	4,006	4,315	13,887	4,302	2,273	1,472	855	695	492	186	172	181	30	3	1
Burkina Faso	45,004	42,187		11,784	8,281	6,861	6,281	3,241	2,477	2,227	2,184	1,956	1,032	591	203	60	30
Cameroon	871	742	393	127	72	30	15	17	19	23	8	5	5	3	0	0	0
Central African Republic							18	9	5	34	26	35	36	nd	nd	nd	nd
Chad				156	1,231	640	149	127	25	3	1	3	0	0	0	0	0
Cote d'Ivoire	1,555	1,360	12,690		8,034	5,061	3,801	2,794	1,254	1,414	476	297	231	198	42	21	10
Ethiopia	3,565	2,333		303	1,120	1,252	514	371	451	366	249	60	29	47	28	17	37
Ghana	179,556	123,793	66,697	33,464	17,918	8,432	8,894	4,877	8,921	5,473	9,027	7,402	4,739	5,611	8,290	7,275	3,981
Kenya	5	6			35	53	23	0	6	7	1	4	8	17	12	7	2
Mali	1,111	884	16,024		12,011	5,581	4,218	2,402	1,099	650	410	290	718	861	829	357	659
Mauritania	447	8,036		1,557	3,533	5,029	1,762	562	388	379	255	136	94	42	13	3	0
Niger	288		32,829	500	21,564	18,562	13,821	2,956	3,030	2,700	1,920	1,166	417	248	293	240	183
Nigeria	640,008	394,082	281,937	183,169	75,752	39,774	16,374	12,282	12,590	13,420	13,237	7,869	5,355	3,820	1,459	495	120
Senegal		38		728	630	195	76	19	4	0	0	0	0	0	0	0	0
Sudan				2,447	2,984	53,271	64,608	118,578	43,596	47,977	66,097	54,890	49,471	41,493	20,299	7,266	5,569
Togo	2,749	3,042	5,118	8,179	10,349	5,044	2,073	1,626	1,762	2,128	1,589	828	1,354	1,502	669	278	73
Uganda	1,309	4,704	1,341	126,369	42,852	10,425	4,810	1,455	1,374	1,061	321	96	55	24	26	4	9
ASIA AND THE MIDDLE EAST																	
India	7,881	4,798	2,185	1,081	755	371	60	9	0	0	0	0					
Pakistan	534	160	106	23	2	0	0	0	0	0	0	0					
Yemen						94	82	62	7	0	0	0					
Number of countries reporting	15	15	11	15	18	19	20	20	20	20	20	20	17	16	16	16	16
Number of Cases	892,055	623,579	423,326	374,202	221,010	164,977	129,852	152,859	77,863	78,557	96,293	75,223	63,716	54,638	32,193	16,026	10,674

Number of Dracunculiasis Cases Reported

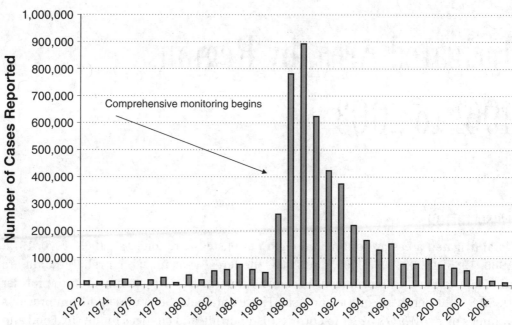

FIGURE DT 15.1   Reported Dracunculiasis cases, 1972-2005.

Irrigated Area, by Region, 1961 to 2003

Description

Total irrigated areas by continental region are listed here for 1961, 1965, 1970, 1975, 1980, 1985, 1990, 1995, 2000, and 2003, the latest year for which reliable data are available. Units are thousands of hectares. After 1990, all irrigated area in the former USSR is split between Europe and Asia. These data have been updated from previous volumes of *The World's Water* to correct misreporting and changes to the UN Food and Agriculture Organization FAOSTAT database.

Figure DT 16.1 shows total irrigated area plotted over this time period. Although irrigated area has continued to increase, it is now increasing at a rate slower than population. As a result, per capita irrigated area is declining.

Limitations

These data depend on in-country surveys, national reports, and estimates by the UN Food and Agriculture Organization. In some regions, multiple cropping may increase the apparent area in production. These data are not reported here. No information is offered about the quality of the land in production. Changes in political borders and the independence of several countries make certain continental time-series comparisons misleading. Data for the Soviet Union, Yugoslavia, and Czechoslovakia are provided through 1990; thereafter, the irrigated areas of the newly independent states are reported. When summing by continental area, however, trends will appear misleading because some of the newly independent states are now included in Asia, whereas others are in Europe. Thus, meaningful time-series trends, by continent, cannot be seen for these areas. The time-series for Africa, North and Central America, South America, and Oceania do not suffer from this problem.

Source

United Nations Food and Agriculture Organization. 2006. FAOSTAT database. http://faostat. fao.org/faostat/collections?subset=agriculture.

DATA TABLE 16   Irrigated Area, by Region, 1961 to 2003

Thousand hectares

Region	1961	1965	1970	1975	1980	1985	1990	1995	2000	2003
Africa	7,389	7,770	8,434	8,943	9,315	10,073	11,000	12,463	13,162	13,370
Asia [1]	89,883	96,713	109,373	121,151	131,659	141,780	155,525	181,767	193,169	193,890
Europe [1]	8,468	9,401	10,583	12,704	14,479	16,018	17,414	26,150	25,341	25,208
North & Central Americas	17,950	19,526	20,939	22,822	27,571	27,454	28,893	30,465	31,426	31,264
Oceania	1,081	1,370	1,590	1,622	1,686	1,959	2,118	2,695	2,682	2,844
South America	4,661	5,070	5,673	6,403	7,382	8,269	9,494	10,155	10,488	10,522
Former Soviet Union [1]	9,400	9,900	11,100	14,500	17,200	19,689	20,800			
Totals	138,832	149,750	167,692	188,145	209,292	225,242	245,244	263,695	276,268	277,098

Notes:

[1] After 1990, all irrigated area in the former USSR is split among Europe and Asia.

299

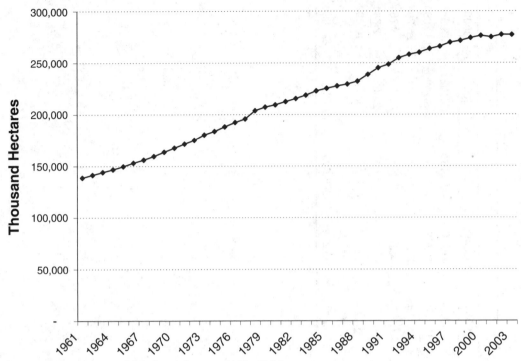

FIGURE DT 16.1   Total irrigated area, 1961–2003.

Irrigated Area, Developed and Developing Countries, 1961 to 2003

Description

Total irrigated areas by UN Food and Agriculture Organization's "developed" and "developing" country categories are listed here for 1961, 1965, 1970, 1975, 1980, 1985, 1990, 1995, 2000, and 2003, the latest year for which reliable data are available. Also shown are the annual percentage changes for developed countries, developing countries, and the world. Units of area are thousands of hectares; units for annual changes are percentages. From 1965 to 2000, annual rates are calculated by dividing five-year differences by five. Total irrigated area worldwide just barely increased from 2000 to 2003—a total of only 170,000 hectares, and the area irrigated in developing countries actually declined for the first time in this period. In addition, the average annual rate of change continues to drop for all regions. In the developed world, total irrigated area has effectively remained the same for nearly a decade. Overall, the rate of increase in irrigated area worldwide has fallen well below 1 percent per year, which means that total per capita irrigated area is actually decreasing. Note that these data have been updated and changed from the previous volume of *The World's Water* for all years reported.

Limitations

These data depend on in-country surveys, national reports, and estimates by the UN Food and Agriculture Organization. In some regions, multiple cropping may increase the apparent area in production, but multicropping is not reported here. No differentiation is made about the quality of the land in production. Recent changes in political borders and the independence of several countries make certain time-series comparisons difficult or inappropriate.

Source

United Nations Food and Agriculture Organization. 2006. FAOSTAT database. http://faostat. fao.org/faostat/collections?subset=agriculture.

DATA TABLE 17 Irrigated Area, Developed and Developing Countries, 1961 to 2003

Thousand hectares

	1961	1965	1970	1975	1980	1985	1990	1995	2000	2003
Developed Countries	37,180	40,232	44,278	50,381	58,926	62,555	66,196	69,240	69,308	69,133
Developing Countries	101,652	109,518	123,414	137,764	150,366	162,687	179,048	194,455	206,960	207,965
World	138,832	149,750	167,692	188,145	209,292	225,242	245,244	263,695	276,268	277,098
Annual Rate of Change: Developed Countries		1.64	2.01	2.76	3.39	1.23	1.16	0.92	0.02	-0.08
Annual Rate of Change: Developing Countries		1.55	2.54	2.33	1.83	1.64	2.01	1.72	1.29	0.16
Annual Rate of Change: World		1.57	2.40	2.44	2.25	1.52	1.78	1.50	0.95	0.10

Notes: From 1965 to 2000, annual rates are calculated by dividing five-year differences by five.

The U.S. Water Industry Revenue (2003) and Growth (2004-2006)

Description

The water industry comprises many diverse companies, products, and operations, ranging from the production of pipes and water meters to desalination plants, sophisticated wastewater treatment systems, water utilities, and water-related consulting. Chapter 6 describes this sector in detail, along with water-related concerns, risks, and strategies. This table provides a summary of the revenue earned by the U.S. water industry, broken into ten categories, as of 2003. That year, the U.S. water industry had total revenues of just under $100 billion. Also shown are projections for growth by sector. Most sectors are projecting growth under 10 percent annually, but specific companies and subsectors may see more rapid growth, or may see regional differences in efforts and funding.

Limitations

Many ways are known to parse the water "industry" that would produce different groups and divisions.

Source

Data compiled from the *Environmental Business Journal,* 2003, as cited in *The Environmental Benchmarker and Strategist Annual Water Issue,* Winter 2005.

DATA TABLE 18   The U.S. Water Industry Revenue (2003) and Growth (2004 to 2006)

(Revenues in Millions)

Business Segment	2003 Revenue	2004-2006 Growth
Drinking water utilities	$32,650	3-4%
Wastewater utilities	$30,780	3-4%
Water treatment equipment	$8,860	4-6%
Delivery equipment	$8,860	2-3%
Consulting/Engineering	$6,090	5-6%
Chemicals	$3,600	0-1%
Contract operations	$2,290	6-10%
Maintenance services	$1,460	3-5%
Instruments/Monitoring	$800	5-7%
Analytic testing	$480	2-4%
Total U.S. Water Industry	$95,870	

Pesticide Occurrence in Streams, Groundwater, Fish, and Sediment in the United States

Description

Pesticides are used for various purposes in the United States, ranging from protecting large-scale agricultural crop production to domestic lawn and garden care. These pesticides, however, can impair water quality and cause adverse human and environmental health impacts. Table 19 presents a very limited range of data available from newly published surveys in the United States of the presence of pesticides in agricultural, urban, and undeveloped areas, and in fish tissue and stream sediments.

These data show that pesticides were detected in 97 percent of all stream samples taken from 1992 to 2001 in both agricultural and urban areas. Even in undeveloped areas, pesticides were detected in 65 percent of stream samples. Slightly lower levels of detection were reported for samples taken from shallow groundwater aquifers, but even in these water sources, more than half of all samples showed pesticide contamination in both urban and agricultural areas.

Similarly, organochlorine pesticides (such as DDT) and their by-products were detected in fish and streambed sediments in very large fractions of agricultural and urban samples. Fish were almost universally contaminated with organochlorine, and a disturbing 80 percent of sediment samples in urban areas contained these substances—even higher than the 57 percent found in agricultural areas. Most organochlorine pesticides have not been used in the United States for several years before the study period, so these data highlight their persistence in the environment.

Pesticides were also found in concentrations that exceeded human-health benchmarks (see the source for more data on these exceedance rates). Streams with pesticide concentrations exceeding human-health benchmarks were typically found in urban and agricultural areas.

Limitations

These results are based on data from a total of 178 streams in the United States, with samples taken over a decade. The results shown do not account for seasonal and annual variations, which can be substantial due to changes in stream flow and the timing of pesticide application. Some sites were sampled in multiple years; others for only one or two years. Mean values were then used to summarize exceedance values. The pesticide findings are based on a time-weighted analysis of 4,380 water samples, adjusted to try to avoid biases caused by differences in sampling intensity among sites and seasons. Far more extensive sampling is needed, not just in the United States, but also in places where water-quality sampling may be limited or even nonexistent.

Source

Gillion, R. J., Barbash, J. E., Crawford, C. G., Hamilton, P. A., Martin, J. D., Nakagaki, N., Nowell, L. H., Scott, J. C., Stackelberg, P. E., Thelin, G. P., Wolock, D. M. 2006. *Pesticides in the nation's streams and ground water, 1992–2001*. 172 pp. United States Geological Survey, Circular 1291.

DATA TABLE 19 Pesticide Occurrence in Streams, Groundwater, Fish, and Sediment in the United States

Percentage of Time or Samples with Pesticide Detection

	Agricultural Areas	Urban Areas	Undeveloped Areas
Streams	97	97	65
Shallow groundwater	61	55	29

Percentage of Samples with One or More Organochlorine Compounds

	Agricultural Areas	Urban Areas	Undeveloped Areas
Fish tissue	92	94	57
Bed sediment	57	80	24

Global Desalination Capacity and Plants—January 1, 2005

Description

Desalination provides fresh water through various processes. Chapter 3 provides an update on the status of desalination worldwide. This table shows the number of plants proposed, in construction, or in operation as a function of capacity, as of January 1, 2005. In total, the plants in this database have a combined capacity exceeding 55 million cubic meters per day. In reality, however, actual operating capacity is far lower than this (see Data Tables 21 and 22). We estimate that actual operating capacity was closer to 32 million cubic meters per day at the beginning of 2005.

More than 50 percent of all desalination capacity, and over 99 percent of the numbers of plants, produce less than 100,000 cubic meters of water per day. Fewer than 70 plants, out of a total of over 10,000, are responsible for more than 45 percent of all desalination capacity, but many of the largest plants in this database are not in operation, or even under construction—only proposed (see Data Table 21).

Limitations

Extreme care should be taken in using these data, because the list includes (but does not clearly identify) plants that have been ordered but never built, built but never operated, operated but then shut down, or are still operating. No separate list of plants in operation only is available, though Table 22 provides a better sense of plants commissioned to operate.

These data were collected from a wide range of sources, from desalting plant suppliers to plant operators, and therefore depend on the accuracy of the information supplied. Plants with capacities less than 500 cubic meters per day are not included. This table includes desalination plants that use seawater, brackish water, river, pure, and other sources of water, not exclusively seawater desalination plants.

Source

Wangnick/GWI. 2005. 2004 *Worldwide desalting plants inventory*. Global Water Intelligence. Oxford, England. (Data provided to the Pacific Institute and used with permission.)

DATA TABLE 20 Global Desalination Plants and Capacity, January 1, 2005

Total Capacity: cubic meters per day	55,377,299
Number of Plants	10,597

Plant Capacity Ranges (m³/d)	Production Capacity (m³/d)	Percent of Total Capacity	Number of Plants
up to 5000 m³/d:	8,458,296	15.3	9,411
5000-10,000	3,538,791	6.4	478
10,000-20,000	4,543,670	8.2	306
20,000-30,000	2,813,661	5.1	112
30,000-40,000	1,807,389	3.3	51
40,000-50,000	2,213,031	4.0	48
50,000-60,000	1,383,728	2.5	25
60,000-70,000	466,750	0.8	7
70,000-80,000	697,300	1.3	9
80000-90,000	1,095,386	2.0	13
90,000-100,000	2,559,099	4.6	27
100,000-150,000	5,279,903	9.5	43
150,000-200,000	2,431,790	4.4	14
200,000-250,000	3,246,650	5.9	14
250,000-300,000	5,584,897	10.1	20
300,000-350,000	1,004,050	1.8	3
350,000-400,000	1,541,508	2.8	4
400,000-450,000	408,600	0.7	1
450,000-500,000	2,725,800	4.9	6
500,000-550,000	0	0.0	0
550,000-600,000	1,167,000	2.1	2
600,000-650,000	0	0.0	0
650,000-700,000	0	0.0	0
700,000-750,000	730,000	1.3	1
750,000-800,000	800,000	1.4	1
800,000-850,000	0	0.0	0
850,000-900,000	880,000	1.6	1
Totals	**55,377,299**	**100**	**10,597**

100 Largest Desalination Plants Planned, in Construction, or in Operation—January 1, 2005

Description

Desalination provides fresh water through various processes. Chapter 3 provides an update on the status of desalination worldwide. This table shows the 100 largest plants proposed, in construction, or in operation as a function of capacity, location, source of water, and estimated construction and operation dates, as of January 1, 2005. These plants represent more than 40 percent of all desalination capacity. This table includes desalination plants that use seawater, brackish water, river, pure, and other sources of water, not exclusively seawater desalination plants. Extreme care should be taken in using these figures, because many of the largest plants in this database are not in operation, or even under construction—only proposed. Indeed, it is likely that many of these plants will not be built, or will be built at a far later date than indicated in this table.

Limitations

These data were collected from a wide range of sources, from desalting plant suppliers to plant operators to urban planners, and therefore depend on the accuracy of the information supplied. This list includes plants that have been ordered but never built, built and never operated, operated but then shut down, or are still operating. For example, one of the largest plants on this list is a plant that is supposed to begin construction in the San Francisco, California, region in 2006 at a capacity of 454,000 cubic meters per day. No such plant is even close to construction anywhere in California (see Chapter 3). A separate list of plants that are in operation is not available, but we estimate that more than half of the plants in the table will not be in operation in 2006.

Source

Wangnick/GWI. 2005. 2004 *Worldwide desalting plants inventory.* Global Water Intelligence. Oxford, England. (Data provided to the Pacific Institute and used with permission.)

DATA TABLE 21   100 Largest Desalination Plants Planned, in Construction, or in Operation, January 1, 2005

Country	Location	Total Capacity (m³/d)	Source of Water	Estimated Construction Start	Planned Operation Year
Saudi Arabia SA	Shuaiba III	880,000	SEA	2004	2007
Saudi Arabia SA	Ras Al-Zour	800,000	SEA	2004	2007
Saudi Arabia SA	Al Jobail II Ex	730,000	SEA	2004	2007
UAE AE	Jebel Ali M	600,000	SEA	2008	2011
Kuwait KW	Al-Zour North	567,000	SEA	2004	2007
UAE AE	Shuweihat	455,000	SEA	2001	2004
UAE AE	Shuweihat 2	454,600	SEA	2004	2006
USA US	CA SanFrancisco	454,200	SEA	2006	2008
UAE AE	Fujairah II	454,000	SEA	2004	2007
UAE AE	Qidfa	454,000	SEA	2004	2006
Saudi Arabia SA	Al Jobail	408,600	SEA	2004	2007
Israel IL	Ashkelon	395,000	SEA	2001	2004
Saudi Arabia SA	Shuaiba III	390,908	SEA	2000	2003
UAE AE	Jebel Ali L-2	363,200	SEA	2004	2007
USA US	TX Pt. Comfort	340,650	SEA	2004	2006
UAE AE	Jebel Ali L-1	317,800	SEA	2003	2005
UAE AE	Jebel Ali N	300,000	SEA	2010	2013
Kuwait KW	Sulaibya	300,000	WASTE	2001	2003
India IN	Minjur Chennai	300,000	SEA	2005	2006
UAE AE	Taweelah B III	295,490	SEA	2005	2008
UAE AE	Fujairah	295,100	SEA	2001	2003
UAE AE	Umm Al Nar	284,125	SEA	2000	2002
USA US	PR Puerto Rico	284,000	SEA	2004	2006
UAE AE	Mirfa	277,000	SEA	2004	2007
Bahrain BH	Hidd 3	272,400	SEA	2004	2006
Saudi Arabia SA	Al Jobail I Ext	272,000	SEA	2004	2007
Saudi Arabia SA	Al Jobail III	272,000	SEA	2004	2006
Saudi Arabia SA	Al Khobar IV	272,000	SEA	2004	2006
Saudi Arabia SA	Shuaiba IV	272,000	SEA	2004	2007
USA US	CA Orange Count	265,000	RIVER	2004	2007
USA US	CA Fountain Val	264,950	WASTE	2004	2006
Libya LY	Tripoli	250,000	SEA	2004	2006
UAE AE	Taweelah A1 Ext	239,680	SEA	2000	2003
UAE AE	Taweelah C RO	227,300	SEA	2003	2006
Saudi Arabia SA	Ras Az Zawr	227,000	SEA	2004	2006
Qatar QA	Ras Laffan	227,000	SEA	2006	2009
Kuwait KW	Subiya 2	227,000	SEA	2004	2006
Kuwait KW	Subiya	227,000	SEA	2003	2007
USA US	CA Carlsbad	189,250	SEA	2005	2008
USA US	CA Huntington B	189,250	SEA	2004	2006
USA US	CA San Diego	189,250	SEA	2004	2006
UAE AE	Jebel Ali K II	182,000	SEA	2000	2002
Saudi Arabia SA	Al Bahah I	182,000	SEA	2004	2006
Qatar QA	Ras Laffan	182,000	SEA	2001	2004
Qatar QA	Ras Laffan 2	181,840	SEA	2004	2007
UAE AE	Fujairah	170,000	SEA	2001	2004

continues

DATA TABLE 21 *continued*

Country	Location	Total Capacity (m³/d)	Source of Water	Estimated Construction Start	Planned Operation Year
Spain ES	Malaga	165,000	BRACK	2001	2003
Kuwait KW	Shuwaikh	163,000	SEA	2004	2006
Saudi Arabia SA	Al Wasia	153,000	BRACK	2002	2004
Israel IL	Negev Arava	152,000	BRACK	2004	2006
USA US	FL Boca Raton	151,400	RIVER	2001	2003
Libya LY	Benghazi South	150,000	SEA	2004	2006
Spain ES	Murcia	147,000	SEA	2003	2004
Jordan JO	Zara Maain	145,344	BRACK	2003	2005
Jordan JO	Zara Maain	145,000	BRACK	2003	2005
China CN	Yantai	143,000	SEA	2004	2006
Israel IL	Ashdod	137,000	SEA	2004	2007
Israel IL	Hadera-Caesarea	136,260	SEA	2004	2007
Singapore SG	Singapore I	136,000	SEA	2003	2005
Iraq IQ		130,000	BRACK	2004	2005
Mexico MX	Hermosillo	128,690	SEA	2001	2004
Australia AU	WA Perth	123,300	SEA	2004	2006
Spain ES	Carboneras	120,000	SEA	2000	2001
Singapore SG	Ulu Pandan	116,000	WASTE	2004	2006
UAE AE	Umm Al Nar IWPP	115,244	SEA	2003	2007
Egypt EG	Sinai	113,650	SEA	2004	2006
Trinidad To. TT	Point Lisas	113,636	SEA	2000	2002
Saudi Arabia SA	Tabuk I	113,636	SEA	2004	2007
USA US	FL Palm Beach 3	113,550	RIVER	2002	2004
UAE AE	Jebel Ali G RO	113,500	SEA	2005	2007
Kuwait KW	Al-Zour North	113,500	SEA	2004	2006
Saudi Arabia SA	Shuqaiq II	109,000	SEA	2004	2006
USA US	TX El Paso	104,088	BRACK	2004	2006
USA US	CA Dana Point	102,195	SEA	2004	2006
UAE AE	Mirfa	102,000	SEA	2000	2001
Algeria DZ	Algiers Djinet	100,000	SEA	2004	2006
Algeria DZ	Algiers Zeralda	100,000	SEA	2004	2006
Algeria DZ	Mostaganem	100,000	SEA	2004	2007
USA US	TX Brownsville	94,625	SEA	2005	2007
USA US	TX Freeport	94,625	SEA	2005	2007
USA US	FL Tampa Bay II	94,625	SEA	2004	2007
USA US	TX Corpus Chris	94,625	SEA	2004	2006
USA US	FL Tampa Bay	94,625	SEA	2001	2003
USA US	FL S. Miami Hei	94,625	RIVER	2004	2006
Pakistan PK	Karachi	94,625	SEA	2004	2006
Pakistan PK	Gwadar	94,625	BRACK	2004	2006
Iran IR	Bandar Imam	93,600	BRACK	2000	2002
Singapore SG	Bedok	92,000	SEA	2004	2006
Saudi Arabia SA	Buraydah	91,000	SEA	2003	2004
Qatar QA	Ras Abu Font B1	91,000	SEA	2004	2007
Oman OM	Barka	90,920	SEA	2000	2003
Algeria DZ	Arzew	88,888	SEA	2003	2005
Algeria DZ	Arzew	88,000	SEA	2001	2003

Country	Location	Total Capacity (m³/d)	Source of Water	Estimated Construction Start	Planned Operation Year
Israel IL	Haifa	83,270	SEA	2004	2006
Israel IL	Palmachin	83,270	SEA	2004	2006
Israel IL	Palmahim	83,270	SEA	2003	2006
Israel IL	Ashdod	82,190	SEA	2003	2006
Israel IL	Shomrad	82,190	SEA	2003	2006
Libya LY	Azzawiya	80,000	SEA	2004	2006
Libya LY	Misurata	80,000	SEA	2004	2006
Total Capacity		**21,404,184**			

Installed Desalination Capacity by Year, Number of Plants, and Total Capacity, 1945 to 2004

Description

Desalination provides fresh water through various processes. Chapter 3 provides an update on the status of desalination worldwide. This table shows the annual increments to installed worldwide capacity and the cumulative capacity in million cubic meters per day. The number of plants added each year is also shown. This table includes desalination plants that use seawater, brackish water, river, pure, and other sources of water, not exclusively seawater desalination plants. Care should be taken in using these figures, because some of these plants may have been constructed, operated, and then shut down. No details are available on plant closures. Figure DT 22.1 shows the increase in desalination capacity over time.

Limitations

These data were collected from a wide range of sources, from desalting plant suppliers to plant operators to urban planners, and therefore depend on the accuracy of the information supplied. This list includes plants that have been built but never operated, operated but then shut down, or are still operating. Excluded from this list are plants that are only planned or under construction. As a result, this table is a better indication of global capacity than Tables 20 and 21, though it still includes plants that may no longer be in operation. Even this list, however, may also include plants scheduled for completion by 2004 that were never completed.

Source

Wangnick/GWI. 2005. 2004 *Worldwide desalting plants inventory*. Global Water Intelligence. Oxford, England. (Data provided to the Pacific Institute and used with permission.)

DATA TABLE 22 Installed Desalination Capacity by Year, Number of Plants, and Total Capacity, 1945 to 2004

Year	Number of Plants	Installed Capacity (m³/d)	Cumulative Capacity (m³/d)
1945	1	326	326
1946	0	-	326
1947	5	2,461	2,787
1948	1	114	2,901
1949	4	2,960	5,861
1950	5	3,000	8,861
1951	2	446	9,307
1952	14	7,295	16,602
1953	7	7,096	23,698
1954	13	15,879	39,577
1955	13	7,113	46,690
1956	26	13,310	60,000
1957	18	8,232	68,232
1958	8	5,758	73,990
1959	21	29,315	103,305
1960	23	19,742	123,047
1961	18	10,055	133,102
1962	20	28,314	161,416
1963	24	40,282	201,698
1964	24	21,761	223,459
1965	27	43,842	267,301
1966	37	38,842	306,143
1967	23	53,760	359,903
1968	53	116,887	476,790
1969	37	179,499	656,289
1970	54	115,358	771,647
1971	78	272,358	1,044,005
1972	70	109,729	1,153,734
1973	160	256,816	1,410,550
1974	166	228,701	1,639,251
1975	176	484,941	2,124,192
1976	191	241,856	2,366,048
1977	257	451,860	2,817,908
1978	224	572,873	3,390,781
1979	280	676,744	4,067,525
1980	304	963,998	5,031,523
1981	235	419,997	5,451,520
1982	307	860,906	6,312,426
1983	284	1,636,511	7,948,937
1984	330	815,495	8,764,432
1985	316	1,118,472	9,882,904
1986	341	619,837	10,502,741
1987	295	633,634	11,136,375
1988	304	1,050,311	12,186,686
1989	319	884,050	13,070,736
1990	324	936,610	14,007,346
1991	275	611,609	14,618,955
1992	336	918,189	15,537,144

continues

DATA TABLE 22   *continued*

Year	Number of Plants	Installed Capacity (m^3/d)	Cumulative Capacity (m^3/d)
1993	308	822,755	16,359,899
1994	372	931,244	17,291,143
1995	476	1,580,061	18,871,204
1996	415	1,277,372	20,148,576
1997	384	1,534,241	21,682,817
1998	400	1,535,182	23,217,999
1999	343	1,290,485	24,508,484
2000	457	1,791,110	26,299,594
2001	409	1,796,573	28,096,167
2002	346	1,644,347	29,740,514
2003	266	2,872,564	32,613,078
2004	176	3,014,296	35,627,374
Total	**10,402**	**35,627,374**	

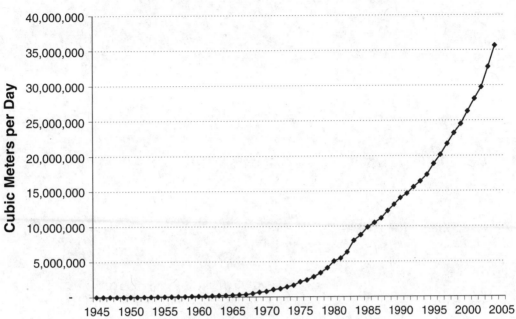

FIGURE DT 22.1   Cumulative installed desalination capacity, 1945–2004.

Water Units, Data Conversions, and Constants

Water experts, managers, scientists, and educators work with a bewildering array of different units and data. These vary with the field of work: engineers may use different water units than hydrologists; urban water agencies may use different units than reservoir operators; academics may use different units than water managers. But they also vary with regions: water agencies in England may use different units than water agencies in France or Africa; hydrologists in the eastern United States often use different units than hydrologists in the western United States. And they vary over time: today's water agency in California may sell water by the acre-foot, but its predecessor a century ago may have sold miner's inches or some other now arcane measure.

These differences are of more than academic interest. Unless a common "language" is used, or a dictionary of translations is available, errors can be made or misunderstandings can ensue. In some disciplines, unit errors can be more than embarrassing; they can be expensive, or deadly. In September 1999, the $125 million Mars Climate Orbiter spacecraft was sent crashing into the face of Mars instead of into its proper safe orbit above the surface because one of the computer programs controlling a portion of the navigational analysis used English units incompatible with the metric units used in all the other systems. The failure to translate English units into metric units was described in the findings of the preliminary investigation as the principal cause of mission failure.

This table is a comprehensive list of water units, data conversions, and constants related to water volumes, flows, pressures, and much more. Most of these units and conversions were compiled by Kent Anderson and initially published in P. H. Gleick, 1993, *Water in Crisis: A Guide to the World's Fresh Water Resources*, Oxford University Press, New York.

Water Units, Data Conversions, and Constants

Prefix (Metric)	Abbreviation	Multiple	Prefix (Metric)	Abbreviation	Multiple
deka-	da	10	deci-	d	0.1
hecto-	h	100	centi-	c	0.01
kilo-	k	1000	milli-	m	0.001
mega-	M	10^6	micro-	µ	10^{-6}
giga-	G	10^9	nano-	n	10^{-9}
tera-	T	10^{12}	pico-	P	10^{-12}
peta-	P	10^{15}	femto-	f	10^{-15}
exa-	E	10^{18}	atto-	a	10^{-18}

LENGTH (L)

1 micron (µ)	$= 1 \times 10^{-3}$ mm	10 hectometers	= 1 kilometer
	$= 1 \times 10^{-6}$ m	1 mil	= 0.0254 mm
	$= 3.3937 \times 10^{-5}$ in		$= 1 \times 10^{-3}$ in
1 millimeter (mm)	= 0.1 cm	1 inch (in)	= 25.4 mm
	$= 1 \times 10^{-3}$ m		= 2.54 cm
	= 0.03937 in		= 0.08333 ft
1 centimeter (cm)	= 10 mm		= 0.0278 yd
	= 0.01 m	1 foot (ft)	= 30.48 cm
	$= 1 \times 10^{-5}$ km		= 0.3048 m
	= 0.3937 in		$= 3.048 \times 10^{-4}$ km
	= 0.03281 ft		= 12 in
	= 0.01094 yd		= 0.3333 yd
1 meter (m)	= 1000 mm		$= 1.89 \times 10^{-4}$ mi
	= 100 cm	1 yard (yd)	= 91.44 cm
	$= 1 \times 10^{-3}$ km		= 0.9144 m
	= 39.37 in		$= 9.144 \times 10^{-4}$ km
	= 3.281 ft		= 36 in
	= 1.094 yd		= 3 ft
	$= 6.21 \times 10^{-4}$ mi		$= 5.68 \times 10^{-4}$ mi
1 kilometer (km)	$= 1 \times 10^5$ cm	1 mile (mi)	= 1609.3 m
	= 1000 m		= 1.609 km
	= 3280.8 ft		= 5280 ft
	= 1093.6 yd		= 1760 yd
	= 0.621 mi	1 fathom (nautical)	= 6 ft
10 millimeters	= 1 centimeter	1 league (nautical)	= 5.556 km
10 centimeters	= 1 decimeter		= 3 nautical miles
10 decimeters (dm)	= 1 meter	1 league (land)	= 4.828 km
10 meters	= 1 dekameter		= 5280 yd
10 dekameters (dam)	= 1 hectometer		= 3 mi
		1 international nautical mile	= 1.852 km
			= 6076.1 ft
			= 1.151 mi

Water Units, Data Conversions, and Constants *(continued)*

AREA (L^2)

1 square centimeter (cm^2)	$= 1 \times 10^{-4} m^2$	**1 square foot (ft^2)**	$= 929.0 \text{ cm}^2$
	$= 0.1550 \text{ in}^2$		$= 0.0929 \text{ m}^2$
	$= 1.076 \times 10^{-3} \text{ ft}^2$		$= 144 \text{ in}^2$
	$= 1.196 \times 10^{-4} \text{ yd}^2$		$= 0.1111 \text{ yd}^2$
1 square meter (m^2)	$= 1 \times 10^{-4}$ hectare		$= 2.296 \times 10^{-5}$ acre
	$= 1 \times 10^{-6} \text{ km}^2$		$= 3.587 \times 10^{-8} \text{ mi}^2$
	$= 1$ centare (French)	**1 square yard (yd^2)**	$= 0.8361 \text{ m}^2$
	$= 0.01$ are		$= 8.361 \times 10^{-5}$ hectare
	$= 1550.0 \text{ in}^2$		$= 1296 \text{ in}^2$
	$= 10.76 \text{ ft}^2$		$= 9 \text{ ft}^2$
	$= 1.196 \text{ yd}^2$		$= 2.066 \times 10^{-4}$ acres
	$= 2.471 \times 10^{-4}$ acre		$= 3.228 \times 10^{-7} \text{ mi}^2$
1 are	$= 100 \text{ m}^2$	**1 acre**	$= 4046.9 \text{ m}^2$
1 hectare (ha)	$= 1 \times 10^4 \text{ m}^2$		$= 0.40469 \text{ ha}$
	$= 100$ are		$= 4.0469 \times 10^{-3} \text{ km}^2$
	$= 0.01 \text{ km}^2$		$= 43,560 \text{ ft}^2$
	$= 1.076 \times 10^5 \text{ ft}^2$		$= 4840 \text{ yd}^2$
	$= 1.196 \times 10^4 \text{ yd}^2$		$= 1.5625 \times 10^{-3} \text{ mi}^2$
	$= 2.471$ acres	**1 square mile (mi^2)**	$= 2.590 \times 10^6 \text{ m}^2$
	$= 3.861 \times 10^{-3} \text{ mi}^2$		$= 259.0$ hectares
1 square kilometer (km^2)	$= 1 \times 10^6 \text{ m}^2$		$= 2.590 \text{ km}^2$
	$= 100$ hectares		$= 2.788 \times 10^7 \text{ ft}^2$
	$= 1.076 \times 10^7 \text{ ft}^2$		$= 3.098 \times 10^6 \text{ yd}^2$
	$= 1.196 \times 10^6 \text{ yd}^2$		$= 640$ acres
	$= 247.1$ acres		$= 1$ section (of land)
	$= 0.3861 \text{ mi}^2$	**1 feddan (Egyptian)**	$= 4200 \text{ m}^2$
1 square inch (in^2)	$= 6.452 \text{ cm}^2$		$= 0.42 \text{ ha}$
	$= 6.452 \times 10^{-4} \text{ m}^2$		$= 1.038$ acres
	$= 6.944 \times 10^{-3} \text{ ft}^2$		
	$= 7.716 \times 10^{-4} \text{ yd}^2$		

(continues)

Water Units, Data Conversions, and Constants *(continued)*

VOLUME (L^3)

1 cubic centimeter	$= 1 \times 10^{-3}$ liter	**1 cubic foot (ft^3)**	$= 2.832 \times 10^4$ cm^3
(cm^3)	$= 1 \times 10^{-6}$ m^3		$= 28.32$ liters
	$= 0.06102$ in^3		$= 0.02832$ m^3
	$= 2.642 \times 10^{-4}$ gal		$= 1728$ in^3
	$= 3.531 \times 10^{-3}$ ft^3		$= 7.481$ gal
1 liter (1)	$= 1000$ cm^3		$= 0.03704$ yd^3
	$= 1 \times 10^{-3}$ m^3	**1 cubic yard (yd^3)**	$= 0.7646$ m^3
	$= 61.02$ in^3		$= 6.198 \times 10^{-4}$
	$= 0.2642$ gal		acre-ft
	$= 0.03531$ ft^3		$= 46656$ in^3
1 cubic meter (m^3)	$= 1 \times 10^6$ cm^3		$= 27$ ft^3
	$= 1000$ liter	**1 acre-foot**	$= 1233.48$ m^3
	$= 1 \times 10^{-9}$ km^3	**(acre-ft or AF)**	$= 3.259 \times 10^5$ gal
	$= 264.2$ gal		$= 43560$ ft^3
	$= 35.31$ ft^3	**1 Imperial gallon**	$= 4.546$ liters
	$= 6.29$ bbl		$= 277.4$ in^3
	$= 1.3078$ yd^3		$= 1.201$ gal
	$= 8.107 \times 10^{-4}$		$= 0.16055$ ft^3
	acre-ft	**1 cfs-day**	$= 1.98$ acre-feet
1 cubic decameter	$= 1000$ m^3		$= 0.0372$ in-mi^2
(dam^3)	$= 1 \times 10^6$ liter	**1 inch-mi^2**	$= 1.738 \times 10^7$ gal
	$= 1 \times 10^{-6}$ km^3		$= 2.323 \times 10^6$ ft^3
	$= 2.642 \times 10^5$ gal		$= 53.3$ acre-ft
	$= 3.531 \times 10^4$ ft^3		$= 26.9$ cfs-days
	$= 1.3078 \times 10^3$ yd^3	**1 barrel (of oil)**	$= 159$ liter
	$= 0.8107$ acre-ft	**(bbl)**	$= 0.159$ m^3
1 cubic hectometer	$= 1 \times 10^6$ m^3		$= 42$ gal
(ha^3)	$= 1 \times 10^3$ dam^3		$= 5.6$ ft^3
	$= 1 \times 10^9$ liter	**1 million gallons**	$= 3.069$ acre-ft
	$= 2.642 \times 10^8$ gal	**1 pint (pt)**	$= 0.473$ liter
	$= 3.531 \times 10^7$ ft^3		$= 28.875$ in^3
	$= 1.3078 \times 10^6$ yd^3		$= 0.5$ qt
	$= 810.7$ acre-ft		$= 16$ fluid ounces
1 cubic kilometer	$= 1 \times 10^{12}$ liter		$= 32$ tablespoons
(km^3)	$= 1 \times 10^9$ m^3		$= 96$ teaspoons
	$= 1 \times 10^6$ dam^3	**1 quart (qt)**	$= 0.946$ liter
	$= 1000$ ha^3		$= 57.75$ in^3
	$= 8.107 \times 10^5$		$= 2$ pt
	acre-ft		$= 0.25$ gal
	$= 0.24$ mi^3	**1 morgen-foot**	$= 2610.7$ m^3
1 cubic inch (in^3)	$= 16.39$ cm^3	**(S. Africa)**	
	$= 0.01639$ liter	**1 board-foot**	$= 2359.8$ cm^3
	$= 4.329 \times 10^{-3}$ gal		$= 144$ in^3
	$= 5.787 \times 10^{-4}$ ft^2		$= 0.0833$ ft^3
1 gallon (gal)	$= 3.785$ liters	**1 cord**	$= 128$ ft^3
	$= 3.785 \times 10^{-3}$ m^3		$= 0.453$ m^3
	$= 231$ in^3		
	$= 0.1337$ ft^3		
	$= 4.951 \times 10^{-3}$ yd^3		

Water Units, Data Conversions, and Constants *(continued)*

VOLUME/AREA (L^3/L^2)

1 inch of rain	= 5.610 gal/yd^2
	= 2.715 × 10^4 gal/acre

1 box of rain	= 3,154.0 lesh

MASS (M)

1 gram (g or gm)	= 0.001 kg	**1 ounce (oz)**	= 28.35 g
	= 15.43 gr		= 437.5 gr
	= 0.03527 oz		= 0.0625 lb
	= 2.205 × 10^{-3} lb	**1 pound (lb)**	= 453.6 g
1 kilogram (kg)	= 1000 g		= 0.45359237 kg
	= 0.001 tonne		= 7000 gr
	= 35.27 oz		= 16 oz
	= 2.205 lb	**1 short ton (ton)**	= 907.2 kg
1 hectogram (hg)	= 100 gm		= 0.9072 tonne
	= 0.1 kg		= 2000 lb
1 metric ton (tonne or te or MT)	= 1000 kg	**1 long ton**	= 1016.0 kg
	= 2204.6 lb		= 1.016 tonne
	= 1. 102 ton	**1 long ton**	= 2240 lb
	= 0.9842 long ton		= 1.12 ton
1 dalton (atomic mass unit)	= 1.6604 × 10^{-24} g	**1 stone (British)**	= 6.35 kg
			= 14 lb
1 grain (gr)	= 2.286 × 10^{-3} oz		
	= 1.429 × 10^{-4} lb		

TIME (T)

1 second (s or sec)	= 0.01667 min	**1 day (d)**	= 24 hr
	= 2.7778 × 10^{-4} hr		= 86400 s
1 minute (min)	= 60 s	**1 year (yr or y)**	= 365 d
	= 0.01667 hr		= 8760 hr
1 hour (hr or h)	= 60 min		= 3.15 × 10^7 s
	= 3600 s		

DENSITY (M/L^3)

1 kilogram per cubic meter (kg/m^3)	= 10^{-3} g/cm^3	**1 metric ton per cubic meter (te/m^3)**	= 1.0 specific gravity
	= 0.062 lb/ft^3		= density of H$_2$O at 4°C
1 gram per cubic centimeter (g/cm^3)	= 1000 kg/m^3		= 8.35 lb/gal
	= 62.43 lb/ft^3	**1 pound per cubic foot (lb/ft^3)**	= 16.02 kg/m^3

(continues)

Water Units, Data Conversions, and Constants *(continued)*

VELOCITY (L/T)

1 meter per second (m/s)	= 3.6 km/hr = 2.237 mph = 3.28 ft/s	**1 foot per second (ft/s)**	= 0.68 mph = 0.3048 m/s
1 kilometer per hour (km/h or kph)	= 0.62 mph = 0.278 m/s	**velocity of light in vacuum (c)**	= 2.9979×10^8 m/s = 186,000 mi/s
		1 knot	= 1.852 km/h = 1 nautical mile/hour
1 mile per hour (mph or mi/h)	= 1.609 km/h = 0.45 m/s = 1.47 ft/s		= 1.151 mph = 1.688 ft/s

VELOCITY OF SOUND IN WATER AND SEAWATER
(assuming atmospheric pressure and sea water salinity of 35,000 ppm)

Temp, °C	Pure water, (meters/sec)	Sea water, (meters/sec)
0	1,400	1,445
10	1,445	1,485
20	1,480	1,520
30	1,505	1,545

FLOW RATE (L³/T)

1 liter per second (1/sec)	= 0.001 m³/sec = 86.4 m³/day = 15.9 gpm = 0.0228 mgd = 0.0353 cfs = 0.0700 AF/day	**1 cubic decameters per day (dam³/day)**	= 11.57 1/sec = 1.157×10^{-2} m³/sec = 1000 m³/day = 1.83×10^6 gpm = 0.264 mgd
1 cubic meter per second (m³/sec)	= 1000 1/sec = 8.64×10^4 m³/day = 1.59×10^4 gpm = 22.8 mgd = 35.3 cfs = 70.0 AF/day		= 0.409 cfs = 0.811 AF/day
		1 gallon per minute (gpm)	= 0.0631 1/sec = 6.31×10^{-5} m³/sec = 1.44×10^{-3} mgd = 2.23×10^{-3} cfs = 4.42×10^{-3} AF/day
1 cubic meter per day (m³/day)	= 0.01157 1/sec = 1.157×10^{-5} m³/sec = 0.183 gpm = 2.64×10^{-4} mgd = 4.09×10^{-4} cfs = 8.11×10^{-4} AF/day	**1 million gallons per day (mgd)**	= 43.8 1/sec = 0.0438 m³/sec = 3785 m³/day = 694 gpm = 1.55 cfs = 3.07 AF/day

Water Units, Data Conversions, and Constants *(continued)*

FLOW RATE (L^3/T) (continued)

1 cubic foot per second (cfs)	= 28.3 l/sec = 0.0283 m^3/sec = 2447 m^3/day = 449 gpm = 0.646 mgd = 1.98 AF/day	**1 miner's inch**	= 0.02 cfs (in Idaho, Kansas, Nebraska, New Mexico, North Dakota, South Dakota, and Utah) = 0.026 cfs (in Colorado) = 0.028 cfs (in British Columbia)
1 acre-foot per day (AF/day)	= 14.3 l/sec = 0.0143 m^3/sec = 1233.48 m^3/day = 226 gpm = 0.326 mgd = 0.504 cfs	**1 weir** **1 quinaria (ancient Rome)**	= 0.02 garcia = 0.47–0.48 l/sec
1 miner's inch	= 0.025 cfs (in Arizona, California, Montana, and Oregon: flow of water through 1 in^2 aperture under 6-inch head)		

ACCELERATION (L/T^2)

standard acceleration of gravity	= 9.8 m/s^2 = 32 ft/s^2

FORCE (ML/T^2 = Mass × Acceleration)

1 newton (N)	= kg-m/s^2 = 10^5 dynes = 0.1020 kg force = 0.2248 lb force	**1 dyne** **1 pound force**	= g·cm/s^2 = 10^{-5} N = lb mass × acceleration of gravity = 4.448 N

(continues)

Water Units, Data Conversions, and Constants *(continued)*

PRESSURE (M/L² = Force/Area)		**1 kilogram per sq.**	$= 14.22 \text{ lb/in}^2$
1 pascal (Pa)	$= \text{N/m}^2$	**centimeter**	
1 bar	$= 1 \times 10^5 \text{ Pa}$	**(kg/cm²)**	
	$= 1 \times 10^6 \text{ dyne/cm}^2$	**1 inch of water**	$= 0.0361 \text{ lb/in}^2$
	$= 1019.7 \text{ g/cm}^2$	**at 62°F**	$= 5.196 \text{ lb/ft}^3$
	$= 10.197 \text{ te/m}^2$		$= 0.0735$ inch of
	$= 0.9869$ atmos-		mercury at 62°F
	phere	**1 foot of water**	$= 0.433 \text{ lb/in}^2$
	$= 14.50 \text{ lb/in}^2$	**at 62°F**	$= 62.36 \text{ lb/ft}^2$
	$= 1000$ millibars		$= 0.833$ inch of
1 atmosphere (atm)	$=$ standard		mercury at 62°F
	pressure		$= 2.950 \times 10^{-2}$
	$= 760$ mm of		atmosphere
	mercury at 0°C	**1 pound per sq.**	$= 2.309$ feet of
	$= 1013.25$ millibars	**inch (psi or**	water at 62°F
	$= 1033 \text{ g/cm}^2$	**lb/in²)**	$= 2.036$ inches of
	$= 1.033 \text{ kg/cm}^2$		mercury at 32°F
	$= 14.7 \text{ lb/in}^2$		$= 0.06804$
	$= 2116 \text{ lb/ft}^2$		atmosphere
	$= 33.95$ feet of		$= 0.07031 \text{ kg/cm}^2$
	water at 62°F	**1 inch of mercury**	$= 0.4192 \text{ lb/in}^2$
	$= 29.92$ inches of	**at 32°F**	$= 1.133$ feet of
	mercury at 32°F		water at 32°F
TEMPERATURE			
degrees Celsius or	$= (°\text{F}-32) \times 5/9$	**degrees Fahrenheit**	$= 32 + (°\text{C} \times 1.8)$
Centigrade (°C)	$= \text{K}-273.16$	**(°F)**	$= 32 + ((°\text{K}-273.16)$
Kelvins (K)	$= 273.16 + °\text{C}$		$\times 1.8)$
	$= 273.16 + ((°\text{F}- 32)$		
	$\times 5/9)$		

Water Units, Data Conversions, and Constants *(continued)*

ENERGY(ML^2/T^2 = Force × Distance)

1 joule (J)	$= 10^7$ ergs	1 kilowatt-hour	$= 3.6 \times 10^6$ J
	$= N \cdot m$	(kWh)	$= 3412$ Btu
	$= W \cdot s$		$= 859.1$ kcal
	$= kg \cdot m^2/s^2$	l quad	$= 10^{15}$ Btu
	$= 0.239$ calories		$= 1.055 \times 10^{18}$ J
	$= 9.48 \times 10^{-4}$ Btu		$= 293 \times 10^9$ kWh
1 calorie (cal)	$= 4.184$ J		$= 0.001$ Q
	$= 3.97 \times 10^{-3}$ Btu		$= 33.45$ GWy
	(raises 1 g H_2O	1 Q	$= 1000$ quads
	$1°C$)		$\approx 10^{21}$ J
1 British thermal	$= 1055$ J	1 foot-pound (ft-lb)	$= 1.356$ J
unit (Btu)	$= 252$ cal (raises		$= 0.324$ cal
	1 lb H_2O $1°F$)	1 therm	$= 10^5$ Btu
	$= 2.93 \times 10^{-4}$ kWh	1 electron-volt (eV)	$= 1.602 \times 10^{-19}$ J
1 erg	$= 10^{-7}$ J	1 kiloton of TNT	$= 4.2 \times 10^{12}$ J
	$= g \cdot cm^2/s^2$	1 10^6 te oil equiv.	$= 7.33 \times 10^6$ bbl oil
	$= dyne \cdot cm$	(Mtoe)	$= 45 \times 10^{15}$ J
1 kilocalorie (kcal)	$= 1000$ cal		$= 0.0425$ quad
	$= 1$ Calorie (food)		

POWER (ML^2/T^3 = rate of flow of energy)

1 watt (W)	$= J/s$	1 horsepower	$= 0.178$ kcal/s
	$= 3600$ J/hr	(H.P. or hp)	$= 6535$ kWh/yr
	$= 3.412$ Btu/hr		$= 33,000$ ft-lb/min
1 TW	$= 10^{12}$ W		$= 550$ ft-lb/sec
	$= 31.5 \times 10^{18}$ J		$= 8760$ H.P.-hr/yr
	$= 30$ quad/yr	H.P. input	$= 1.34 \times$ kW input
1 kilowatt (kW)	$= 1000$W		to motor
	$= 1.341$ horsepower		$=$ horsepower
	$= 0.239$ kcal/s		input to motor
	$= 3412$ Btu/hr	Water H.P.	$=$ H.P. required to
10^6 bbl (oil) /day	≈ 2 quads/yr		lift water at a
(Mb/d)	≈ 70 GW		definite rate to
1 quad/yr	$= 33.45$ GW		a given distance
	≈ 0.5 Mb/d		assuming 100%
1 horsepower	$= 745.7$W		efficiency
(H.R or hp)	$= 0.7457$ kW		$=$ gpm × total head
			(in feet)/3960

(continues)

Water Units, Data Conversions, and Constants *(continued)*

EXPRESSIONS OF HARDNESS[a]

1 grain per gallon	= 1 grain $CaCO_3$ per U.S. gallon	**1 French degree**	= 1 part $CaCO_3$ per 100,000 parts water
1 part per million	= 1 part $CaCO_3$ per 1,000,000 parts water	**1 German degree**	= 1 part CaO per 100,000 parts water
1 English, or Clark, degree	= 1 grain $CaCO_3$ per Imperial gallon		

CONVERSIONS OF HARDNESS

1 grain per U.S. gallon	= 17.1 ppm, as $CaCO_3$	**1 French degree**	= 10 ppm, as $CaCO_3$
1 English degree	= 14.3 ppm, as $CaCO_3$	**1 German degree**	= 17.9 ppm, as $CaCO_3$

WEIGHT OF WATER

1 cubic inch	= 0.0361 lb	**1 imperial gallon**	= 10.0 lb
1 cubic foot	= 62.4 lb	**1 cubic meter**	= 1 tonne
1 gallon	= 8.34 lb		

DENSITY OF WATER[a]

Temperature		Density
°C	°F	gm/cm³
0	32	0.99987
1.667	35	0.99996
4.000	39.2	1.00000
4.444	40	0.99999
10.000	50	0.99975
15.556	60	0.99907
21.111	70	0.99802
26.667	80	0.99669
32.222	90	0.99510
37.778	100	0.99318
48.889	120	0.98870
60.000	140	0.98338
71.111	160	0.97729
82.222	180	0.97056
93.333	200	0.96333
100.000	212	0.95865

Note: Density of Sea Water: approximately 1.025 gm/cm³ at 15°C.

[a]*Source:* van der Leeden, F., Troise, F. L., and Todd, D. K., 1990. *The Water Encyclopedia*, 2d edition. Lewis Publishers, Inc., Chelsea, Michigan.

Comprehensive Table of Contents

Volume 1
The World's Water 1998-1999: The Biennial Report on Freshwater Resources

Foreword by Anne H. and Paul R. Erhlich   ix

Acknowledgments   xi

Introduction   1

ONE   The Changing Water Paradigm   5

Twentieth-Century Water-Resources Development   6
The Changing Nature of Demand   10
Economics of Major Water Projects   16
Meeting Water Demands in the Next Century   18
Summary: New Thinking, New Actions   32
References   33

TWO   Water and Human Health   39

Water Supply and Sanitation: Falling Behind   39
Basic Human Needs for Water   42
Water-Related Diseases   47
Update on Dracunculiasis (Guinea Worm)   50
Update on Cholera   56
Summary   63
References   64

THREE   The Status of Large Dams: The End of an Era?   69

Environmental and Social Impacts of
  Large Dams   75
New Developments in the Dam Debate   80
The Three Gorges Project, Yangtze
  River, China   84
The Lesotho Highlands Project, Senqu River
  Basin, Lesotho   93
References   101

FOUR   Conflict and Cooperation Over Fresh Water 105

Conflicts Over Shared Water Resources   107
Reducing the Risk of Water-Related Conflict   113
The Israel-Jordan Peace Treaty of 1994   115
The Ganges-Brahmaputra Rivers: Conflict and Agreement 118
Water Disputes in Southern Africa   119
Summary   124
Appendix A
    Chronology of Conflict Over Waters in the Legends,
    Myths, and History of the Ancient Middle East   125
Appendix B
    Chronology of Conflict Over Water:
    1500 to the Present   128
References   132

FIVE   Climate Change and Water Resources:
        What Does the Future Hold? 137

What Do We Know?   138
Hydrologic Effects of Climate Change   139
Societal Impacts of Changes in Water Resources   144
Is the Hydrologic System Showing Signs of Change?   145
Recommendations and Conclusions   148
References   150

SIX   New Water Laws, New Water Institutions 155

Water Law and Policy in New South Africa: A Move Toward
    Equity   156
The Global Water Partnership   165
The World Water Council   172
The World Commission on Dams   175
References   180

SEVEN   Moving Toward a Sustainable Vision for the Earth's
        Fresh Water 183

Introduction   183
A Vision for 2050: Sustaining Our Waters   185

WATER BRIEFS

The Best and Worst of Science: Small Comets and the New
    Debate Over the Origin of Water on Earth   193
Water Bag Technology   200
Treaty Between the Government of the Republic of India and
    the Government of the People's Republic of Bangladesh
    on Sharing of the Ganga/Ganges Waters at Farakka   206

United Nations Conventions on the Law of the Non-
Navigational Uses of International Watercourses   210
Water-Related Web Sites   231

DATA SECTION

Table 1   Total Renewable Freshwater Supply by Country   235
Table 2   Freshwater Withdrawal by Country and Sector   241
Table 3   Summary of Estimated Water Use in the United States, 1900
to 1995   245
Table 4   Total and Urban Population by Country, 1975 to 1995   246
Table 5   Percentage of Population with Access to Safe Drinking Water
by Country, 1970 to 1994   251
Table 6   Percentage of Population with Access to Sanitation by
Country, 1970 to 1994   256
Table 7   Access to Safe Drinking Water in Developing Countries by
Region, 1980 to 1994   261
Table 8   Access to Sanitation in Developing Countries by Region,
1980 to 1994   263
Table 9   Reported Cholera Cases and Deaths by Region,
1950 to 1997   265
Table 10   Reported Cholera Cases and Deaths by Country, 1996 and
1997   268
Table 11   Reported Cholera Cases and Deaths in the Americas, 1991 to
1997   271
Table 12   Reported Cases of Dracunculiasis by Country,
1972 to 1996   272
Table 13   Waterborne Disease Outbreaks in the United States by Type
of Water Supply System, 1971 to 1994   274
Table 14   Hydroelectric Capacity and Production by Country,
1996   276
Table 15   Populations Displaced as a Consequence of Dam
Construction, 1930 to 1996   281
Table 16   Desalination Capacity by Country (January 1, 1996)   288
Table 17   Desalination Capacity by Process (January 1, 1996)   290
Table 18   Threatened Reptiles, Amphibians, and Freshwater
Fish, 1997   291
Table 19   Irrigated Area by Country and Region, 1961 to 1994   297

Index   303

Volume 2
The World's Water 2000-2001: The Biennial Report on Freshwater Resources

Foreword by Timothy E. Wirth   xiii

Acknowledgments   xv

Introduction   xvii

ONE   The Human Right to Water   1

Is There a Human Right to Water?   2
Existing Human Rights Laws, Covenants, and Declarations   4
Defining and Meeting a Human Right to Water   9
Conclusions   15
References   15

TWO   How Much Water Is There and Whose Is It? The World's Stocks
and Flows of Water and International River Basins   19

How Much Water Is There? The Basic Hydrologic Cycle   20
International River Basins: A New Assessment   27
The Geopolitics of International River Basins   35
Summary   36
References   37

THREE   Pictures of the Future: A Review of Global Water Resources
Projections   39

Data Constraints   40
Forty Years of Water Scenarios and Projections   42
Analysis and Conclusions   58
References   59

FOUR   Water for Food: How Much Will Be Needed?   63

Feeding the World Today   63
Feeding the World in the Future: Pieces of the Puzzle   65
How Much Water Will Be Needed to Grow Food?   78
Conclusions   88
Referencess   89

FIVE   Desalination: Straw into Gold or Gold into Water   93

History of Desalination and Current Status   94
Desalination Technologies   98
Other Aspects of Desalination   106
The Tampa Bay Desalination Plant   108
Summary   109
References   110

SIX   The Removal of Dams: A New Dimension to an Old Debate   113

Economics of Dam Removal   118
Dam Removal Case Studies: Some Completed Removals   120
Some Proposed Dam Removals or Decommissionings   126
Conclusion   134
References   134

SEVEN   Water Reclamation and Reuse: Waste Not, Want Not   137

Wastewater Uses   139
Direct and Indirect Potable Water Reuse   151
Health Issues   152
Wastewater Reuse in Namibia   155
Wastewater Reclamation and Reuse in Japan   158
Wastewater Costs   159
Summary   159
References   161

WATER BRIEFS

Arsenic in the Groundwater of Bangladesh and West Bengal,
    India   165
Fog Collection as a Source of Water   175
Environment and Security: Water Conflict Chronology—
    Version 2000   182
Water-Related Web Sites   192

DATA SECTION

Table 1   Total Renewable Freshwater Supply, by Country   197
Table 2   Freshwater Withdrawal, by Country and Sector   203
Table 3   World Population, Year 0 to A.D. 2050   212
Table 4   Population, by Continent, 1750 to 2050   213
Table 5   Renewable Water Resources and Water Availability, by
           Continent   215
Table 6   Dynamics of Water Resources, Selected Countries,
           1921 to 1985   218
Table 7   International River Basins of the World   219
Table 8   Fraction of a Country's Area in International
           River Basins   239
Table 9   International River Basins, by Country   247
Table 10   Irrigated Area, by Country and Region, 1961 to 1997   255
Table 11   Irrigated Area, by Continent, 1961 to 1997   264
Table 12   Human-Induced Soil Degradation, by Type and Cause, Late
           1980s   266
Table 13   Continental Distribution of Human-Induced Salinization
           268
Table 14   Salinization, by Country, Late 1980s   269
Table 15   Total Number of Reservoirs, by Continent and Volume   270

Table 16   Number of Reservoirs Larger than 0.1 km^3, by Continent,
                 Time Series   271

Table 17   Volume of Reservoirs Larger than 0.1 km^3, by Continent,
                 Time Series   273

Table 18   Dams Removed or Decommissioned in the United States,
                 1912 to Present   275

Table 19   Desalination Capacity, by Country, January 1999   287

Table 20   Total Desalination Capacity, by Process, June 1999   289

Table 21   Desalination Capacity, by Source of Water, June 1999   290

Table 22   Number of Threatened Species, by Country/Area, by Group,
                 1997   291

Table 23   Countries with the Largest Number of Fish Species   298

Table 24   Countries with the Largest Number of Fish Species per Unit
                 Area   299

Table 25   Water Units, Data Conversions, and Constants   300

Index   311

Volume 3
The World's Water 2002-2003: The Biennial Report on Freshwater Resources

Foreword by Amory B. Lovins   xiii

Acknowledgments   xv

Introduction   xvii

ONE   The Soft Path for Water   1
          by Gary Wolff and Peter H. Gleick

          A Better Way   3
          Dominance of the Hard Path in the Twentieth Century   7
          Myths about the Soft Path   9
          One Dimension of the Soft Path: Efficiency of Use   16
          Moving Forward on the Soft Path   25
          Conclusions   30
          References   30

TWO   Globalization and International Trade of Water   33
          *by Peter H. Gleick, Gary Wolff, Elizabeth L. Chalecki, and
              Rachel Reyes*

          The Nature and Economics of Water, Background and
              Definitions   34
          Water Managed as Both a Social and Economic Good   38
          The Globalization of Water: International Trade   41
          The Current Trade in Water   42
          The Rules: International Trading Regimes   47
          References   54

THREE   The Privatization of Water and Water Systems   57
          *by Peter H. Gleick, Gary Wolff, Elizabeth L. Chalecki, and
              Rachel Reyes*

          Drivers of Water Privatization   58
          History of Privatization   59
          The Players   61
          Forms of Privatization   63
          Risks of Privatization: Can and Will They Be Managed?   67
          Principles and Standards for Privatization   79
          Conclusions   82
          References   83

FOUR   Measuring Water Well-Being: Water Indicators and Indices   87
         by Peter H. Gleick, Elizabeth L. Chalecki, and Arlene Wong

         Quality-of-Life Indicators and Why We Develop Them   88
         Limitations to Indicators and Indices   92
         Examples of Single-Factor or Weighted Water Measures   96
         Multifactor Indicators   101
         Conclusions   111
         References   112

FIVE   Pacific Island Developing Country Water Resources and
         Climate Change   113
         by William C.G. Burns

         PIDCs and Freshwater Sources   113
         Climate Change and PIDC Freshwater Resources   119
         Potential Impacts of Climate Change on PIDC Freshwater
             Resources   124
         Recommendations and Conclusions   125
         References   127

SIX   Managing Across Boundaries: The Case of the
         Colorado River Delta   133
         by Michael Cohen

         The Colorado River   134
         The Colorado River Delta   139
         Conclusions   144
         References   145

SEVEN   The World Commission on Dams Report: What Next?   149
         by Katherine Kao Cushing

         The WCD Organization   149
         Findings and Recommendations   151
         Strategic Priorities, Criteria, and Guidelines 153
         Reaction to the WCD Report   155
         References 172

WATER BRIEFS
         The Texts of the Ministerial Declarations from The Hague
             March 2000) and Bonn (December 2001)   173
         The Southeastern Anatolia (GAP) Project and Archaeology   181
         by Amar S. Mann

Water Conflict Chronology   194
by Peter S. Gleick

Water and Space   209
by Elizabeth L. Chalecki

Water-Related Web Sites   225

DATA SECTION   237

Table 1   Total Renewable Freshwater Supply, by Country
            (2002 Update)   237
Table 2   Fresh Water Withdrawals, by Country and Sector
            (2002 Update)   243
Table 3   Access to Safe Drinking Water by Country,
            1970 to 2000   252
Table 4   Access to Sanitation by Country, 1970 to 2000   261
Table 5   Access to Water Supply and Sanitation by Region, 1990 and
            2000   270
Table 6   Reported Cases of Dracunculiasis by Country,
            1972 to 2000   273
Table 7   Reported Cases of Dracunculiasis Cases, Eradication
            Progress, 2000   276
Table 8   National Standards for Arsenic in Drinking Water   278
Table 9   United States National Primary Drinking Water
            Regulations   280
Table 10   Irrigated Area, by Region, 1961 to 1999   289
Table 11   Irrigated Area, Developed and Developing Countries, 1960
            to 1999   290
Table 12   Number of Dams, by Continent and Country   291
Table 13   Number of Dams, by Country   296
Table 14   Regional Statistics on Large Dams   300
Table 15   Commissioning of Large Dams in the 20th Century, by
            Decade   301
Table 16   Water System Rate Structures   303
Table 17   Water Prices for Various Households   304
Table 18   Unaccounted-for Water   305
Table 19   United States Population and Water Withdrawals, 1900 to
            1995   308
Table 20   United States GNP and Water Withdrawals,
            1900 to 1996   310
Table 21   Hong Kong GDP and Water Withdrawls, 1952 to 2000   313
Table 22   China GDP and Water Withdrawls, 1952 to 2000   316

Water Units, Data Conversions, and Constants   318

Index   329

Volume 4
The World's Water 2004–2005: The Biennial Report on Freshwater Resources

Foreword   xiii

Introduction   xv

ONE   The Millennium Development Goals for Water: Crucial
           Objectives, Inadequate Commitments   1
           by Peter H. Gleick

           Setting Water and Sanitation Goals   2
           Commitments to Achieving the MDGs for Water   2
           Consequences: Water-Related Diseases   7
           Measures of Illness from Water-Related Diseases   9
           Scenarios of Future Deaths from Water-Related
               Diseases   10
           Conclusions   14
           References   14

TWO   The Myth and Reality of Bottled Water   17
           by Peter H. Gleick

           Bottled Water Use History and Trends   18
           The Price and Cost of Bottled Water   22
           The Flavor and Taste of Water   23
           Bottled Water Quality   25
           Regulating Bottled Water   26
           Comparison of U.S. Standards for Bottled Water and
               Tap Water   36
           Other Concerns Associated with Bottled Water   37
           Conclusions   41
           References   42

THREE   Water Privatization Principles and Practices   45
           *by Meena Palaniappan, Peter H. Gleick, Catherine Hunt,
               Veena Srinivasan*

           Update on Privatization   46
           Principles and Standards for Water   47
           Can the Principles Be Met?   48
           Conclusions   73
           References   74

FOUR   Groundwater: The Challenge ofMonitoring and
         Management 79
         by Marcus Moench

         Conceptual Foundations   80
         Challenges in Assessment   80
         Extraction and Use   81
         Groundwater in Agriculture   88
         The Analytical Dilemma   90
         A Way Forward: Simple Data as a Catalyst for Effective
             Management   97
         References   98

FIVE   Urban Water Conservation: A Case Study of Residential
         Water Use in California   101
         by Peter H. Gleick, Dana Haasz, Gary Wolff

         The Debate over California's Water   102
         Defining Water "Conservation" and "Efficiency"   103
         Current Urban Water Use in California   105
         A Word About Agricultural Water Use   107
         Economics of Water Savings   107
         Data and Information Gaps   108
         Indoor Residential Water Use   109
         Indoor Residential Water Conservation: Methods and
             Assumptions   112
         Indoor Residential Summary   118
         Outdoor Residential Water Use   118
         Current Outdoor Residential Water Use   119
         Existing Outdoor Conservation Efforts and Approaches   120
         Outdoor Residential Water Conservation: Methods and
             Assumptions   121
         Residential Outdoor Water Use Summary   125
         Conclusions   126
         Abbreviations and Acronyms   126
         References   127

SIX   Urban Water Conservation: A Case Study of Commercial and
         Industrial Water Use in California   131
         *by Peter H. Gleick, Veena Srinivasan, Christine Henges-Jeck,
             Gary Wolff*

         Background to CII Water Use   132
         Current California Water Use in the CII Sectors   133
         Estimated CII Water Use in California in 2000   138

    Data Challenges   139

    The Potential for CII Water Conservation and Efficiency
       Improvements: Methods and Assumptions   140

    Methods for Estimating CII Water Use and Conservation
       Potential   143

    Calculation of Conservation Potential   145

    Data Constraints and Conclusions   148

    Recommendations for CII Water Conservation   150

    Conclusions   153

    References   154

SEVEN   Climate Change and California Water Resources   157

    by Michael Kiparsky and Peter H. Gleick

    The State of the Science   158

    Climate Change and Impacts on Managed Water-Resource
       Systems   172

    Moving From Climate Science to Water Policy   175

    Conclusions   183

    References   184

WATER BRIEFS

One   3rd World Water Forum in Kyoto: Disappointment and
       Possibility   189
    by Nicholas L. Cain

    Ministerial Declaration of the 3rd World Water Forum:
       Message from the Lake Biwa and Yodo River Basin   198

    NGO Statement of the 3rd World Water Forum   202

Two   The Human Right to Water: Two Steps Forward, One Step
       Back   204
    by Peter H. Gleick

    Substantive Issues Arising in the Implementation of the Inter-
       national Covenant on Economic, Social, and Cultural Rights.
       United Nations General Comment No. 15 (2002)   213

Three   The Water and Climate Bibliography   228
    by Peter H. Gleick and Michael Kiparksy

Four   Environment and Security: Water Conflict Chronology
       Version 2004–2005   234
    by Peter H. Gleick

    Water Conflict Chronology   236
    by Peter H. Gleick

DATA SECTION

Data Table 1   Total Renewable Freshwater Supply by Country
       (2004 Update)   257

Data Table 2   Freshwater Withdrawals by Country and Sector
       (2004 Update)   263

Data Table 3   Deaths and DALYs from Selected Water-Related
                Diseases   272

Data Table 4   Official Development Assistance Indicators   274

Data Table 5   Aid to Water Supply and Sanitation by Donor,
                1996 to 2001   278

Data Table 6   Bottled Water Consumption by Country, 1997 to 2002   280

Data Table 7   Global Bottled Water Consumption by Region,
                1997 to 2002   283

Data Table 8   Bottled Water Consumption, Share by Region,
                1997 to 2002   285

Data Table 9   Per-capita Bottled Water Consumption by Region,
                1997 to 2002   286

Data Table 10   United States Bottled Water Sales, 1991 to 2001   288

Data Table 11   Types of Packaging Used for Bottled Water in Various
                Countries, 1999   289

Data Table 12   Irrigated Area by Region, 1961 to 2001   291

Data Table 13   Irrigated Area, Developed and Developing Countries,
                1961 to 2001   293

Data Table 14   Global Production and Yields of Major Cereal Crops, 1961
                to 2002   295

Data Table 15   Global Reported Flood Deaths, 1900 to 2002   298

Data Table 16   United States Flood Damage by Fiscal Year,
                1926 to 2001   301

Data Table 17   Total Outbreaks of Drinking Water-Related Disease,
                United States, 1973 to 2000   304

Data Table 18
                18.a Extinction Rate Estimates for Continental North
                American Fauna (percent loss per decade)   309
                18.b Imperiled Species for North American Fauna   309

Data Table 19   Proportion of Species at Risk, United States   311

Data Table 20   United States Population and Water Withdrawals
                1900 to 2000   313

Data Table 21   United States Economic Productivity of Water,
                1900 to 2000   317

WATER UNITS, DATA CONVERSIONS, AND CONSTANTS   321

COMPREHENSIVE TABLE OF CONTENTS   331

Volume 1:   The World's Water 1998–1999: The Biennial Report on Fresh-
            water Resources   331

Volume 2:   The World's Water 2000–2001: The Biennial Report on Fresh-
            water Resources   334

Volume 3:   The World's Water 2002–2003: The Biennial Report on Fresh-
            water Resources   337

COMPREHENSIVE INDEX   341

Comprehensive Index

KEY (book volume in boldface numerals)

1: The World's Water 1998–1999: The Biennial Report on Freshwater Resources

2: The World's Water 2000–2001: The Biennial Report on Freshwater Resources

3: The World's Water 2002–2003: The Biennial Report on Freshwater Resources

4: The World's Water 2004–2005: The Biennial Report on Freshwater Resources

5: The World's Water 2006–2007: The Biennial Report on Freshwater Resources

ABB, **1:**85

Abbasids, **3:**184

Abi-Eshuh, **1:**69, **5:**5

Abou Ali ibn Sina, **1:**51

Abu-Zeid, Mahmoud, **1:**174, **4:**193

Acceleration, measuring, **2:**306, **3:**324, **4:**331

Access to water and environmental justice, **5:**124–25, 127

see also Conflict/cooperation concerning freshwater; Drinking water, access to clean; Environmental flow; Human right to water; Renewable freshwater supply; Sanitation services; Stocks and flows of freshwater; Withdrawals, water

Acres International, **1:**85

Adams, Dennis, **1:**196

Adaptive capacity, **4:**236

Adaptive management and environmental flows, **5:**45

Adriatic Sea, **3:**47

Adulteration, food, **4:**32–33

Advanced Radio Interferometry (ARISE), **3:**221

AES Tiete, **5:**152

Africa:

availability, water, **2:**217

bottled water, **4:**288, 289, 291, **5:**281, 283

cholera, **1:**57, 59, 61–63, 266, 269, **3:**2, **5:**289, 290

climate change, **1:**148

conflict/cooperation concerning freshwater, **1:**119–24, **5:**7, 9

dams, **3:**292, **5:**151

desalination, **2:**94, 97

dracunculiasis, **1:**52–55, 272, **3:**274, **5:**295, 296

drinking water, **1:**252, 262, **3:**254–55, **5:**240, 241

droughts, **5:**93, 100

economic development derailed, **1:**42

environmental flow, **5:**32

fog collection as a source of water, **2:**175

Global Water Partnership, **1:**169

groundwater, **4:**85–86

human right to water, **4:**211

hydroelectric production, **1:**71, 277–78

irrigation, **1:**298–99, **2:**80, 85, 256–57, 265, **3:**289, **4:**296, **5:**299

Millennium Development Goals, **4:**7

needs, basic human, **1:**47

population data/issues, **1:**247, **2:**214

reclaimed water, **1:**28, **2:**139

renewable freshwater supply, **1:**237–38, **2:**199–200, 217, **3:**239–40, **4:**263–64, **5:**223, 224

reservoirs, **2:**270, 272, 274

salinization, **2:**268

sanitation services, **1:**257, 264, **3:**263–65, 271, **5:**249–50, 259

threatened/at risk species, **1:**292–93

well-being, measuring water scarcity and, **3:**96

withdrawals, water, **1:**242, **2:**205–7, **3:**245–47, **4:**269–70, **5:**230–31

African Development Bank (AfDB), **1:**95–96, 173, **3:**162–63

Agreements, international, *see* Law/legal instruments/regulatory bodies; United Nations

Agriculture:

Alcamo et al. water assessment, **2:**56–57

by region, **3:**289

cereal production, **2:**64, **4:**299–301

conflict/cooperation concerning freshwater, **1:**111

data problems, **3:**93

droughts, **5:**92, 94, 98, 103

Falkenmark and Lindh water assessment, **2:**47–49

floods, **5:**109

Gleick water assessment, **2:**54–55

groundwater, **4:**83, 87, 88–90

inefficient and wasteful water use, **4:**107

irrigation

business/industry, water risks, **5:**162

by continent, **2:**264–65

by country and region, **1:**297–301, **2:**255–63, **4:**295–96, **5:**298–300

climate change, **4:**174–75

conflict/cooperation concerning freshwater, **1:**110

developing countries, **1:**24, **3:**290, **5:**301–2

drip, **1:**23–24, **2:**82, 84

Edwards Aquifer, **3:**74

efficiency improvement opportunities, **3:**20

expanding water-resources infrastructure, **1:**6

government involvement in water projects, **1:**8

hard path for meeting water-related needs, **3:**2

projections, review of global water resources, **2:**45

reclaimed water, **2:**139, 142, 145–46

343

Seckler et al. water assessment, **2:**57–58
Southeastern Anatolia Project, **3:**182
total irrigated areas, **4:**297–98
Kalinin and Shiklomanov and De Mare water assessment, **2:**46–47
pesticides and fertilizer runoff, **5:**128, 305–7
pricing, water, **1:**117
projections, review of global water resources, **2:**45–46
Raskin et al. water assessment, **2:**55–56
reclaimed water, **1:**28, 29
subsidies, **1:**24–25
sustainable vision for the earth's freshwater, **1:**187–88
water-use efficiency, **3:**4, 19–20
well-being, measuring water scarcity and, **3:**99
World Water Forum (2003), **4:**203
see also Food needs for current/future populations
Aguas Argentinas, **3:**78, **4:**47
Aguas de Barcelona, **3:**63
Aguas del Aconquija, **3:**70
Al-Qaida, **5:**15
Albania, **1:**71, **3:**47
Albright, Madeleine K., **1:**106
Altamonte Springs (FL), **2:**146
American Association for the Advancement of Science, **1:**149, **4:**176
American Convention on Human Rights (1969), **2:**4, 8
American Cyanamid, **1:**52
American Fare Premium Water, **4:**39
American Fisheries Society, **2:**113, 133
American Geophysical Union (AGU), **1:**197–98
American Rivers and Trout Unlimited (AR & TU), **2:**118, 123
American Society of Civil Engineers, **5:**24
American Water Works, **3:**61
American Water Works Association (AWWA), **2:**41, **3:**59, **4:**176, **5:**24
American Water/Pridesa, **5:**62
Amoebiasis, **1:**48
Amount of water, *see* Stocks and flows of freshwater
Amphibians, threatened, **1:**291–96
Anatolia region, *see* Southeastern Anatolia Project
Angola, **1:**119, 121, **2:**175, **3:**49, **5:**9

Anheuser-Busch, **5:**150
Ankara University, **3:**183
Antiochus I, **3:**184–85
Apamea-Seleucia archaeological site, **3:**185–87
Apartheid, **1:**158–59
Appleton, Albert, **4:**52–53
Aqua Vie Beverage Corporation, **4:**38
Aquaculture, **2:**79
Aquafina, **4:**21
Aquarius Water Trading and Transportation, Ltd., **1:**201–2, 204
Aral Sea, **1:**24, **3:**3, 39–41, 77
Archaeology, *see* Southeastern Anatolia Project
Area, measuring, **2:**302, **3:**320, **4:**327
Argentina, **3:**13, 60, 70, **4:**40, 47
Arizona, **3:**20, 138
Army Corps of Engineers, U.S. (ACoE), **1:**7–8, **2:**132, **3:**137
Arrowhead, **4:**21
Arsenic in groundwater, **2:**165–73, **3:**278–79, **4:**87, **5:**20
Artemis Society, **3:**218
Artesian water/well water, **4:**28
Artuqids, **3:**188–89
Ascension Island, **2:**175
Asia:
agriculture, **4:**88–90
availability, water, **2:**24, 217
bottled water, **4:**18, 41, 291, **5:**163, 281, 283
cholera, **1:**56, 58, 61, 266, 269–70, **3:**2, **5:**289, 290
climate change, **1:**147
conflict/cooperation concerning freshwater, **1:**111
dams, **3:**294–95
dracunculiasis, **1:**272–73, **3:**274–75, **5:**295, 296
drinking water, **1:**253–54, 262, **3:**257–59, **5:**243–45
environmental flow, **5:**32–33
floods, **5:**108
food needs for current/future populations, **2:**75, 79
Global Water Partnership, **1:**169
groundwater, **4:**84, 88–90, 96
hydroelectric production, **1:**71, 278–79
international river basin, **2:**30
irrigation, **1:**299–300, **2:**80, 86, 259–61, 265, **3:**289, **4:**296, **5:**299
needs, basic human, **1:**47
population data/issues, **1:**248–49, **2:**214
pricing, water, **1:**24, **3:**69
privatization, **3:**61
renewable freshwater supply, **1:**238–39, **2:**217, **3:**240–41, **4:**264–65, **5:**225, 226
reservoirs, **2:**270, 272, 274

runoff, **2:**23
salinization, **2:**268
sanitation services, **1:**258–59, 264, **3:**266–68, 271, **5:**252–54, 258–61
supply systems, ancient, **1:**40
threatened/at risk species, **1:**293–94
well-being, measuring water scarcity and, **3:**96
withdrawals, water, **1:**243–44, **2:**208–9, **3:**248–49, **4:**272–73, **5:**233–34
Asian Development Bank (ADB), **1:**17, 173, **3:**118, 163, 169, **4:**7, **5:**125
Asmal, Kader, **1:**160, **3:**169
Assessments:
Aquastat database, **4:**81–82
Colorado River Severe Sustained Drought study (CRSSD), **4:**166
Comprehensive Assessment of the Freshwater Resources of the World, **2:**10, **3:**90
Dow Jones Indexes, **3:**167
High Efficiency Laundry Metering and Marketing Analysis project, The (THELMA), **4:**115
Human Development Index, **2:**165
Human Development Report, **4:**7
Human Poverty Index, **3:**87, 89, 90, 109–11, **5:**125
hydrologic cycle and accurate quantifications, **4:**92–96
International Journal on Hydropower and Dams, **1:**70
International Rice Research Institute (IRRI), **2:**75, 76
international river basins, **2:**27–35
measurements, **2:**300–309, **3:**318–27, **4:**325–34
National Assessment on the Potential Consequences of Climate Variability and Change, **4:**176
Palmer Drought Severity Index, **5:**93
Standard Industrial Classification (SIC), **4:**132
Standard Precipitation Index, **5:**93
Stockholm Water Symposiums (1995/1997), **1:**165, 170
Third Assessment Report, **3:**121–23
Water in Crisis: A Guide to the World's Fresh Water Resources, **2:**300, **3:**318

Water-Global Assessment and Prognosis (WaterGAP), **2:**56
World Health Reports, **4:**7
World Resources Reports, **3:**88
see also Data issues/problems; monitoring/management problems *under* Groundwater; Projections, review of global water resources; Stocks and flows of freshwater; Well-being, measuring water scarcity and
Association of Southeast Asian Nations (ASEAN), **1:**169
Assyrians, **3:**184
Atlanta (GA), **3:**62, **4:**46–47
Atlantic Salmon Federation, **2:**123
Atmosphere, harvesting water from the, **2:**175–81
see also Outer space, search for water in
Austin (TX), **1:**22
Australia:
availability, water, **2:**24, 217
bottled water, **4:**26
conflict/cooperation concerning freshwater, **5:**10
desalination, **5:**69
environmental flow, **5:**33, 35, 42
globalization and international trade of water, **3:**45, 46
irrigation, **2:**85
Millennium Development Goals, **4:**7
privatization, **3:**60, 61
renewable freshwater supply, **2:**217
reservoirs, **2:**270, 272, 274
terrorism, **5:**16
Austria, **3:**47
Availability, water, **2:**24–27, 215–17
see also Conflict/cooperation concerning freshwater; Drinking water, access to clean; Environmental flow; Human right to water; Renewable freshwater supply; Sanitation services; Stocks and flows of freshwater; Withdrawals, water
Azov Sea, **1:**77

Babbitt, Bruce, **2:**124–25, 128
Babylon, ancient, **1:**109, 110
Bag technology, water, **1:**200–205
Baker, James, **1:**106
Bakersfield (CA), **1:**29
Balfour Beatty, **3:**166
Bangladesh:
agriculture, **4:**88
arsenic in groundwater, **2:**165–73

Arsenic Mitigation/Water Supply Project, **2:**172
conflict/cooperation concerning freshwater, **1:**107, 109, 118–19, 206–9
drinking water, **3:**2
floods, **5:**106
groundwater, **4:**88
Rural Advancement Committee, **2:**168
Banks, Harvey, **1:**9
Barlow, Nadine, **3:**215
Basic water requirement (BWR), **1:**44–46, **2:**10–13, **3:**101–3
Basin irrigation, **2:**82
Bass, **2:**123
Bath Iron Works, **2:**124
Batman Begins, **5:**14
BC Hydro, **1:**88
Beard, Dan, **2:**129
Bechtel, **3:**63, 70
Belgium, **5:**10
Benin, **1:**55
Best available technology (BAT), **3:**18, 22–23, **4:**104, **5:**157
Best practicable technology (BPT), **3:**18, **4:**104
Beverage Marketing Corporation (BMC), **4:**18
Biologic/chemical attacks, vulnerability to, **5:**16–22
Birds affected by dams, **1:**90, 98
Black Sea, **1:**77
Block pricing, **1:**26, **4:**56
Boiling and desalination, **2:**99
Bolivia, **3:**68, 69–72, **4:**54, 56–57
Books and portrayals of terrorism, **5:**14
Border irrigation, **2:**82
Boron, **5:**75
Bosnia, **1:**71
Botswana, **1:**119, 122–24
Bottled water:
brands, leading, **4:**21–22
business/industry, water risks that face, **5:**163
consumption
by country, **4:**284–86, **5:**276–79, 284–86
by region, **4:**287–88, **5:**169–70, 280–81
per-capita by region, **4:**290–91, **5:**171, 282–83
share by region, **4:**289
developing countries, **3:**44, 45
environmental issues, **4:**41
flavor and taste, **4:**23–24
history and trends, **4:**18–22
industry associations: standards/rules, **4:**34–35
international standards, **4:**35–36
overview, **4:**xvi, 17
packaging used, **4:**293–94
poor, selling water to the, **4:**40–41

price and cost, **4:**22–23
quality, **4:**17, 25–26, 31–32, 37–40
recalls, **4:**37–40, **5:**171–74
sales, global, **3:**43, **5:**169
standards/regulations, **4:**26–27, 34–35, **5:**171, 174
summary/conclusions, **4:**41
tap water and bottled water, comparing U.S. standards for, **4:**36–37
United States federal regulations
adulteration, food, **4:**32–33
enforcement/regulatory action, **4:**34
good manufacturing practices, **4:**32
identity standards, **4:**27–31
quality, **4:**31–32
sampling/testing/FDA inspections, **4:**33–34
United States, sales and imports in, **3:**44, 343, **4:**292
Boundaries, managing across, **3:**133–34
see also Colorado River; International river basins
Brazil:
bottled water, **4:**40, **5:**170
business/industry, water risks, **5:**149, 151–52
cholera, **1:**59
conflict/cooperation concerning freshwater, **1:**107
dams, **1:**16, **5:**134
environmental flow, **5:**34
hydroelectric production, **1:**71
monitoring and privatization, **3:**76–77
needs, basic human, **1:**46
privatization, **3:**76–78
runoff, **2:**23
sanitation services, **3:**6
Three Gorges Dam project, **1:**89
Brine and desalination, **2:**107, **5:**77–80
British Columbia Hydro International, **1:**85
British Geological Survey (BGS), **2:**169
British Medical Association, **4:**63–64
Bruce Banks Sails, **1:**202
Bruvold, William, **4:**24
Burkina Faso, **1:**55, **4:**211
Burns, William, **3:**xiv
Burundi, **1:**62
Business for Social Responsibility, **5:**156

Business sector, **3:**22–24, 169–70
see also Privatization of
water/water systems;
Urban water conserva-
tion: commercial/indus-
trial water use in
California
Business/industry, water risks:
Anheuser-Busch, **5:**150
availability/reliability of
supply, **5:**146
China, **5:**147, 149, 160, 165
climate change, **5:**152
developing countries, **5:**150
energy and water links, **5:**150,
151–52
India, **5:**146, 147, 165
management
best available technology,
5:157
companies, review of
specific, **5:**153–55
continuous improvement,
commit to, **5:**158
hydrological/social/econo
mic/political
factors, **5:**153, 156
partnerships, form
strategic, **5:**158
policies/goals/targets,
establishing,
5:156–57
report performance,
measure and,
5:157–58
risks factored into
decisions, **5:**157
stakeholders, consulting,
5:145
supply chain, working
collaboratively
with, **5:**156
overview, **5:**145–46
privatization, public
opposition to,
5:152–53
public's role in water
policy, **5:**150–51
quality, water, **5:**146–49
summary/conclusions,
5:163–65
supply-chain vulnerabil-
ity, **5:**149
water industry, overview
bottled water, **5:**163
desalination, **5:**161
disinfection/purifica-
tion of drinking
water, **5:**159
distribution, infra-
structure for,
5:160–61
efficiency, improving
water use,
5:161–62
high-quality water,
processes
requiring,
5:147–48, 160
irrigation, **5:**162

overview, **5:**158–59
revenue and growth in
U.S., **5:**303–4
utilities, water,
5:162–63
wastewater treatment,
5:153, 159–60
Bussi, Antonio, **3:**70
Byzantines, **3:**184

Calgon, **3:**61
California:
agriculture, **4:**89
bottled water, **4:**24
business/industry, water
risks, **5:**158, 162
California Bay-Delta
Authority, **4:**181
California Central Valley
Project, **4:**173
California Department of
Water Resources, **1:**9,
29, **3:**11, **4:**170, 232
California Energy Commis-
sion, **4:**157, 176, 232
California Regional Assessment Group, **4:**176
climate change, **1:**144–45,
4:232
conflict/cooperation con-
cerning freshwater,
1:109
dams, **1:**75, **2:**120–23
desalination, **1:**30, 32, **5:**51,
52, 63–69, 71, 73, 74
East Bay Municipal Utilities
District, **1:**29
economics of water projects,
1:16
efficiency, improving water-
use, **1:**19
environmental justice,
5:122–23
floods, **5:**111
fog collection as a source of
water, **2:**175
food needs for current/future
populations, **2:**87
groundwater, **4:**89
industrial water use, **1:**20–21
Irvine Ranch Water District,
4:124–25
Metropolitan Water District
of Southern California,
1:22
privatization, **3:**73
projections, review of global
water resources, **2:**43
reclaimed water
agriculture, **1:**29, **2:**142–46
drinking water, **2:**152
first state to attempt,
2:137–38
groundwater recharge,
2:151
health issues, **2:**154–55
Irvine Ranch Water
District, **2:**147
Kelly Farm marsh, **2:**149

San Jose/Santa Clara
Wastewater
Pollution Control
Plant, **2:**149–50
uses of, **2:**141–45
West Basin Municipal
Water District,
2:148–49
soft path for meeting water-
related needs, **3:**20–22,
24–25
subsidies, **1:**24–25
toilets, energy-efficient,
1:22
twentieth-century water-
resources develop-
ment, **1:**9
Western Canal Water District,
2:121
see also Climate change: California; Urban water
listings
California-American Water
Company (Cal AM), **5:**74
Camdessus, Michael, **4:**195–96
Cameroon, **1:**55, **5:**32
Canada:
availability, water, **2:**217
bottled water, **4:**25, 26, 39,
288, 289, 291, **5:**281,
283
Canadian International
Development Agency,
2:14
cholera, **1:**266, 270
climate change, **1:**147, 148
conflict/cooperation con-
cerning freshwater,
5:6, 8
dams, **1:**75, **3:**293
data problems, **3:**93
dracunculiasis, **1:**52
drinking water, **1:**253, **3:**256,
5:242
environmental flow, **5:**34
fog collection as a source of
water, **2:**179
General Agreement on Tariffs
and Trade, **3:**50
groundwater, **4:**86
hydroelectric production,
1:71, 74, 278
international river basin,
2:33
irrigation, **1:**299, **2:**265, **3:**289,
4:296, **5:**299
North American Free Trade
Agreement, **3:**51–54
population data/issues,
1:247, **2:**214
renewable freshwater supply,
1:238, **2:**200, 217,
3:240, **4:**264
reservoirs, **2:**270, 272, 274
runoff, **2:**23
salinization, **2:**268
sanitation services, **1:**258,
3:265, 272
threatened/at risk species,
1:293

withdrawals, water, **1:**242, **2:**207, **3:**247, **4:**271
World Water Council, **1:**172
Canary Islands, **3:**46
Cap and trade market and environmental flows, **5:**42
Cape Verde Islands, **2:**175
Carbon dioxide, **1:**138, 139, **3:**120, 215, **4:**160, 164
Caribbean:
  dams, **3:**293–94
  drinking water, **1:**253, 262, **3:**255–57, **5:**241, 242
  groundwater, **4:**86
  hydroelectric production, **1:**278
  population data/issues, **1:**247–48, **2:**214
  sanitation services, **1:**258, **3:**265–66, 271, **5:**250, 251, 259
  threatened/at risk species, **1:**293
Caribbean National Forest, **5:**34
Carolina Power and Light Company, **2:**126
Carp, **2:**118
Caspian Sea, **1:**77
Cat's Cradle (Vonnegut), **5:**14
Catley-Carlson, Margaret, **4:**xii–xiv
Cellatex, **5:**15
Center for Environmental Systems Research, **2:**56
Centers for Disease Control and Prevention (CDC), **1:**52, 55, 57
Central America:
  availability, water, **2:**217
  cholera, **1:**266, 270, 271
  dams, **3:**293–94
  drinking water, **1:**253, **3:**255, 256, **5:**241–42
  environmental flow, **5:**34
  groundwater, **4:**86
  hydroelectric production, **1:**278
  irrigation, **1:**299, **2:**265, **4:**296
  population data, total/urban, **1:**247–48
  renewable freshwater supply, **1:**238, **2:**200, 217, **3:**240, **4:**264, **5:**224
  reservoirs, **2:**270, 272, 274
  salinization, **2:**268
  sanitation services, **1:**258, **3:**265–66, **5:**250–51
  threatened/at risk species, **1:**293
  withdrawals, water, **1:**242–43, **2:**207, **3:**247, **4:**271, **5:**231–32
  World Commission on Dams, **3:**165
  see also Latin America
Centre for Ecology and Hydrology, **3:**110–11
Centro de Investigaciones Sociales Alternativas, **2:**179

Cereal production, **2:**64, **4:**299–301
Chad, **1:**55
Chakraborti, Dipankar, **2:**167
Chalecki, Elizabeth L., **3:**xiv
Chanute (KS), **2:**152
Chemical/biologic attacks, vulnerability to, **5:**16–22
Chiang Kai-shek, **5:**5
Chile:
  cholera, **1:**59
  environmental flow, **5:**34, 37
  fog collection as a source of water, **2:**177–78
  General Agreement on Tariffs and Trade, **3:**49
  privatization, **3:**60, 66, 78
  subsidies, **4:**57–58
China:
  agriculture, **4:**88, 90
  basic water requirement, **2:**13
  bottled water, **4:**21, 40, **5:**170
  business/industry, water risks that face, **5:**147, 149, 160, 165
  dams, **1:**69, 70, 77, 78, 81, **5:**15–16, 133, 134
  displaced people, dams and, **1:**78
  droughts, **5:**97
  economics of water projects, **1:**16
  environmental flow, **5:**32
  floods, **5:**106
  food needs for current/future populations, **2:**74
  globalization and international trade of water, **3:**46
  groundwater, **2:**87, **3:**2, 50, **4:**79, 82, 83, 88, 90, 96–97, **5:**125
  hydroelectric production, **1:**71
  industrial water use, **5:**125
  irrigation, **2:**86
  needs, basic human, **1:**46
  pricing, water, **1:**25
  privatization, **3:**59, 60
  sanitation services, **5:**124, 147
  Three Gorges Dam project, **1:**85, 88, 89
  use, average domestic water, **1:**46
  withdrawals, water, **3:**316–17
  World Commission on Dams, **3:**170–71
  see also Three Gorges Dam project
Chipko movement in India, **5:**124
Chitale, Madhav, **1:**174
Chlorination, **1:**47, 60, **4:**39, **5:**159
Cholera, **1:**48, 56–63, 265–71, **3:**2, **5:**287–92
Christie Malry's Own Double Entry, **5:**14
CII sectors (commercial/industrial sectors), *see* Urban water conservation: commercial/industrial water use in California

Cincinnati Enquirer, **5:**171
CIPM Yangtze Joint Venture, **1:**85
Clams, **3:**142–43
Clark Atlanta University, **3:**62
Clean water, *see* Drinking water, access to clean; Health, water and human
Clementine spacecraft, **1:**197, **3:**212–13
Climate change:
  Bibliography, The Water & Climate, **4:**xvii, 232–37
  business/industry, water risks, **5:**152
  changes occurring yet?, **1:**145–48
  desalination, **5:**80–81
  droughts, **5:**112–13
  environmental flow, **5:**45
  environmental justice, **5:**136–37
  floods, **5:**112–13
  food needs for current/future populations, **2:**87–88
  hydrologic cycle, **1:**139–43, **5:**117
  overview, **1:**137–39
  recommendations and conclusions, **1:**148–50
  reliability, water-supply, **5:**74
  societal impacts, **1:**144–45
  sustainable vision for the earth's freshwater, **1:**191
Climate change: California
  overview, **4:**xvii, 157–58
  policy, moving from science to demand management/conservation/efficiency, **4:**179–80
  economics/pricing/markets, **4:**180–81
  information gathering/reducing uncertainty, **4:**182–83
  infrastructure, existing, **4:**175–77
  institutions/institutional behaviors, new, **4:**181–82
  monitoring, hydrologic and environmental, **4:**183
  new supply options, **4:**178–79
  overview, **4:**175
  planning and assessment, **4:**178
  reports recommending integration of science/water policy, **4:**176
  science, the state of the
    ecosystems, **4:**171–72
    evaporation and transpiration, **4:**159–60
    groundwater, **4:**170
    lake levels and conditions, **4:**168–69

overview, **4:**158–59,
    172–73
precipitation, **4:**159
quality, water, **4:**167–68
runoff, large area/regional
    river, **4:**163–67
sea level, **4:**169–70
snowpack, **4:**160–61
soil moisture, **4:**167
storms/extreme events
    and variability,
    4:161–63
temperature, **4:**159
summary/conclusions,
    4:183–84
systems, managed water
    resource
    agriculture, **4:**174–75
    hydropower and thermal
        power generation,
        4:173–74
    infrastructure, water
        supply, **4:**173
Clinton, Bill, **2:**127, 134
Clothes washing, emerging
    technologies for, **5:**219,
    220
Coastal development and desali-
    nation, **5:**80
Coastal floods, **5:**104, 108
Coca-Cola, **4:**21, 38, **5:**127, 146,
    163
Cogeneration systems and
    desalination, **2:**107
Colombia, **1:**59, **3:**60, **5:**12
Colorado, **4:**95–96
Colorado River:
    climate change, **1:**142, 144,
        4:165–67
    conflict/cooperation con-
        cerning freshwater,
        1:109, 111
    delta characteristics,
        3:139–43
    fisheries, **1:**77
    Glen Canyon Dam, **1:**75–76,
        2:128–31
    Hoover Dam, **1:**69
    hydrology, **3:**135–37
    institutional control of,
        3:134
    Las Vegas attempting to
        lessen dependence on,
        5:74
    legal framework ("Law of the
        River"), **3:**137–39
    plants/vegetation, **3:**139–42
    resilience,
        vegetation/wildlife,
        3:134
    restoration opportunities,
        3:143–44
    summary/conclusions,
        3:144–45
Columbia River Alliance, **2:**133
Comets (small) and origins of
    water on earth, **1:**193–98,
    3:209–10, 219–20
Commagenian kings, **3:**185

Commercial sector, **3:**22–24,
    169–70
    see also Business/industry,
        water risks; Privatization;
        Urban water conserva-
        tion: commercial/indus-
        trial water use in
        California
Commissions, *see* International
    listings; Law/legal instru-
    ments/regulatory bodies;
    United Nations; World
    listings
Commodification, **3:**35
Commonwealth Development
    Corporation, **1:**96
Compaore, Blaise, **1:**52
Concession models, privatiza-
    tion and, **3:**66–67
Conferences/meetings, interna-
    tional, *see* International
    listings; Law/legal instru-
    ments/regulatory bodies;
    United Nations; World
    listings
Conflict/cooperation concerning
    freshwater:
    Africa, **1:**119–24
    chronology of disputes over
        course of history,
        1:108–9, 125–30, **2:**35,
        182–89, **3:**194–206,
        4:xvii–xviii, 238–56,
        5:5–15, 190–213
    Convention on the Non-Nav-
        igational Uses of Inter-
        national Watercourses,
        1:107, 114, 124, 210–30
    droughts, **5:**99
    environmental deficiencies
        and resource scarci-
        ties, **1:**105–6
    geopolitics of international
        river basins, **2:**35–36
    human right to water, **2:**3
    India-Bangladesh, **1:**107, 109,
        118–19, 206–9
    inequities in water distribu-
        tion/use/develop-
        ment, **1:**111–13
    instrument/tool of conflict,
        water as an, **1:**109–10
    Israel-Jordan, **1:**107, 115–16
    Middle East, other water
        issues in the, **1:**117–18
    military/political goal, water
        as a, **1:**108–9
    overview, **1:**107, **3:**2–3,
        5:189–90
    privatization, **3:**xviii, 70–71,
        79, **4:**54, 67
    reducing the risk of conflict,
        1:113–15
    security analysis, shift in
        international, **1:**105
    summary/conclusions, **1:**124
    sustainable vision for the
        earth's freshwater,
        1:190

    systems as targets of conflict,
        1:110–11
    see also Terrorism
Congo, Democratic Republic of,
    5:9
Conservation, *see* Environmental
    flow; Soft path for
    meeting water-related
    needs; Sustainable vision
    for the Earth's freshwater;
    Twenty-first century
    water-resources develop-
    ment; Urban water *listings*
Consumption/consumptive use,
    1:12, 13, **3:**103
    see also Projections, review of
        global water resources;
        Withdrawals, water
Contracts, privatization and,
    4:65–67
Conventions, international
    legal/law, *see* Interna-
    tional *listings;* Law/legal
    instruments/regulatory
    bodies; United Nations;
    World *listings*
Cook Islands, **3:**118
Cooling and CII water use, **4:**135
Cost effectiveness, **4:**105
    see also Economy/economic
        issues
Costa Rica, **5:**34
Côte d'Ivoire, **1:**55, **4:**65–67
Councils, *see* Law/legal instru-
    ments/regulatory bodies
Court decisions and
    conflict/cooperation con-
    cerning freshwater, **1:**109,
    120
    see also Law/legal instru-
        ments/regulatory bodies;
        Legislation
Covenant, the Sword, and the
    Arm of the Lord (CSA),
    5:21–22
Covenants, *see* International
    Covenant on Economic,
    Social, and Cultural
    Rights; Law/legal instru-
    ments/regulatory bodies;
    United Nations
Crane, Siberian, **1:**90
Crisis management, droughts
    and, **5:**99, 111
Critical Trends, **3:**88
Croatia, **1:**71
Crop yields and food needs for
    current/future popula-
    tions, **2:**74–76
Cropping intensity, **2:**76
Cryptosporidium, **1:**48, **2:**157,
    4:52, **5:**2, 159
Cucapá people, **3:**139
Cultural importance of water,
    3:40
Curacao, **2:**94–95
Current good manufacturing
    practice (CGMP), **4:**27
CW Leonis, **3:**219–20

Cyanide, **5**:20
Cyberterrorism, **5**:16
Cyprus, **1**:202–4, **2**:108
Cyrus the Great, **1**:109

da Vinci, Leonardo, **1**:109
Dams:
    business/industry, water
        risks, **5**:151
    by continent and country,
        3:291–99
    debate, new developments in
        the, **1**:80–83
    displaced people, **1**:77–80, 85,
        90, 97–98, 281–87
    economic issues, **1**:16,
        2:117–19, 122–24, 127,
        129–30
    environmental flow, **5**:32–34
    environmental justice,
        5:133–36
    environmental/social
        impacts, **1**:15, 75–80,
        83
    floods, **5**:106
    Gabcikovo-Nagymaros
        project, **1**:109, 120
    grandiose water-transfer
        plans, **1**:74–75
    history of large, **1**:69
    hydroelectric production,
        1:70–75
    Korean peninsula, **1**:109–10
    large, U.S. begins construc-
        tion of, **1**:69–70
    opposition to, **1**:80–82
    private-sector funding, **1**:82
    runoff, humanity appropriat-
        ing half of the world's,
        5:29
    schistosomiasis, **1**:49
    terrorism risks, **5**:15–16
    threatened/at risk species,
        3:3
    twentieth-century water-
        resources develop-
        ment, **1**:6
    World Water Forum (2003),
        4:193
    see also World Commission
        on Dams
Dams, removing/decommission-
    ing:
    1912 to present, **2**:275–86
    case studies: completed
        removals
        Edwards, **2**:xix, 123–25
        Maisons-Rouges and
            Saint-Etienne-du-
            Vigan, **2**:125
        Newport No. 11, **2**:125–26
        Quaker Neck, **2**:126
        Sacramento River valley,
            2:120–23
    economics, **2**:118–19
    hydroelectric production,
        1:83
    overview, **2**:xix, 113–14

proposed
    Elwha and Glines Canyon,
        2:127–28
    Glen Canyon, **2**:128–31
    Pacific Northwest,
        network of dams
        in, **2**:131–34
    Peterson, **2**:128
    Savage Rapids, **2**:128
    Scotts Peak, **2**:126–27
purpose for being built no
    longer valid, **2**:117
renewal of federal
    hydropower licenses,
    2:114–15
safety issues, **2**:130
states taking action, **2**:117–18
summary/conclusions, **2**:134
twentieth century by decade,
    3:301–2
Dams, specific:
    Ataturk, **1**:110, **3**:182, 184, 185
    Auburn, **1**:16
    Bakun, **1**:16
    Balbina, **5**:134
    Banqiao, **5**:15–16
    Batman, **3**:187
    Birecik, **3**:185, 186–87
    Bonneville, **1**:69
    Chixoy, **3**:13
    Cizre, **3**:189–90
    Condit, **2**:119
    Edwards, **1**:83, **2**:xix, 119, 123
    Elwha, **1**:83
    Farakka Barrage, **1**:118–19
    Fort Peck, **1**:69
    Fort Randall, **1**:70
    Garrison, **1**:70, **5**:123
    Glen Canyon, **1**:76
    Glines Canyon, **1**:83
    Gorges, **5**:133
    Grand Coulee, **1**:69
    Hetch Hetchy, **1**:15, 80–81
    Hoa Binh, **5**:134
    Hoover, **1**:69, **3**:137
    Ice Harbor, **2**:131–34
    Ilisu, **3**:187–89, 191
    Itaipu, **1**:16
    Kalabagh, **1**:16
    Karakaya, **3**:182, 184
    Kariba, **5**:134
    Katse, **1**:93, 95
    Keban, **3**:184
    Koyna, **1**:77
    Laguna, **3**:136
    Little Goose, **2**:131–34
    Lower Granite, **2**:131–34
    Lower Monumental, **2**:131–34
    Manitowoc Rapids, **2**:119
    Morelos, **3**:138, 142
    Nam Theun I, **1**:16
    Nurek, **1**:70
    Oahe, **1**:70
    Pak Mun, **5**:134
    Peterson, **2**:128
    Quaker Neck, **2**:126
    Sadd el-Kafara, **1**:69
    Salling, **2**:119
    Sandstone, **2**:119

Sapta Koshi High, **1**:16
Sardar Sarovar, **5**:133
Savage Rapids, **2**:119, 128
Shasta, **1**:69, **5**:123
Shimantan, **5**:15–16
Snake River, **2**:131–33
St. Francis, **5**:16
Ta Bu, **5**:134
Welch, **2**:118
Woolen Mills, **2**:117–19
Yacyreta, **3**:13
see also Lesotho Highlands
    project; Southeastern
    Anatolia Project; Three
    Gorges Dam
Dasani, **4**:21
Data issues/problems:
    climate change, **4**:183
    constraints on data and pro-
        jections of global water
        resources, **2**:40–42
    conversions/water
        units/constants,
        5:319–28
    conversions/water units/
        constants, data,
        2:300–309, **3**:318–27,
        4:325–34
    groundwater, **4**:97–98
    open access to information,
        4:70–73
    urban commercial/industrial
        water use in California,
        4:139–40, 148–50,
        152–53
    urban residential water use
        in California, **4**:108–9
    well-being, measuring water
        scarcity and, **3**:93
de Melo, Carlos, **3**:6
Decision making, open/demo-
    cratic, **4**:70–73
Declarations, *see* Law/legal
    instruments/regulatory
    bodies; United Nations
Deer Park, **4**:21
Deltas, river, **3**:xix
    see also Colorado River; Inter-
        national river basins
Demand management, **4**:179–80
    see also Withdrawals, water;
        Water-use efficiency
Dengue fever, **3**:2
Denmark, **1**:52
Density, measuring, **2**:304, **3**:327,
    4:329, 334
Desalination:
    advantages and disadvan-
        tages, **5**:66–76
    business/industry, water
        risks, **5**:161
    California, **1**:30, 32, **5**:51–52,
        63–69, 71, 73, 74
    capacity by country/process/
        source of water, **1**:131,
        288–90, **2**:287–90,
        5:58–60
    capacity statistics, **5**:56–57,
        59–60, 308–17

climate change, **5:**80–81
concentrate disposal, **2:**107
economic issues, **1:**30, **2:**95, 105–9, **5:**62–63, 66, 68–73
energy use/reuse, **2:**107, **5:**69–71, 75–76
environmental effects of, **5:**76–80
global status of, **5:**55–58
health, water quality and, **5:**74–75
history and current status, **2:**94–98, **5:**54
intakes, water: impingement/ entrainment, **5:**76–77
membrane processes
electrodialysis, **2:**101–2
overview, **2:**101
reverse osmosis, **2:**96, 102–3
Nauru, **3:**118
oversight process, regulatory and, **5:**81–82
overview, **1:**29–30, **2:**93–94, **5:**51–53
plants, capacity of actual/planned, **5:**308–17
processes, other
freezing, **2:**104
ion-exchange methods, **2:**104
membrane distillation, **2:**104–5
overview, **2:**103–4
solar and wind-driven systems, **2:**105–6
reliability value of, **5:**73–74
salt concentrations of different waters, **2:**94 **5:**53
source of water/process, capacity by, **5:**56–57, 59–60
summary/conclusions/rec-ommendations, **2:**109–10, **5:**82–86
Tampa Bay desalination plant, **2:**108–9, **5:**61–63
technologies used, **5:**54–55
thermal processes
multiple effect distillation, **2:**99–100
multistage flash distilla-tion, **2:**100
overview, **2:**98–99
vapor compression distil-lation, **2:**100–101
United States, **5:**58–63
Deutsche Morgan Grenfell, **1:**96
Developing countries:
agriculture, **1:**24, **3:**290
bottled water, **4:**40–41
business/industry, water risks, **5:**150
cholera, **1:**56
dams, **1:**82
diseases, water-related, **5:**117
dracunculiasis, **1:**51

drinking water, **1:**261–62
economic development derailed, **1:**42
efficiency, improving water-use, **1:**19
food needs for current/future populations, **2:**69
industrial water use, **1:**21
irrigation, **3:**290, **4:**297–98, **5:**301–2
Pacific Islands, **5:**136
population increases and lack of basic water services, **3:**2
privatization, **3:**79, **4:**46
sanitation services, **1:**263–64
toilets, energy-efficient, **1:**22
unaccounted for water, **4:**59
see also, Environmental justice; Pacific Island developing countries
Development organizations and WCD report, interna-tional, **3:**164–65
Development, the right to, **2:**8–9
Diarrhea, **1:**48, **4:**8, 11
Direct uses of water, **3:**18–19
Disability adjusted life year (DALY), **4:**9, 276–77
Discrimination, environmental, **5:**118–19
Diseases, water-related, **1:**186–87
see also under Health, water and human *and* Millen-nium Development Goals
Dishwashers, **4:**109, 116
Displaced people, dams and, **1:**77–80, 85, 90, 97–98, 281–87, **5:**134, 151
Dolphins, **1:**77, 90, **3:**49, 50
Dow Jones Indexes, **3:**167
Downstream users and the human right to water, **5:**37
Dracunculiasis, **1:**39, 48–56, 272–73, **3:**273–77, **5:**293–97
Dressler, Alexander, **1:**194
Drinking water, access:
arsenic in groundwater, **2:**165–73, **3:**278–79
Bolivia, **3:**71
by country, **1:**251–55, **3:**252–60, **5:**237–46
defining terms, **4:**28
developing countries, **1:**261–62
disinfection and purification, **5:**159
Edwards Aquifer, **3:**74
excreta, diseases and water contaminated by, **1:**47–48
fluoride, **4:**87
international organizations, recommendations by, **2:**10–11
reclaimed water, **2:**151–52
twentieth-century water-resources develop-ment, **3:**2

United States, primary drinking water regula-tions in, **3:**280–88
well-being, measuring water scarcity and, **3:**96–98
World Health Organization, **4:**208
World Water Forum (2003), **4:**202
see also Health, water and human; Millennium Development Goals; Human right to water; Soft path for meeting water-related needs; Well-being, measuring water scarcity and
Drip irrigation, **1:**23, **2:**82, 84
Droughts, **1:**142, 143, **4:**163, 203–4
beginning of, determining the, **5:**93
causes of, **5:**95–96
characteristics vary by region/water-use activity, **5:**92
defining terms, **5:**92
disturbances promoting ecosystem diversity, **5:**91
economy/economic issues, **5:**91–92, 98, 103
effects of, **5:**95–99
future of, **5:**112–13
management
crisis management, **5:**99
impact and vulnerability assessment, **5:**100–101
mitigation and response, **5:**101–3
monitoring and early warning, **5:**99–100
risk management, **5:**99
National Drought Mitigation Center, **5:**94
overview, **5:**91–92
short-lived or persistent, **5:**93
summary/conclusions, **5:**113–14
United States, areas of extreme drought in, **5:**93
DuPont, **1:**52
Dust Bowl (1930s), **5:**93
Dutch Water Line strategy, **5:**5
Dynamics, water, **2:**218
Dysentery, **1:**42, **4:**64

Early Warning Monitoring to Detect Hazardous Events in Water Supplies, **5:**2, 20
Earth Water, **5:**163
Earthquakes and floods, **5:**106
Earthquakes caused by filling reservoirs, **1:**77, 97
Earths' water, origins of the, **1:**93–98, **3:**209–12
East Bay Municipal Utilities District (EBMUD), **5:**73

East Timor, **5:**9

Economies of scale and hard
    path for meeting water-
    related needs, **3:**8

Economy/economic issues:
    access to water, **5:**125
    bag technology, water, **1:**200,
        204–5
    bottled water, **4:**17, 22–23
    cholera, **1:**60
    Colorado River, **3:**143
    conflict/cooperation con-
        cerning freshwater,
        5:15
    dams, **1:**82, **2:**117–19, 122–24,
        127, 129–30, 132
    desalination, **1:**30, **2:**95,
        105–9, **5:**62–63, 66,
        68–73
    developing countries, **1:**42
    droughts, **5:**91–92, 98, 103
    economic good, treating
        water as an, **3:**xviii,
        33–34, 37–38, 58,
        4:45
    efficiency, economic, **4:**104,
        105
    environmental flow, **5:**32,
        40–43
    fishing, **2:**117
    floods, **5:**91–92, 108, 109
    Global Water Partnership,
        1:171
    human right to water,
        2:13–14
    industrial water use, **1:**21
    international water meetings,
        5:183
    Lesotho Highlands project,
        1:95–97, 99
    Millennium Development
        Goals, **4:**6–7
    needs, basic human,
        1:46–47
    official development assis-
        tance, **4:**6, 278–83,
        5:262–72
    overruns, water-supply
        project, **3:**13
    Pacific Island developing
        countries, **3:**127
    privatization, **3:**77, **4:**53–60
    productivity of water, U.S.,
        4:321–24
    reclaimed water, **2:**156, 159
    sanitation services, **5:**273–75
    soft path for meeting water-
        related needs, **3:**5, 6–7,
        12–15, 23–25
    Southeastern Anatolia
        Project, **3:**182, 187,
        190–91
    subsidies
        agriculture, **1:**117
        desalination, **5:**69
        engineering projects,
            large-scale, **1:**8
        government and interna-
            tional organiza-
            tions, **1:**17

    privatization, **3:**70–72,
        4:50, 53–60
    twenty-first century
        water-resources
        development,
        1:24–25
    supply-side solutions, **1:**6
    terrorism, **5:**20
    Three Gorges Dam project,
        1:86–89
    twentieth-century water-
        resources develop-
        ment, end of, **1:**16–17
    urban commercial/industrial
        water use in California,
        4:140–43, 151–52
    urban residential water use
        in California, **4:**107–8
    water industry
        revenue/growth,
        5:303–4
    withdrawals, water, **3:**310–17
    World Commission on Dams,
        3:158, 162–64, 167–69
    World Water Forum (2003),
        4:195–96
    see also Business/industry,
        water risks; Globalization
        and international trade of
        water; Environmental
        justice; Pricing, water; Pri-
        vatization; World Bank

Ecosystems, healthy:
    Aral Sea, degradation of the,
        3:39
    climate change, **4:**171–72
    dams and water withdrawals
        destroying, **3:**3
    environmental flow, **5:**30–31
    metals in aquatic ecosystems,
        4:168
    privatization, **4:**51–53
    reclaimed water, **2:**149–50
    World Water Forum (2003),
        4:203
    see also Environmental flow;
        Environmental issues;
        Fish; Soft path for
        meeting water-related
        needs; Sustainable vision
        for the Earth's freshwater

Ecuador, **1:**59, 71, **2:**178–79, **3:**49,
    5:124

Edwards Aquifer, **3:**74–75, **4:**60–61

Edwards Manufacturing
    Company, **2:**124

Eels, **2:**123

Egypt, **1:**118, **2:**26, 33, **4:**18, 40,
    5:163

Eighteen District Towns project,
    2:167

El Niño/Southern Oscillation
    (ENSO), **1:**139, 143, 147,
    3:119, **4:**162, **5:**95, 106

Electricity, *see* Hydroelectric pro-
    duction

Electrodialysis and desalination,
    2:101–2

End use of water as a social
    concern, **3:**7, 8–9

Energy issues:
    business/industry, water
        risks, **5:**150, 151–52
    desalination and energy
        use/reuse, **2:**107,
        5:69–71, 75–76
    droughts, **5:**98
    efficiency, **3:**xiii
    Energy Department, U.S.,
        1:23, **4:**115
    measuring energy, **2:**308,
        3:326, **4:**333

Engineering projects, large-scale,
    see Twentieth-century
    water-resources
    development

England:
    cholera, **1:**56
    desalination, **2:**94
    droughts, **5:**92
    human right to water, **4:**212
    Lesotho Highlands project,
        1:96
    Office of Water Services,
        4:64–65
    privatization, **3:**58, 60, 61, 78,
        4:62–65
    sanitation services, **5:**129–31
    Southeastern Anatolia
        Project, **3:**191
    World Commission on Dams,
        3:162, 170

Enron, **3:**63

Entrainment and desalination,
    5:77

Environment, **2:**182

Environmental flow:
    characteristics of hydrologic
        regimes, **5:**31
    economics/finance, **5:**40–43
    General Accounting Office
        (GAO), **5:**119
    legal framework, **5:**34–37
    policy implementation,
        5:43–45
    projects in practice, **5:**32–34
    science of determining,
        5:38–40
    summary/conclusions,
        5:45–46
    water quality link, **5:**40
    World Commission on Dams'
        recommendations,
        5:30

Environmental issues:
    bottled water, **4:**41
    change, global, **1:**1
    conflict/cooperation con-
        cerning freshwater,
        1:105–6
    dams/reservoirs, **1:**15, 75–80,
        83, 91
    desalination, **5:**76–80
    environmental flow, **5:**32–34
    environmental justice,
        5:117–142
    Lesotho Highlands project,
        1:98
    pesticides, **5:**305–7
    reclaimed water, **2:**149–50

shrimp/tuna and turtle/
dolphin disputes,
3:49, 50
sustainable vision for the
Earth's freshwater,
1:188–90
Three Gorges Dam project,
1:89–92
twentieth-century water-
resources develop-
ment, end of, **1:**12,
15–16
wetland pollution, **1:**6
see also Ecosystems, healthy;
Fish; Environmental flow;
Law/legal instruments/
regulatory bodies; Threat-
ened/at risk species
Environmental justice:
climate change, **5:**136–37
Coca-Cola, **5:**127
dams, **5:**133–36
discrimination, environmental, **5:**118–19
environmentalism of the
poor, **5:**123–24
Environmental Justice
Coalition for Water,
5:122–23
good governance, **5:**138–41
history of movement in U.S.,
5:119–20, 122–23
international context, **5:**118
human right to water, recog-
nition/implementa-
tion, **5:**137–38
overview, **5:**117–18
principles of, **5:**120–22
privatization, **5:**131–33
sanitation services, **5:**127–31
summary/conclusions,
5:141–42
water access, **5:**124–25, 127
water quality, **5:**127–29
women and water, **5:**126, 134
Eritrea, **5:**95
Ethiopia, **1:**55, **4:**211, **5:**95, 97
Ethos Water, **5:**163
Europa Orbiter, **3:**218
Europe:
availability, water, **2:**217
bottled water, **4:**22, 25, 288,
289, 291, **5:**163, 281, 283
cholera, **1:**56, 267, 270, **5:**289,
290
dams, **3:**292–93
drinking water, **5:**159, 245–46
environmental flow, **5:**33
European Union Develop-
ment Fund, **1:**96
food needs for current/future
populations, **2:**69
Global Water Partnership,
1:169
globalization and international trade of water,
3:45, 47
groundwater, **4:**84–85
human right to water, **4:**209,
214

hydroelectric production,
1:71, 279–80
international river basin,
2:29–31
irrigation, **1:**301, **2:**261–62,
265, **3:**289
Lesotho Highlands project,
1:96
Mars, water on, **5:**178
population data/issues,
1:249–50, **2:**214
privatization, **3:**58, 60, 61
renewable freshwater supply,
1:239–40, **2:**201–2, 217,
3:241–42, **4:**265–66,
5:225–26
reservoirs, **2:**270, 274
salinization, **2:**268
sanitation services, **3:**272,
5:255
threatened/at risk species,
1:294–95
waterborne diseases, **1:**48
withdrawals, water, **1:**244,
2:209–11, **3:**249–50,
4:273–75, **5:**235–36
World Commission on Dams,
3:165
Evaporation of water into atmosphere, **1:**141, **2:**20, 22, 83,
4:159–60
Evapotranspiration, **2:**83
Ex-Im Bank, **1:**88–89, **3:**163
Excreta, drinking water contaminated by, **1:**47–48
Export credit agencies (ECAs),
3:191

Faucets, **4:**117
Fedchenko, Alexei P., **1:**51
Fertilizer runoff, **5:**128
Field flooding, **2:**82, 86
Fiji, **3:**45, 46, 118
Filtration, water, **1:**47
Finn, Kathy, **4:**69–70
Fires and droughts, **5:**98, 102
Fish:
aquaculture, **2:**79
climate change, **4:**168
Colorado River, **3:**142
dams, removing/decommis-
sioning
Edwards, **2:**123
Peterson, **2:**128
Quaker Neck, **2:**126
Savage Rapids, **2:**128
Snake River, **2:**131–33
Woolen Mills, **2:**118
dams/reservoirs affecting,
1:77, 83, 90, 98, **2:**117
droughts, **5:**98, 102
floods, **5:**109
food needs for current/future
populations, **2:**79
largest number of species,
countries with,
2:298–99
pesticides, **5:**305–7
Sacramento River valley,
2:120–21

threatened/at risk species,
1:291–96, **3:**3, 39–40,
142
Flash floods, **5:**104
Floods, **1:**142–43, **4:**162–63,
203–4, 302–7, **5:**104–109
causes of, **5:**106
control, **5:**110–112
definition, **5:**104–5
disturbances promoting
ecosystem diversity,
5:91
economy/economic issues,
5:91–92
effects of, **5:**106–10
frequency, calculating flood,
5:105
future of, **5:**112–13
Johnston Flood of 1889,
5:16
management, **5:**110–12
overview, **5:**91–92
summary/conclusions,
5:113–14
Florida and desalination,
5:60–63, 69
Flow rates, **2:**305–6, **3:**104, 324,
4:168, 172, 331
see also Hydrologic cycle;
Stocks and flows of freshwater
Fluoride, **4:**87
Fog collection as a source of
water, **2:**175–81
Fondo Ecuatoriano Canadiense
de Desarrollo, **2:**179
Food needs for current/future
populations:
climate change, **2:**87–88
crop yields, **2:**74–76
cropping intensity, **2:**76
eaten by humans, fraction of
crop production,
2:76–77
inequalities in food distribu-
tion/consumption,
2:64, 67–70
kind of food will people eat,
what, **2:**68–70
land availability/quality,
2:70–71, 73–74
need and want to eat, how
much food will people,
2:67–68
overview, **2:**65
people to feed, how many,
2:66–67
production may be unable to
keep pace with future
needs, **2:**64
progress in feeding Earth's
population, **2:**63–64
summary/conclusions,
2:88
water needed to grow food,
how much
crop type, **2:**78–80
irrigation, **2:**80–81, 81–87
regional diets, **2:**64–66
water quality, **2:**87

Force, measuring, **2:**306, **3:**324, **4:**331
Foreign Affairs, **3:**xiii
France:
  conflict/cooperation concerning freshwater, **5:**15
  dracunculiasis, **1:**52
  Global Water Partnership, **1:**171
  globalization and international trade of water, **3:**45
  human right to water, **4:**209
  Lesotho Highlands project, **1:**96
  privatization, **3:**60, 61
  Three Gorges Dam project, **1:**89
  World Commission on Dams, **3:**159
Frank, Louis A., **1:**194–98, **3:**209–10
Freezing and desalination, **2:**104
French Polynesia, **3:**118
Freshwater, *see* Access to water and environmental justice; Rivers
Furrow irrigation, **2:**82, 86
Future, the, *see* Projections, review of global water resources; Soft path for meeting water-related needs; Sustainable vision for the Earth's freshwater; Twenty-first century water-resources development

Galileo spacecraft, **3:**217, 218
Gambia, **4:**211
Gap, **5:**156
Gardens, energy efficient ways of maintaining, **1:**23
  see also indoor water use *under* Urban water conservation: residential water use in California
Gaziantep Museum, **3:**186, 187
GE Infrastructure, **5:**159–61
GEC Alsthom, **1:**85
General circulation models (GCMs), **3:**121–23, **4:**158, 159, 162, 167
General Electric Canada, **1:**85
Geophysical Research Letters, **1:**194
Germani, Gianfranco, **1:**203
Germany:
  conflict/cooperation concerning freshwater, **5:**5, 7
  international river basin, **2:**29
  Lesotho Highlands project, **1:**96
  privatization, **3:**61
  terrorism, **5:**20
  Three Gorges Dam project, **1:**88, 89
  World Commission on Dams, **3:**159, 170

Ghana, **1:**46, 52, 54
Giardia, **1:**48, **2:**157, **4:**52
Giardiasis, **1:**48
Gleick, Peter, **3:**xiii–xiv, **5:**124
Glen Canyon Institute, **2:**130
Global Environmental Outlook, **3:**88
Global Reporting Initiative (GRI), **5:**158
Globalization and international trade of water:
  business/industry, water risks, **5:**151
  defining terms
    commodification, **3:**35
    economic good, **3:**37–38
    globalization, **3:**34–35
    private/public goods, **3:**34
    privatization, **3:**35
    social good, **3:**36–37
  General Agreement on Tariffs and Trade, 48–51
  North American Free Trade Agreement, **3:**51–54
  overview, **3:**33–34, 41–42
  raw or value-added resource, **3:**42–47
  rules: international trading regimes, **3:**47–48
  social and economic good, water managed as both, **3:**38–40
  World Water Forum (2003), **4:**192, 193
Goddess of the Gorges, The, **1:**84
Goh Chok Tong, **1:**110
Good manufacturing practice (GMP), **4:**32, 33
Goodland, Robert, **1:**77
Gorbachev, Mikhail, **1:**106
Gorton, Slade, **2:**134
Government/politics:
  droughts, **5:**92, 94–95, 103
  environmental justice, **5:**138–41
  human right to water, **2:**3
  irrigation, **1:**8
  military/political goal, water as a, **1:**108–9
  privatization usurping responsibilities of, **3:**68
  privatization, maintaining strong regulation of, **4:**60–73
  subsidies, **1:**17
  twentieth-century water-resources development, **1:**7–8, 17
  WCD report, **3:**170–71
  see also Conflict/cooperation concerning freshwater; Law/legal instruments/regulatory bodies; Legislation; policy, moving from science to *under* Climate change: California; Human right to water; Stocks and flows of freshwater; *individual countries*

Grain production, **2:**64, 299–301
Grand Banks, **1:**77
Grand Canyon, **1:**15, **2:**138, 146
Granite State Artesian, **4:**39
Grants Pass Irrigation District (GPID), **2:**128
Gre Dimse archaeological site, **3:**189
Great Lakes, **1:**111, **3:**50
Greece:
  ancient water systems, **2:**137
  bag technology, water, **1:**202, 204, 205
  conflict/cooperation concerning freshwater, **5:**5
  hydroelectric production, **1:**71
  supply systems, ancient, **1:**40
Greenhouse effect, **1:**137, 138, **3:**126, **4:**171
  see also Climate change *listings*
Gross national product (GNP) and water withdrawals, **3:**310–17
Groundwater:
  arsenic in, **2:**165–73, **3:**278–79, **4:**87
  climate change, **4:**170
  data problems, **3:**93
  Edwards Aquifer, **3:**74–75, **4:**60–61
  food needs for current/future populations, **2:**87
  General Agreement on Tariffs and Trade, **3:**49–50
  hard path for meeting water-related needs, **3:**2
  monitoring/management problems
    agriculture, **4:**88–90
    analytical dilemma, **4:**90–97
    challenges in assessments, **4:**80–81
    conceptual foundations of assessments, **4:**80
    data and effective management, **4:**97–98
    extraction and use, **4:**81–88
    overview, **4:**79
  overextraction, **5:**125, 128
  Pacific Island developing countries, **3:**116–18
  pesticides, **5:**307
  privatization, **3:**77, **4:**60–61
  public ownership rights and privatization, **3:**74
  quality concerns, **4:**83, 87
  reclaimed water, **2:**150–51
  reliability, water-supply, **5:**74
  stocks and flows of freshwater, **2:**20
  well-being, measuring water scarcity and, **3:**104
Guatemala, **3:**13
Guidelines for Drinking-Water Quality, **4:**26–27, 31

Guinea, **3:**76
Guinea worm, **1:**39, 48–56,
    272–73, **3:**273–77
Gulf of California, **3:**141, 142
Gulf of Mexico, **1:**77
Gwembe Tonga people, **5:**134
Habitat loss and droughts, **5:**98,
    103, 109
Habitat simulation as environ-
    mental flow methodology,
    5:39
Haiti, **1:**46
Hallan Cemi archaeological site,
    3:187
Halogen Occultation Experiment
    (HALOE), **1:**196
Hamidi, Ahmed Z., **1:**110
Harcourt, Mike, **1:**88
Hard path for meeting water-
    related needs, **3:**xviii, 2
    see also Soft path for meeting
        water-related needs;
        Twentieth-century water-
        resources development
Hardness, measuring, **2:**309,
    3:327, **4:**334
Harran, **3:**185
Harvard School of Public Health,
    4:9
Harvesting technology and food
    needs for current/future
    populations, **2:**77
Hasankeyf archaeological site,
    3:187–89, 191
Hazardous waste landfills, **5:**119
Health and Human Services
    (HHS), U.S. Department
    of, **5:**24
Health, water and human:
    arsenic in groundwater,
        2:165–73, **3:**278–79,
        4:87
    desalination, **5:**74–75
    diseases, water-related
        cholera, **1:**48, 56–63,
            265–71, **3:**2,
            5:287–92
        dams, removing/decom-
            missioning, **2:**130
        deaths and disability
            adjusted life year,
            4:276–77
        dengue fever, **3:**2
        diarrhea, **1:**48, **4:**8, 11
        dracunculiasis, **1:**39,
            48–56, 272–73,
            3:273–77, **5:**293–97
        dysentery, **1:**42, **4:**64
        emerging diseases, **2:**155
        environmental justice,
            5:128–29
        failure, **3:**2, **5:**117
        malaria, **1:**49–50
        outbreaks in U.S.,
            4:308–12
        overview, **1:**274–75,
            47–50
        schistosomiasis, **1:**48,
            49
    droughts, **5:**102
    floods, **5:**109

fluoride, **4:**87
    needs for water, basic
        human, **1:**42–47
    privatization, **4:**47
    reclaimed water, **2:**152–56
    summary/conclusions,
        1:63–64
    see also diseases, water-
        related *under* Millennium
        Development Goals;
        Drinking water, access;
        Environmental justice;
        Sanitation services; Well-
        being, measuring water
        scarcity and
Heat is On, The (Gelbspan),
    5:136
Helmut Kaiser, **5:**161
Herodotus, **1:**109
Herring, **2:**123
Historic flow as environmental
    flow methodology, **5:**39
Hittites, **3:**184
Hoecker, James, **2:**124
Holistic approaches to environ-
    mental flow methodology,
    5:39
Holland, **5:**5
Holmberg, Johan, **1:**166, 175
Honduras, **1:**71, **4:**54–55
Hong Kong, **3:**46, 313–15
Hoppa, Gregory, **3:**217
Human rights and international
    law, **2:**4–9
Human right to water:
    barriers to, **4:**212–13
    defining terms, **2:**9–13
    environmental flow, **5:**37
    failure to meet, consequences
        of the, **2:**14–15
    is there a right?, **2:**2–3
    laws/covenants/declarations,
        2:4–9
    obligations, translating rights
        into legal, **2:**3, 13–14
    overview, **4:**207–8
    Prior Appropriation Doctrine,
        5:37
    progress toward acknowledg-
        ing, **4:**208–11
    services, access to basic
        water, **2:**1–2
    summary/conclusions, **2:**15
    why bother?, **4:**214
    see also International
        Covenant on Economic,
        Social & Cultural Rights;
        Environmental justice
Hungary, **1:**109, 120
Hunger, estimates of world, **2:**70
Hurrian Kingdom, **3:**183
Hurricane Katrina, **5:**24, 110
Hydraulic geometry as environ-
    mental flow methodology,
    5:39
Hydro Equipment Association
    (HEA), **3:**166–67
Hydro-Quebec International,
    1:85
Hydroelectric production:
    by region, **1:**70–71

California, **4:**173–74
capacity, countries with
    largest installed, **1:**72,
    276–80
Colorado River, **4:**165
dams, removing/decommis-
    sioning, **1:**83
Glen Canyon Dam, **2:**129–30
grandiose water-transfer
    schemes, **1:**74–75
percentage of electricity
    generated with
    hydropower, **1:**73–74
production, countries with
    highest, **1:**72, 276–80
Snake River, **2:**132–33
Southeastern Anatolia
    Project, **3:**182
Three Gorges Dam project,
    1:84
well-being, measuring water
    scarcity and, **3:**103
Hydrologic cycle:
    climate change, **1:**139–43,
        4:183, **5:**117
    desalination, **2:**95, **5:**52
    droughts, **5:**94
    quantifications, accurate,
        4:92–96
    stocks and flows of freshwa-
        ter, **2:**20–27
    see also Environmental flow

Iceland, **1:**71
Idaho Rivers United, **2:**133
Identity standards and bottled
    water, **4:**27–31
Impingement and desalination,
    5:76–77
India:
    agriculture, **4:**88, 89
    arsenic in groundwater,
        2:165–73, **4:**87
    basic water requirement, **2:**13
    bottled water, **4:**22, 25, 40
    business/industry, water risks
        that face, **5:**146, 147,
        165
    cholera, **1:**61
    conflict/cooperation con-
        cerning freshwater,
        1:107, 109, 118–19,
        206–9, **5:**13, 15
    dams, **1:**70, 78, 81, **5:**133
    Dhaka Community Hospital,
        2:170
    displaced people, dams and,
        1:78
    dracunculiasis, **1:**53, 55
    economics of water projects,
        1:16, 17
    environmental justice, **5:**124,
        127
    floods, **5:**106
    groundwater, **3:**2, 50, **4:**82, 83,
        88–90, 92–95, **5:**125,
        128
    human right to water, **4:**211
    industrial water use, **1:**21
    international river basin,
        2:27

irrigation, **2:**85, 86
sanitation services, **5:**128
use, domestic water, **1:**46
World Commission on Dams,
    3:159, 170
Indicators/indices, water-
    related, **3:**87
    see also Well-being,
        measuring water scarcity
        and
Indigenous populations, **5:**123,
    124
    see also Environmental
        justice
Indirect uses of water, **3:**18–19
Indonesia:
    bottled water, **5:**170
    cholera, **1:**58
    climate change, **1:**147
    General Agreement on Tariffs
        and Trade, **3:**49
    needs, basic human, **1:**46
    pricing, water, **1:**25, **3:**69
Industrial sculptures, **5:**219, 220
Industrial water treatment, **5:**160
Industrial water use, **1:**20–21,
    5:124–25
    see also Business/industry,
        water risks; Projections,
        review of global water
        resources; Urban water
        conservation: commer-
        cial/industrial water use
        in California
Industry/trade associations and
    WCD report, **3:**169–70
Infant brands of bottled water,
    4:28
Infrared Space Observatory,
    3:220
Insect vectors, diseases associ-
    ated with water-related,
    1:49–50, **4:**8–9
Institute of Marine Aerodynam-
    ics, **1:**202
Institutions, new water, *see*
    Law/legal
        instruments/regulatory
        bodies
Instream water allocations, pre-
    serving/restoring, **5:**29–30
    see also Environmental flow
Integrated water planning, **1:**17,
    3:21
    see also Global Water Partner-
        ship
Intensity, water, **3:**17–19
Inter-American Development
    Bank, **3:**163
Interferometry, **3:**221
International alliances/confer-
    ences/meetings, time to
    rethink large, **5:**182–85
    see also Law/legal instru-
        ments/regulatory bodies
International Association of
    Hydrological Sciences
    (IAHS), **5:**183
International Bottled Water
    Association (IBWA), **4:**26,
    34, **5:**174

International Commission on
    Irrigation and Drainage
    (ICID), **5:**183
International Commission on
    Large Dams (ICOLD), **1:**70
International Conference on
    Water and the Environ-
    ment (1992), **3:**37
International Council of Bottled
    Water Association
    (ICBWA), **4:**26
International Covenant on
    Economic, Social, and
    Cultural Rights
    (1966/1976), **2:**4, **4:**208
    actors other than states, obli-
        gations of, **4:**231
    Article 2 (1), **2:**6–7
    Articles 11 and 12, **2:**7
    Declaration on the Right to
        Development, **2:**9
    implementation at national
        level, **4:**228–30
    introduction, **4:**216–18
    normative content of the
        right of water,
        4:218–21
    special topics of broad appli-
        cation, **4:**220–21
    states parties' obligations,
        4:222–26
    violations, **4:**226–28
International development
    organizations and WCD
    report, **3:**164–65
International Drinking Water
    Supply and Sanitation
    Decade (1981-90), **3:**37
International Food Policy
    Research Institute (IFPRI),
    2:64
International Freshwater Confer-
    ence in Bonn (2001),
    3:xviii
International Hydrological
    Program (IHP), **5:**183
International Joint Commission
    between U.S. and Canada,
    3:50
International Law Association,
    5:35
International Law Commission,
    1:107
International Maize and Wheat
    Improvement Center, **2:**75
International river basins:
    a new assessment, **2:**27–35
    by country, **2:**247–54
    fraction of a country's area in,
        2:239–46
    of the world, **2:**219–38
    see also Colorado River
International Water Association
    (IWA), **5:**182
International Water Ltd., **3:**70
International Water Manage-
    ment Institute (IWMI),
    3:197, **4:**88, 108
International Water Resources
    Association (IWRA), **1:**172,
    5:183

Internet, the, **1:**231–34, **2:**192–96,
    3:225–35
Ion-exchange desalination
    methods, **2:**104
Iran, **1:**58, **5:**8
Iraq, **1:**59, 110–11, 118, **5:**13,
    15–16
Irrigation for gardens/lawns,
    4:122
    see also irrigation *under* Agri-
        culture
Israel:
    conflict/cooperation con-
        cerning freshwater,
        1:107, 109, 110–11,
        115–16, **5:**6, 7, 10,
        14–15
    desalination, **5:**51, 69, 71,
        72
    drip irrigation, **1:**23
    environmental flow, **5:**33
    globalization and interna-
        tional trade of water,
        3:45
    reclaimed water, **1:**25, 29,
        2:138, 142
    terrorism, **5:**21
    well-being, measuring water
        scarcity and, **3:**98
Italy, **3:**47, 61, **5:**11

Japan:
    conflict/cooperation con-
        cerning freshwater, **5:**5
    dracunculiasis, **1:**52, 53
    environmental flow, **5:**42
    industrial water use, **1:**20
    reclaimed water, **2:**139, 140,
        158–59
    soft path for meeting water-
        related needs, **3:**23
    World Commission on Dams,
        3:159
Jarboe, James F., **5:**4
Jefferson, Thomas, **2:**94
Jerusalem Post, **5:**71
Jolly, Richard, **2:**3, **4:**196
Jordan, **1:**107, 109, 115–16, **2:**33
    conflict/cooperation con-
        cerning freshwater,
        5:12
    environmental flow, **5:**33
Jupiter's moons, search for water
    on, **3:**217–18

Kantor, Mickey, **3:**51–52
Kelp beds and desalination, **5:**79
Kennebec Coalition, **2:**123
Kennebec Hydro Developers,
    2:124
Kennedy, John F., **2:**95
Kenya:
    dracunculiasis, **1:**53, 55
    droughts, **5:**92, 99
    fog collection as a source of
        water, **2:**175
    food needs for current/future
        populations, **2:**76–77
    sanitation services, **1:**42
Khan, Akhtar H., **1:**39, **4:**71
King, Angus, **2:**123

Kiribati, **3:**118

Kitchens and CII water use,
      4:135, 137

Kokh, Peter, **3:**218

Korean peninsula, **1:**53, 109–10

Korean War, **1:**110

Kosovo, **5:**9

Kruger National Park, **1:**120–23,
      5:32

Kurdish Workers' Party (PKK),
      5:22

Kuwait, **1:**111, **2:**94, 97, **4:**18,
      5:69, 160, 163

La Niña, **3:**119, **5:**95

La Paz-El Alto, **3:**68, 69–72

Labeling and bottled water,
      4:28–31

Lagash-Umma border dispute,
      5:5

Lake levels/conditions and
      climate change, **4:**168–69

Lakes, specific:
      Chad, **1:**111, 148
      Chapala, **3:**77
      Kostonjärvi, **5:**33
      Mead, **3:**137, 140
      Mono, **5:**37, 41
      Oulujärvi, **5:**33
      Powell, **1:**76, **2:**129, 130

Land availability/quality and
      food needs for
      current/future popula-
      tions, **2:**70–71, 73–74

Land-use management and
      floods, **5:**111–12

Landscape design, **1:**23,
      4:122–23, 135, 137–38

Laos, **1:**16, 71

Las Vegas (NV) and water supply
      reliability, **5:**74

Laser leveling, agriculture and,
      3:20

Latin America:
      bottled water, **4:**40
      cholera, **1:**56, 57, 59–61, **3:**2
      climate change, **1:**147
      dams, **1:**77, 81
      drinking water, **1:**262
      hydroelectric production, **1:**71
      irrigation, **2:**86
      needs, basic human, **1:**47
      population, **2:**214
      sanitation services, **1:**264,
            3:271, **5:**259
      see also Central America;
            South America

Laundry water and CII water
      use, **4:**138

Lavelin International, **1:**85

Law/legal instruments/regula-
      tory bodies:
      Agenda 21, **5:**34
      Agreement on Technical
            Barriers to Trade
            (TBT), **4:**35
      Agreement on the Applica-
            tion of Sanitary and
            Phytosanitary
            Measures (SPS), **4:**35

Beijing Platform of Action,
      2:8

Berlin Conference Report
      (2004), **5:**35

Bern Geneva Convention
      (1877), **1:**114

Bureau of Government
      Research (BGR),
      4:69–70

Bureau of Reclamation, U.S.
      (BoR), **1:**7, 69, 88,
      2:128, **3:**137, **4:**10

Cairo Programme of Action,
      2:8

California Coastal Commis-
      sion (CCC), **5:**2

California Energy Commis-
      sion, **4:**157, **5:**76

Climate and Water Panel,
      1:149

Code of Federal Regulations
      (CFR), **4:**31–32

Codex Alimentarius Commis-
      sion (CAC), **4:**26, 35–36

Colorado River, **3:**137–39

Consortium for Energy Effi-
      ciency (CEE), **4:**115

Consultative Group on Inter-
      national Agricultural
      Research (CGIAR), **3:**90

Convention of the Law of the
      Non-Navigational Uses
      of International Water-
      courses (1997), **5:**35

Convention of the Rights of
      the Child (1989), **2:**4, 9,
      4:209

Convention on Biological
      Diversity (CBD), **3:**166,
      5:34

Copenhagen Declaration, **2:**8

Corporate Industrial Water
      Management Group,
      5:157

Declaration on the Right to
      Development (1986),
      2:4

Dublin Conference (1992),
      1:24, 165–66, 169, **3:**37,
      58, 101, **5:**34

Earth Summit (1992), **3:**38,
      88, 101, **5:**137

Emergency Management and
      Emergency Prepared-
      ness Office, **5:**24

environmental flow, **5:**34–37

Environmental Modification
      Convention (1977),
      1:114, **5:**4

Environmental Protection
      Agency (EPA), **2:**123,
      152, **4:**37, 52, **5:**23, 24

European Convention on
      Human Rights (1950),
      2:4, 8

Federal Bureau of Investiga-
      tion (FBI), **5:**24

Federal Emergency Manage-
      ment Agency (FEMA),
      5:24, 96–97

Federal Energy Regulatory
      Commission (FERC),
      1:83, **2:**123–24, 126,
      5:36

First National People of Color
      Environmental Lead-
      ership Summit (1991),
      5:120

Fish and Wildlife Service, U.S.
      (FWS), **2:**126, 128

Food and Agricultural Orga-
      nization (FAO), **5:**126

Food and Drug Administra-
      tion (FDA), **4:**26, 37,
      40, **5:**171
      see also United States federal
            regulations *under* Bottled
            water

Ganges Water Agreement
      (1977), **1:**119

General Agreement on Tariffs
      and Trade (GATT),
      3:47–52

Geneva Conventions, **5:**4

Global Reporting Initiative
      (GRI), **5:**158

Global Water Partnership
      (GWP), **1:**165–72, 175,
      176, **5:**183

groundwater, **4:**95–96

Hague Declaration (2000),
      5:139, 140

Helsinki Rules (1966), **1:**114

Homeland Security, U.S.
      Department of, **5:**23

Housing and Urban Develop-
      ment, U.S. Depart-
      ment of (HUD), **4:**117

human rights and interna-
      tional law, **2:**4–9

India-Bangladesh, **1:**107, 119,
      206–9

Intergovernmental Panel on
      Climate Change
      (IPCC), **1:**137, 138, 140,
      145, 149, **3:**120–23,
      5:81, 136

international water meetings,
      5:182–85

Interior Department, U.S.,
      2:123, 127

Israel-Jordan Peace Treaty
      (1994), **1:**107, 115–16

Joint Declaration to Enhance
      Cooperation in the
      Colorado River Delta,
      3:144

Kyoto Protocol, **5:**137

Kyoto Third World Water
      Forum (2003), **5:**183

Mar del Plata Conference
      (1977), **1:**40, 42,
      2:8, 10, 47, **4:**209,
      5:183, 185

Massachusetts Water
      Resources Authority,
      3:20

Mekong River Commission
      (MRC), **1:**82,
      3:165, **5:**35

Ministerial Declaration from Bonn (2001), **3:**173, 178–80

Ministerial Declaration from the Hague (2000), **3:**173–77, **4:**2

Minute 306, **3:**144–45

Natural Resources Council of Maine, **2:**123

North American Free Trade Agreement (NAFTA), **3:**47–48, 51–54

North American Water and Power Alliance (NAWAPA), **1:**74

Okavango River Basin Commission (OKACOM), **1:**122, 124

Organisation for African Unity (OAU), **1:**120

Organization for Economic Cooperation and Development (OECD), **2:**56, **3:**90, 91, 164, **4:**6

overview, **1:**155

Ramsar Convention, **3:**166, **5:**34

role of international law, **1:**114–15

Secretariat, Global Water Partnership's, **1:**170–71

Snake River Dam Removal Economics Working Group, **2:**132

South African Department of Water Affairs and Forestry, **1:**96

South Asian Association for Regional Cooperation (SAARC), **1:**118

Southern Africa Development Community (SADC), **1:**156–58, 169, **3:**165

Southwest Florida Water Management District (SWFWMD), **2:**108, **5:**62

Surface Water Treatment Rule (SWTR), **4:**52

Swedish International Development Agency (SIDA), **1:**165, 170, 171, **2:**14, **3:**162

Third World Centre for Water Management, **5:**139

U.S. Agency for International Development (USAID), **1:**44, **2:**10, 14

U.S. Mexico Treaty on the Utilization of the Colorado and Tijuana Rivers, **3:**138

U.S. National Primary Drinking Water Regulation (NPDWR), **3:**280–88

Upper Occoquan Sewage Authority, **2:**152

Vienna Declaration, **2:**8

Water Aid and Water for People, **2:**14

Water Environment Federation (WEF), **3:**78, **4:**62, **5:**182–83

Water Forum, Global Water Partnership's, **1:**170–71

Water Law Review Conference (1996), **1:**161

Water Sentinel Initiative, **5:**23

Water Supply and Sanitation Collaborative Council (WSSCC), **5:**138

see also American *listings;* International *listings;* Environmental justice; National *listings;* World *listings*

Lawns, energy efficient ways of maintaining, **1:**23

see also indoor water use *under* Urban water conservation: residential water use in California

Le Moigne, Guy, **1:**172, 174

Leak rates, **4:**109, 117–18

Leases and environmental flows, **5:**42

Leasing contracts and privatization, **3:**66

Lebanon, **1:**115

Lechwe, Kafue, **5:**32

Lecornu, Jacques, **1:**174

Legislation:

Bioterrorism Act of 2002, **5:**23

Clean Air Act, **4:**68

Clean Water Act, **1:**15, **4:**68, **5:**36

Electric Consumers Protection Act, **5:**36

Elwha River Ecosystem and Fisheries Restoration Act of 1992, **2:**127

Endangered Species Act (ESA) of 1973, **1:**15, **5:**37

Energy Policy Act of 1992, **1:**21

Federal Food, Drug, and Cosmetic Act, **4:**27

Federal Power Act, **5:**36

Federal Reclamation Act of 1902, **1:**8

Federal Wild and Scenic Rivers Act of 1966, **1:**15

Flood Control Act of 1936, **1:**16

National Environmental Policy Act (NEPA) of 1969/1970, **2:**130, **5:**36

National Wild and Scenic Rivers Act of 1997, **2:**120

Nutrition Labeling and Education Act of 1990, **4:**28

Public Health, Security, and Bioterrorism Preparedness and Response Act of 2002, **5:**23

Safe Drinking Water Act of 1974, **1:**15, **3:**280, **4:**27

Saline Water Conversion Act of 1952, **2:**94, 95

Water Desalination Act, **2:**95

Wild and Scenic Rivers Act, **5:**36

California

Central Valley Project Improvement Act, **5:**36

Coastal Act, **5:**80

Water Conservation in Landscaping Act of 1990, **4:**120–21

Israel

Water Law of 1959, **5:**35

Japan

River Law of 1997, **5:**35

South Africa

Act 54 of 1956, **1:**93, 160

National Water Law of 1998, **5:**35, 37

Switzerland

Water Protection Act of 1991, **5:**35

Length, measuring, **2:**301, **3:**319, **4:**326

Lesotho, **3:**159–60

Lesotho Highlands project:

chronology of events, **1:**100

components of, **1:**93, **95**

displaced people, **1:**97–98

economic issues, **1:**16

financing the, **1:**95–97

impacts of the, **1:**97–99

Kingdom of Lesotho, geographical characteristics of, **1:**93, 94

Lesotho Highlands Development Authority, **1:**96, 98

management team, **1:**93

opposition to, **1:**81, 98, 99

update, project, **1:**99–101

Levees and flood management, **5:**111

Levi Strauss, **5:**156

Li Bai, **1:**84

Liberia, **1:**63

Licenses for hydropower dams, **2:**114–15, 123–24

Linnaeus, **1:**51

Lovins, Amory B., **3:**xiii–xiv

Low-energy precision application (LEPA), **1:**23

Lunar Prospector spacecraft, **1:**197, **3:**213

Macedonia, **1:**71, **5:**10

Machiavelli, **1:**109

Malaria, **1:**49–50

Malawi, **4:**22

Malaysia:

conflict/cooperation concerning freshwater, **1:**110

data, strict access to water,
2:41
economics of water projects,
1:16
floods, 5:106
globalization and international trade of water,
3:46
hydroelectric production,
1:71
prices, water, 3:69
privatization, 3:61
Singapore, water disputes
with, 1:22
Maldives, 5:106
Mali, 1:52–53, 55
Mallorca, 3:45, 46
Manila Water Company, 4:46
Mao Tse-tung, 1:85
Mariner 4, 5:175, 177
Marion Pepsi-Cola Bottling Co.,
4:38
Market approach to water
pricing, 1:27
Mars Climate Orbiter, 2:300
Mars Express, 5:178
Mars Global Surveyor (MGS),
3:214, 5:177
Mars Odyssey, 3:xx
Mars Orbital Camera (MOC),
3:215, 5:178
Mars Reconnaissance Orbiter,
5:180
Mars, water on, 3:214–17, 5:175
future Mars missions, 5:180
history of, 5:178–80
instrumental analyses, 5:178
missions to Mars, 5:175–78
overview, 5:175
visual evidence of, 5:177–78
Marshall Islands, 3:118, 5:136
Mass, measuring, 2:304
Mauritania, 1:55, 4:82
Maximum available savings
(MAS), 3:18, 4:105
Maximum cost-effective savings
(MCES), 3:18, 24, 4:105
Maximum practical savings
(MPS), 3:18, 24, 4:105
Maytag Corporation, 1:23
McKernan, John, 2:123
McPhee, John, 2:113
Measurements, water, 2:25,
300–309, 3:318–27,
4:325–34
see also Assessments; Well-
being, measuring water
scarcity and
Meat consumption, 2:68–69, 72,
79–80
Media portrayals of terrorism,
5:14
Mediterranean Sea, 1:75, 77
Medusa Corporation, 1:204
Meetings/conferences, interna-
tional, *see* International
listings; Law/legal instru-
ments/regulatory bodies;
United Nations; World
listings

Membrane technologies and
desalination, 2:104–5,
5:54–55, 64
Metals in aquatic ecosystems,
4:168
Meteorites, water-bearing,
3:210–12
Meteorological droughts, 5:94
Methane, 1:138, 139
Mexico:
bottled water, 4:40, 5:170
Colorado River, 3:134, 137,
138, 141, 144–45
efficiency, improving water-
use, 1:19
environmental flow, 5:37
groundwater, 4:82, 83, 96,
5:125
hydroelectric production,
1:71
irrigation, 3:289
monitoring efforts, 3:76–77
North American Free Trade
Agreement, 3:51–54
privatization, 3:60
sanitation services, 3:272
Microfiltration and desalination,
5:55
Middle East:
bottled water, 4:289, 291,
5:281, 283
conflict/cooperation con-
cerning freshwater,
1:107, 109, 111,
115–18, 2:182, 5:15
consumption, 4:288
desalination, 2:94, 97, 5:54,
55, 57, 58, 68–69
dracunculiasis, 1:272–73,
3:274–75, 5:295,
296
environmental flow, 5:33
groundwater, 4:82, 85, 5:125
irrigation, 2:87
reclaimed water, 1:28, 2:139
see also Southeastern
Anatolia Project; *individual countries*
Military/political goal, water as
a, 1:108–9
see also Conflict/cooperation
concerning freshwater;
Terrorism
Millennium Development Goals
(MDGs):
commitments to achieving
the goals, 4:2, 6–7
diseases, water-related
classes of, four, 4:8–9
future deaths from, 4:10,
12–13
measures of illness/death,
4:9–11
mortality from, 4:9–10
overview, 4:7–8
overview, 4:xv, 1
projections for meeting,
4:13–14
summary/conclusions, 4:14
water/sanitation goals, 4:2–5

Mineral water, 4:29
Ministerial statements/declara-
tions at global water con-
ferences, 5:184–85
see also Law/legal instru-
ments/regulatory bodies
Minoan civilization, 1:40, 2:137
Mohamad, Mahathir, 1:110
Mokaba, Peter, 1:123
Moldavia, 5:8
Monitoring and privatization,
3:75–77, 81–82, 4:59–60,
62–65
see also monitoring/manage-
ment problems *under*
Groundwater
Monkey Wrench Gang, The
(Abbey), 2:129, 5:14
Monterey County (CA), 2:151
Moon, search for water on the,
3:212–14
Mothers of East Los Angeles,
1:22
Mount Pelion, 4:39
Movies and portrayals of
terrorism, 5:14
Mozambique, 1:63, 119–21, 5:7
Mueller, Robert, 5:4
Muir, John, 1:80–81
Multiple-effect distillation
desalination process,
2:99–100
Multistage flash distillation
(MSF) desalination
process, 2:96, 100
Municipal water, 4:29
see also Urban *listings*

Nahkla meteorite, 3:216
Nalco, 3:61
Namibia:
conflict/cooperation con-
cerning freshwater,
1:119, 122–24
fog collection as a source of
water, 2:175
Lesotho Highlands project,
1:98–99
reclaimed water, 1:28, 2:152,
156–58
Nanofiltration and desalination,
5:55
Narmada Project, 1:17
National Academy of Sciences,
1:28, 2:155, 166
National Aeronautics and Space
Administration (NASA),
5:96, 178
see also Outer space, search
for water in
National Arsenic Mitigation
Information Centre,
2:172
National Council of Women of
Canada, 3:78, 4:62
National Drought Policy Commission, 5:95
National Environmental Protection Agency of China,
1:92

National Fish and Wildlife Foundation, **2:**124
National Geographic, **3:**89
National Institute of Preventative and Social Medicine, **2:**167
National Marine Fisheries Service, **2:**128, 132
National Park Service, U.S. (NPS), **2:**117, 127
National Radio Astronomy Observatory, **3:**221
Native Americans, **5:**123
Natural label on bottled water, **4:**29
Natural Springs, **4:**38
Nauru, **3:**45, 46, 118
Needs, basic water, **1:**185–86, **2:**10–13, **4:**49–51
    see also Drinking water, access to clean; Health, water and human; Human right to water; Sanitation services; Well-being, measuring water scarcity and
Negev desert, **4:**89
Nepal:
    arsenic in groundwater, **4:**87
    bottled water, **4:**22–23
    conflict/cooperation concerning freshwater, **5:**11
    dams, **1:**16, 71, 81
    irrigation, **2:**86
    World Commission on Dams, **3:**160
Nestle, **4:**21, 40, 41, **5:**163
Netherlands:
    arsenic in groundwater, **2:**167
    climate change, **4:**158, 232
    dracunculiasis, **1:**52
    Global Water Partnership, **1:**169
    Millennium Development Goals, **4:**7
    open access to information, **4:**72–73
    public-private partnerships, **3:**66
Neufeld, David, **1:**197
Neutron Spectrometer, **3:**213
Nevali Cori archaeological site, **3:**185
New Economy of Water, The (Gleick), **4:**47
New Hampshire, **2:**118
New Orleans (LA), **4:**67–70
New Orleans City Business, **4:**69
New Orleans Times Picayune, **4:**69
New York (NY), **4:**51–53
New York Times, **2:**113, **5:**20
New Yorker, **2:**113
New Zealand:
    bottled water, **4:**26
    climate change, **5:**136
    environmental flow, **5:**33
    globalization and international trade of water, **3:**45, 46

privatization, **3:**61
    reservoirs, **2:**270, 272, 274
Newton Valley Water, **4:**38
Nigeria, **1:**52, 55, **5:**41
Nike, **5:**156
Nitrous oxide, **1:**138, 139
Nongovernmental organizations (NGOs), **1:**81, **3:**157, **4:**198, **205–6**
    see also individual organizations
Nordic Water Supply Company, **1:**202–5
Norway, **1:**52, 89, **3:**160
Nutrient cycling/loading, **1:**77, **4:**172

Oak Ridge Laboratory, **1:**23, **4:**115
Oberti Olives, **3:**22–23
Oceania:
    availability, water, **2:**24, 217
    bottled water, **4:**289, 291, **5:**281, 283
    cholera, **1:**267, 270, **5:**289, 290
    consumption, **4:**288
    dams, **3:**295
    drinking water, **1:**254–55, 262, **3:**259–60, **5:**245
    groundwater, **4:**86
    hydroelectric production, **1:**71, 280
    irrigation, **1:**301, **2:**263, 265, **3:**289, **4:**296, **5:**299
    population data/issues, **1:**250, **2:**214
    privatization, **3:**61
    renewable freshwater supply, **1:**240, **2:**202, 217, **3:**242, **4:**266, **5:**227
    salinization, **2:**268
    sanitation services, **1:**259–60, 264, **3:**268–69, 272, **5:**254–55, 261
    threatened/at risk species, **1:**295–96
    withdrawals, water, **1:**244, **2:**211, **3:**251, **4:**275, **5:**236
    see also Pacific Island developing countries
Official development assistance (ODA), **4:**6, 278–83, **5:**262–72
Ogallala Aquifer, **3:**50
Ogoni people, **5:**124
Olivero, John, **1:**196
Oman, the Sultanate of, **2:**179–80
Ontario Hydro, **1:**88
Orange County (CA), **2:**152
Order of the Rising Sun, **5:**20
Orion Nebula, **3:**220
Outer space, search for water in:
    clouds, interstellar, **3:**219–20
    earth's water, origin of, **3:**209–12
    exploration plans, **3:**216–17
    Jupiter's moons, **3:**217–18
    Mars, **3:**214–17

moon, the, **3:**212–14
    solar system, beyond our, **3:**218–19
    summary/conclusions, **3:**221–22
    universe, on the other side of the, **3:**220–21
    see also Mars, water on
Ozguc, Nimet, **3:**185

Pacific Institute:
    bottled water, **4:**24
    climate change, **4:**232, 234, 236
    conflict/cooperation concerning freshwater, **2:**182, **4:**238
    privatization, **4:**45–46, 193–94, **5:**132–33
    urban residential water use in California, **4:**105
Pacific Island developing countries (PIDCs):
    climate change
        overview, **3:**xix
        precipitation, **3:**124–25
        projections for 21st century, **3:**121–23
        science overview, **3:**119–21
        sea-level rise, **3:**124
        severe impacts of, **5:**136
        storms and temperatures, **3:**125
    freshwater resources
        description and status of, **3:**115–18
        overview, **3:**113–14
        threats to, **3:**118–19
    profile of, **3:**115
    summary/conclusions, **3:**125–27
    terrain of, **3:**116
    see also Oceania
Packard Humanities Institute, **3:**187
Pakistan:
    agriculture, **4:**88, 89
    bottled water, **4:**40
    conflict/cooperation concerning freshwater, **5:**10, 13, 15
    dracunculiasis, **1:**53
    groundwater, **4:**82, 88, 89
    Orangi Pilot Project, **4:**71–72
    World Commission on Dams, **3:**160
Palau, **3:**118
Palestinians, **1:**109, 118, **5:**6, 10, 13–15
Panama, **3:**160
Paraguay, **3:**13
Parasites, *see* diseases, water-related *under* Health, water and human
Peak-load pricing, **1:**26
Pennsylvania, **2:**118
Pepsi, **4:**21, **5:**146
Permitting, wastewater, **4:**150
Perrier, **4:**21, 38, 40, 41, **5:**151

Persian Gulf War, **1:**110, 111
Persians, **3:**184
Peru, **1:**46, 59–60, **2:**178–79
Pesticides, **5:**20, 128, 305–7
PET, **4:**39, 41
Pets, purchased food going to feed, **2:**77
Philippines:
    bottled water, **4:**40
    cholera, **1:**63
    conflict/cooperation concerning freshwater, **5:**11
    dams, **1:**71
    prices, water, **3:**69
    privatization, **3:**60, 61, 66, **4:**46
    World Commission on Dams, **3:**161
Pinchot, Gifford, **1:**80–81
Pluto, **3:**220
Poland, **3:**66, 160–61, **4:**38
Poland Spring, **4:**21
Polar satellite, **3:**209–10
Politics, *see* Government/politics
Pollution, *see* Environmental issues
Pomona (CA), **2:**138
Poor/poverty, **4:**40–41, **5:**123–24
    See also Developing countries; Environmental justice
Population issues:
    0 to AD 2050, world population from, **2:**212
    by continent, **2:**213–14
    dams and displaced people, **1:**77–80
    developing countries lacking basic water services, **3:**2
    diseases, projected deaths from water-related, **4:**12–13
    expanding water-resources infrastructure, **1:**6
    food needs, **2:**63–64, 66–67
    growth between 2000 and 2020, **4:**10
    Pacific Island developing countries, **3:**118
    total and urban population data, **1:**246–50
    withdrawals, water, **1:**10, 12, 13, **3:**308–9
Portugal, **1:**71
Poseidon Resources, **2:**108
Poseidon Water Resources, **5:**61
Postel, Sandra, **1:**111
Power, measuring, **2:**308, **3:**326, **4:**333
Precipitation:
    climate change, **1:**140–41, 146–47, **4:**159, 166
    Pacific Island developing countries, **3:**115–16, 124–25
    stocks and flows of freshwater, **2:**20, 22
Precision Fabrics Group, **1:**52

Pressure, measuring, **2:**307, **3:**325, **4:**332
Preston, Guy, **4:**50
Pricing, water:
    bottled water, **4:**22–23
    climate change, **4:**180–81
    households in different/cities/countries, **3:**304
    Jordan, **1:**117
    privatization, **3:**69–71, 73, **4:**53–55
    rate structures, water system, **3:**303
    twentieth-century water-resources development, **1:**24–28
    urban commercial/industrial water use in California, **4:**150
    urban residential water use in California, **4:**124–25
    see also subsidies *under* Economy/economic issues
Private goods, **3:**34
Privatization:
    business/industry, water risks that face, **5:**152–53
    conflict/cooperation concerning freshwater, **3:**xviii, 70–71, 79, **4:**54, 67
    defining terms, **3:**35
    drivers behind, **3:**58–59
    economic issues, **3:**70–72, **4:**50, 53–60
    environmental justice, **5:**131–33
    failed privatizations, **3:**70
    forms of, **3:**63–67, **4:**47, 48
    history of, **3:**59–61
    opposition to, **3:**58
    overview, **3:**57–58, **4:**xvi, 45–46
    players involved, **3:**61–63
    principles and standards
        can the principles be met, **4:**48–49
        economics, use sound, **3:**80–81, **4:**53–60
        overview, **3:**79, **4:**47–48
        regulation and public oversight, government, **3:**81–82, **4:**60–73
        social good, manage water as a, **3:**80, **4:**49–53
    risks involved
        affordability questions, pricing and, **3:**69–73
        dispute-resolution process, weak, **3:**79
        ecosystems and downstream water users, **3:**77
        efficiency, water, **3:**77–78

        government, usurping responsibilities of, **3:**68
        irreversible, privatization may be, **3:**79
        local communities, transferring assets out of, **3:**79
        monitoring, lack of, **3:**75–77
        overview, **3:**67–68
        public ownership, failing to protect, **3:**74–75
        quality, water, **3:**78
        underrepresented communities, bypassing, **3:**68
    sanitation services, **5:**273–75
    summary/conclusions, **3:**82–83, **4:**73–74
    update on, **4:**46–47
    World Water Forum (2003), **4:**192, 193–94
Process water use and CII water use, **4:**134–36
Procter & Gamble, **5:**157
Productivity, water, **3:**17–19
Projections, review of global water resources:
    Alcamo et al. (1997), **2:**56–57
    analysis and conclusions, **2:**58–59
    data constraints, **2:**40–42
    defining terms, **2:**41
    Falkenmark and Lindh (1974), **2:**47–49
    Gleick (1997), **2:**54–55
    in 2002, **3:**xvii–xviii
    inaccuracy of past projections, **2:**43–44
    Kalinin and Shiklomanov (1974) and De Mare (1976), **2:**46–47
    L'vovich (1974), **2:**44–47
    Nikitopoulos (1962, 1967), **2:**44
    overview, **2:**39–40
    Raskin et al. (1997, 1998), **2:**55–56
    Seckler et al. (1988), **2:**57–58
    Shiklomanov (1993, 1998), **2:**50–53
    World Resources Institute (1990) and Belyaev (1990), **2:**49–50
    see also Sustainable vision for the earth's freshwater
Public Citizen, **4:**69
Public goods, **3:**34
Public Limited Companies (PLC), **4:**72–73
Public participation in water decision making, **5:**150–51
Public Trust Doctrine, **5:**37
Public-private partnerships, **3:**74–75, **4:**60–73, 193–94
Puerto Rico, **3:**46, **5:**34
Pupfish, desert, **3:**142
Pure Life, **4:**40
Purified water, **4:**29

Quantitative measures of water availability/use, **2:**25

Race and environmental discrimination, **5:**118
  see also Environmental justice
Radiative forcing, **3:**120
Rail, Yuma clapper, **3:**142
Rand Water, **1:**95, 96
Raw or value-added resource, water traded as a, **3:**42–47
Reagan, Ronald, **2:**95
Recalls, bottled water, **4:**37–40, **5:**171–74
Reclaimed water:
  agricultural water use, **2:**139, 142, 145–46
  costs, wastewater, **2:**159
  defining terms, **2:**139
  environmental and ecosystem restoration, **2:**149–50
  food needs for current/future populations, **2:**87
  groundwater recharge, **2:**150–51
  health issues, **2:**152–56
  Israel, **1:**25, 29, **2:**138, 142
  Japan, **2:**139, 140, 158–59
  Namibia, **2:**152, 156–58
  overview, **1:**28–29, **2:**137–38
  potable water reuse, direct/indirect, **2:**151–52
  primary/secondary/tertiary treatment, **2:**138
  processes involved, **2:**140
  summary/conclusions, **2:**159–61
  urban areas, **1:**25, **2:**146–49, **4:**151
  uses, wastewater, **2:**139, 141–42
  see also under California
Regulatory bodies
  see also Law/legal instruments/regulatory bodies
Rehydration therapy, cholera and, **1:**57
Reliability, desalination and water-supply, **5:**73–74
Religious importance of water, **3:**40
Renewable freshwater supply:
  by continent, **2:**215–17
  by country, **1:**235–40, **2:**197–202, **3:**237–42; **4:**261–266, **5:**221–27
  globalization and international trade of water, **3:**39
Reptiles, threatened, **1:**291–96
Reservoirs:
  U.S. capacity, **1:**70
  built per year, number, **2:**116
  climate change, **1:**142, 144–45
  earthquakes caused by filling, **1:**77, 97
  environmental issues, **1:**75, 77, 91

Mars, **3:**215–16
number larger than 0.1km by continent/time series, **2:**271–72
orbit of earth affected by, **1:**70
sediment, **1:**91, **2:**127, **3:**136, 139, **4:**169
total number by continent/volume, **2:**270
twentieth-century water-resources development, **1:**6
volume larger than 0.1km by continent/time series, **2:**273–74
volume of U.S., **2:**116
  see also Dams
Reservoirs, specific:
  Diamond Valley, **3:**13
  Imperial, **3:**136
  Itaipú, **1:**75
  Mesohora, **1:**144
  Occoquan, **2:**152
Residential water use, **1:**21–23
  see also Urban water conservation: residential water use in California
Restrooms and CII water use, **4:**135, 136
Reuse, water, *see* Reclaimed water
Revelle, Roger, **1:**149
Reverse osmosis (RO) and desalination, **2:**96, 102–3, **5:**51, 55, 57, 58, 60, 72
Rhodesia, **5:**7
Rice and food needs for current/future populations, **2:**74–79
Right to water:
  see also Human right to water
Risk management, droughts and, **5:**99
Rivers:
  climate change, **1:**142, 143, 145, 148
  dams' ecological impact on, **1:**77, 91
  deltas, **3:**xix
  Federal Wild and Scenic Rivers Act, **1:**15
  floods, **5:**104
  flow rates, **2:**305–6, **3:**104, 324, **4:**168, 172, 331
  pollution and large-scale engineering projects, **1:**6
  restoring, movement towards, **2:**xix
  runoff, **1:**142, 143, 148, **2:**222–24, **4:**163–67, **5:**29
  see also Environmental flow; International river basins; law/legal instruments *under* South Africa; Stock and flows of freshwater; Sustainable vision for the earth's freshwater

Rivers, specific:
  Abang Xi, **1:**69
  Allier, **2:**125
  Amazon, **1:**75, 111, **2:**32
  American, **1:**16
  Amu Darya, **3:**3, 39–40
  Apple, **2:**118
  AuSable, **2:**119
  Bhagirathi, **1:**16
  Brahmaputra, **1:**107, 111, 118–19, 206–9
  Butte Creek, **2:**120–23
  Carmel, **5:**74
  Cauvery, **1:**109, **2:**27
  Clyde, **2:**125–26
  Colorado, *see* Colorado River
  Columbia, **3:**3
  Congo, **1:**75, 111, 156, **2:**31, 32–33
  Crocodile, **1:**123
  Danube, **1:**109, **2:**31, **5:**33, 111
  Elwha, **2:**117, 127–28
  Euphrates, **1:**109–11, 118, **2:**33, **3:**182, 183–87
  Ganges, **1:**107, 111, 118–19, 206–9, **4:**81
  Gila, **3:**139, 140
  Gordon, **2:**126–27
  Han, **1:**109–10
  Incomati, **1:**120–23
  Indus, **1:**16, 77, 111
  Jordan, **1:**107, 109, 111, 115–16, **2:**31, **5:**33
  Kennebec, **2:**117, 123–25, **5:**34
  Kettle, **2:**119
  Kissimmee, **5:**32
  Kromme, **5:**32
  Laguna Salada, **3:**139
  Lamoille, **2:**128
  Lerma, **3:**77
  Letaba, **1:**123
  Limpopo, **1:**120–23
  Logone, **5:**32
  Loire, **2:**117, 125
  Lower Snake, **2:**131–34
  Luvuvhu, **1:**123
  Mahaweli Ganga, **1:**16, **5:**32
  Malibamats'o, **1:**93
  Manavgat, **1:**203, 204–5, **3:**45, 47
  Manitowoc, **2:**119
  Maputo, **1:**121
  McCloud, **5:**123
  Mekong, **1:**111
  Merrimack, **2:**118
  Meuse, **5:**15
  Milwaukee, **2:**117–19
  Mississippi, **2:**32, 33, **3:**13, **5:**110, 111
  Missouri, **3:**13
  Mooi, **4:**51
  Murray-Darling, **5:**33, 42
  Narmada, **5:**133
  Neuse, **2:**126
  Niger, **1:**55, 111, **2:**31, 85
  Nile, **1:**77, 111, **2:**26, 32–33, **3:**10–11, **5:**111
  Okavango, **1:**111, 119, 121–24
  Olifants, **1:**123

Orange, **1:**93, 98–99, 111
Orontes, **1:**111, 115
Pamehac, **5:**34
Paraná, **1:**111, **3:**13
Patauxent, **5:**34
Po, **5:**111
Prairie, **2:**118
Rhine, **5:**111
Rhone, **3:**45, 47
Rio Grande, **1:**111, **5:**41
Rogue, **2:**119, 128
Sabie, **5:**32
Sacramento, **2:**120–23, **4:**164, 167, 169, **5:**34, 111
San Joaquin, **4:**164, 169
Senegal, **1:**111, **2:**85
Shingwedzi, **1:**123
Sierra Nevada, **4:**164
Snake, **2:**131–34, **3:**3
Spöl, **5:**33
Suzhou, **5:**32
Syr Darya, **3:**3, 39–40
Tarim, **5:**32
Temuka, **5:**33
Theodosia, **5:**34
Tigris, **1:**69, 111, 118, **2:**33, **3:**182, 187–90
Vaal, **1:**95
Volga, **1:**77
Wadi Mujib, **5:**33
Waitaki, **5:**33
White Salmon, **2:**119
Yahara, **2:**118
Yangtze, **5:**133
Yarmouk, **1:**109, 115–16
Yellow, **5:**5, 15–16, 111
Zambezi, **1:**111
  see also Lesotho Highlands project; Three Gorges Dam
Road not Taken, The (Frost), **3:**1
Roaring Springs/Global Beverage Systems, **4:**39
Rodenticides, **5:**20
Rome, ancient, **1:**40, **2:**137, **3:**184
Roome, John, **1:**99
Roosevelt, Franklin, **1:**69
Runoff, river, **1:**142, 143, 148, **2:**22–24, **4:**163–67, **5:**29
Rural development and World Water Forum (2003), **4:**203
Russell, James M., III, **1:**196
Russia:
  aquatic species threatened by dams, **1:**77
  dams, **1:**75, 77
  groundwater, **5:**125
  hydroelectric production, **1:**71
  irrigation, **4:**296
  see also Soviet Union, former
Rwanda, **1:**62
RWE/Thames, **5:**162

S&W Water, LLC, **5:**61
Safeway Water, **4:**39
Salat Tepe archaeological site, **3:**189

Salinization:
  by country, **2:**269
  climate change, **4:**168, 169–70
  continental distribution, **2:**268
  groundwater, **4:**87
  salt concentrations of different waters, **2:**21, 94
  soil fertility, **2:**73–74
  see also Desalination
Salmon, **1:**77, **2:**117, 120–21, 123, 128, 132, 133, **3:**3
Samoa, **3:**118
Samosata, **3:**184–85
Samsat, **3:**184–85
San Francisco Bay, **3:**77, **4:**169, 183, **5:**73
San Francisco Chronicle, **4:**24
San Jose/Santa Clara Wastewater Pollution Control Plant, **2:**149–50
San Pellegrino, **4:**21
Sanitation services:
  by country, **1:**256–60, **3:**261–69, **5:**247–55
  by region, **3:**270–72, **5:**256–61
  China, **5:**124, 147
  developing countries, **1:**263–64
  environmental justice, **5:**127–31
  falling behind, **1:**39–42, **5:**117, 124
  Honduras, **4:**54–55
  international organizations, recommendations by, **2:**10–11
  investment in infrastructure projects with private participation, **5:**273–75
  official development assistance, **4:**282–83
  Pakistan, **4:**71–72
  twentieth-century water-resources development, **3:**2
  well-being, measuring water scarcity and, **3:**96–98
  women and access to water, **5:**126
  World Health Organization, **4:**208
  World Water Forum (2003), **4:**202, 205
  see also Health, water and human; Millennium Development Goals; Human right to water; Soft path for meeting water-related needs; Well-being, measuring water scarcity and
Santa Barbara (CA), **5:**63–64
Santa Rosa (CA), **2:**145–46
Sapir, Eddie, **4:**69
Sargon of Assyria, **1:**110
Saudi Arabia:
  desalination, **2:**94, 97
  dracunculiasis, **1:**52

groundwater, **3:**50
  international river basin, **2:**33
  pricing, water, **1:**24
Save the Children Fund, **4:**63
Save Water and Energy Education Program (SWEEP), **4:**114
Schistosomiasis, **1:**48, 49
School of Environmental Studies (SOES), **2:**167, 171
Scientific American, **3:**89
Sea-level rise, **3:**124, **4:**169–70
Seagram Company, **3:**61
Seasonal pricing, **1:**26
Sedimentation, dams/reservoirs and, **1:**91, **2:**127, **3:**136, 139, **4:**169
Seismicity, reservoir-induced, **1:**77, 97
Seljuk Turks, **3:**184, 188
Senegal, **1:**55
Serageldin, Ismail, **1:**166
Services, basic water, *see* Drinking water, access to clean; Health, water and human; Human right to water; Sanitation services
Servicio Nacional de Meterologia e Hidrologia, **2:**179
Sewer systems, condominial, **3:**6
  see also Sanitation services
Shad, **2:**123
Shady, Aly M., **1:**174
Shaping the 21st Century project, **3:**91
Shoemaker, Eugene, **1:**196
Showerheads, **4:**109, 114
Shrimp, **3:**49, 50, 141, 142
Sierra Club, **1:**81
Singapore:
  conflict/cooperation concerning freshwater, **1:**110
  data, strict access to water, **2:**41
  desalination, **2:**108, **5:**51
  toilets, energy-efficient, **1:**22
  water-use efficiency, **4:**58–60
Skanska, **3:**167
Slovakia, **1:**109, 120
Slovenia, **1:**71
SNC, **1:**85
Snow, John, **1:**56–57
Snowfall/snowmelt, changes in, **1:**142, 147, **4:**160–61
Social goods and services, **3:**36–37, 80, **4:**49–53
Société de distribution d'eau de la Côte d'Ivoire (SODECI), **4:**66
Société pour l'aménagement urbain et rural (SAUR), **4:**66
Socioeconomic droughts, **5:**94
Soft Energy Paths (Gleick), **3:**xiii
Soft path for water:
  dominance of hard path in twentieth century, **3:**7–9
  economies of scale in collection/distribution, **3:**8

efficiency of use: definitions/concepts
  agriculture, **3:**19–20
  businesses, **3:**22–24
  conservation and water-use efficiency, **3:**17
  maximum practical/cost-effective savings, **3:**23
  municipal scale, **3:**20–22
  overview, **3:**16–17
  poem, **5:**219
  productivity and intensity, water, **3:**17–19
  social objectives, establishing, **3:**17
emerging technologies, **5:**23–24
end-use technology, simple, **3:**8–9
hard and soft path, differences between, **3:**3, 5–7
how much water is really needed, **3:**4
moving forward
  overview, **3:**25–26
  step 1: identifying the potential, **3:**26–27
  step 2: identifying barriers, **3:**27–28
  step 3: making social choices, **3:**28–29
  step 4: implementing demand management programs, **3:**29
myths about
  cost-effective, efficiency improvements are not, **3:**12–15
  demand management is too complicated, **3:**15–16
  market forces, water demand is unaffected by, **3:**9
  opportunities are small, efficiency, **3:**9
  real, conserved water is not, **3:**10–11
  risky, efficiency improvements are, **3:**11–12
overview, **3:**xviii
redefining the energy problem, **3:**xiii
sewer systems, condominial, **3:**6
shared water resources *vs.* individual water use, **3:**7
summary/conclusions, **3:**30
user participation, **3:**5, 6
see also Sustainable vision for the earth's freshwater; Urban water listings
Soil:
  climate change and moisture changes, **1:**141–42, 148, **4:**167
  degradation by type/cause, **2:**266–67

food needs for current/future populations, **2:**71, 73–74
hard path for meeting water-related needs, **3:**2
Solar energy and desalination, **2:**105–6
Solar radiation powering climate, **1:**138
Solon, **5:**5
Somalia, **5:**106
Sonoran Desert, **3:**142
South Africa:
  bottled water, **4:**22
  conflict/cooperation concerning freshwater, **1:**107, 119–21, 123–24, **5:**7, 9
  dams, **1:**81
  Development Bank of South Africa, **1:**95, 96
  environmental flow, **5:**32, 35, 37, 42
  General Agreement on Tariffs and Trade, **3:**49
  human right to water, **2:**9, **4:**211
  law/legal instruments
    Apartheid Equal Rights Amendment (ERA), **1:**158–59
    Constitution and Bill of Rights, **1:**159–60, **2:**9
    hydrology of Southern Africa, **1:**156–58
    National Water Conservation Campaign, **1:**164–65
    review process, legal/policy, **1:**160–64
    White Paper on Water Supply, **1:**160
  privatization, **3:**60, **4:**49–51
  South African Department of Water Affairs and Forestry, **1:**96
  World Commission on Dams, **3:**161
  see also Lesotho Highlands project
South America:
  availability, water, **2:**217
  bottled water, **4:**18, 289, 291, **5:**163, 281, 283
  cholera, **1:**266, 270, 271
  dams, **1:**75, **3:**293
  drinking water, **1:**253, **3:**257, **5:**243
  environmental flow, **5:**34
  groundwater, **4:**86
  hydroelectric production, **1:**278
  irrigation, **1:**299, **2:**80, 259, 265, **3:**289, **4:**296, **5:**299
  population data, total/urban, **1:**248
  privatization, **3:**60
  renewable freshwater supply, **1:**238, **2:**200–201, 217, **3:**240, **4:**264, **5:**224

reservoirs, **2:**270, 272, 274
runoff, **2:**23
salinization, **2:**268
sanitation services, **1:**258, **3:**266, **5:**252
threatened/at risk species, **1:**293
withdrawals, water, **1:**243, **2:**207–8, **3:**247–48, **4:**271–72, **5:**232–33
see also Latin America
Southeastern Anatolia Project (GAP):
  archaeology in the region, **3:**183
  Euphrates River, developments on the, **3:**183–87
  overview, **3:**181–83
  summary/conclusions, **3:**190–91
  Tigris River, developments along the, **3:**187–90
Southern Bottled Water Company, **4:**39
Soviet Union, former:
  cholera, **1:**58
  climate change, **1:**147
  dams, **1:**70, **3:**293
  environmental movement, **1:**15
  international river basin, **2:**29, 31
  irrigation, **1:**301, **2:**263, 265, **3:**289, **4:**296
  renewable freshwater supply, **4:**266
  withdrawals, water, **1:**244, **2:**211, **3:**250–51
  see also Russia
Spain:
  agriculture, **4:**89
  conflict/cooperation concerning freshwater, **5:**5
  dams, **1:**71
  environmental flow, **5:**33
  globalization and international trade of water, **3:**45, 47
  groundwater, **4:**89
  Three Gorges Dam project, **1:**89
  World Commission on Dams, **3:**161
Sparkling water, **4:**30
Spectrometer, neutron, **3:**213
Spectroscopy, telescopic, **5:**175
Spragg, Terry, **1:**203–5
Spring water, **4:**30
Sprinkler systems for irrigation, **2:**82
Sri Lanka, **1:**69, **2:**86, **3:**161–62
  dams, **5:**134
  environmental flow, **5:**32
  floods, **5:**106
St. Petersburg (FL), **2:**146–47
Stakeholder participation in water management, **5:**140–41
Starbucks, **5:**163

Stocks and flows of freshwater:
    flows of freshwater, **2:**22–24
    geopolitics of international
        river basins, **2:**35–36
    hydrologic cycle, **2:**20–27
    international river basins: a
        new assessment,
        2:27–35
    major stocks of water on
        earth, **2:**21–22
    overview, **2:**19–20
    summary/conclusions,
        2:36–37
Stone & Webster Company,
    2:108, **5:**61
Storage volume relative to
    renewable supply (S/O),
    3:102
Storm frequency/intensity,
    changes in, **1:**142–43,
    4:161–63
Stream sediments and pesti-
    cides, **5:**305–7
Strong, Maurice, **1:**88
*Structure of Scientific Revolu-
    tions* (Kuhn), **1:**193
Sturgeon, **1:**77, 90, **2:**123
*Submillimeter Wave Astronomy
    Satellite* (SWAS), **3:**219–20
Subsidies, *see under* Economy/
    economic issues
Sudan, **1:**55, **2:**26, **5:**7, 13
Suez Lyonnaise des Eaux,
    3:61–63, **4:**46
Supervisory Control and Data
    Acquisition (SCADA), **5:**16
Supply-side development, *see*
    Twentieth-century water-
    resources development
Surface irrigation, **2:**82, 84
Sustainable Asset Management
    (SAM) Group, **3:**167
Sustainable vision for the Earth's
    freshwater:
    agriculture, **1:**187–88
    climate change, **1:**191
    conflict/cooperation con-
        cerning freshwater,
        1:190
    criteria, sustainability,
        1:17–18
    diseases, water-related,
        1:186–87
    ecosystems water needs
        identified and met,
        1:188–90
    introduction, **1:**183–84
    needs, basic human,
        1:185–86
    public participation, **1:**82
    see also Soft path for meeting
        water-related needs;
        Twenty-first century
        water-resources develop-
        ment; Urban water *listings*
Swaziland, **1:**121
Sweden, **1:**52, 96, **3:**162
Switzerland, **1:**89, 171, **3:**162,
    5:33, 35
Sydney Morning Herald, **5:**66

Synthesis Report (2001), **5:**136
Syria, **1:**109, 110–11, 116, 118
Syriana, **5:**14

Taenia solium, **4:**8
Tahoe-Truckee Sanitation
    Agency, **2:**152
Tajikistan, **5:**9
Tampa Bay (FL), **2:**108–9,
    5:61–63
Tanzania, **1:**63
Taskun Kale archaeological site,
    3:184
Tear Fund, **5:**131
Technical efficiency, **4:**103–4
Temperature rise, global, **1:**138,
    145, **3:**120–23, **4:**159, 166
    see also Climate change
    listings
Temperature, measuring, **2:**307,
    3:325, **4:**332
Tennant Method and environ-
    mental flow, **5:**38
Tennessee Valley Authority,
    1:69–70, 145
Terrorism, **2:**35, **5:**1–3
    chemical/biologic attacks,
        vulnerability to,
        5:16–22
    defining terms, **5:**3–5
    detection and protection
        challenges, **5:**23
    early warning systems,
        5:23–24
    environmental terrorism,
        5:3–5
    history of water-related
        conflict, **5:**5–15
    infrastructure attacks, vulner-
        ability to, **5:**15–16
    overview, **5:**1–2
    physical access, protection by
        denying, **5:**22–23
    policy in the U.S., security,
        5:23, 25
    public perception/response,
        5:2–3
    response plans, emergency,
        5:24–25
    safeguards (water treatment)
        reducing vulnerability,
        5:2
    summary/conclusions,
        5:25–26
Texas, **3:**74–75
Thailand, **4:**40, **5:**33, 106, 134
Thames Water, **3:**63, **5:**62
Thatcher, Margaret, **1:**106, **3:**61
Thirsty for Justice, **5:**122
Threatened/at risk species:
    by country, **2:**291–97
    by region, **1:**291–96
    Colorado River, **3:**134, 142
    dams, **1:**77, 83, 90, **2:**120, 123
    extinction rates for fauna
        from continental North
        America, **4:**313–14
    proportion of species at risk
        in U.S., **4:**313–16

twentieth-century water-
    resources develop-
    ment, **3:**3
    water transfers, **3:**39–40
Threats to water systems, *see*
    Terrorism
Three Affiliated Tribes, **5:**123
Three Gorges Dam:
    chronology of events, **1:**85–87
    displaced people, **1:**78, 85, 90,
        5:134, 151
    economic issues, **1:**16
    financing the, **1:**86–89
    hydroelectric production, **1:**84
    impacts of the, **1:**89–92
    largest most powerful ever
        built, **1:**84
    opposition to, **1:**91–93
Tier pricing, **1:**26
Tille Hoyuk archaeological site,
    3:185
Time, measuring, **2:**304, **4:**329
Togo, **1:**55
Toilets, **1:**21–22, **3:**4, 118, **4:**104,
    109, 113–14
Tonga, **3:**46, 118
Touré, A. T., **1:**53
Toxic waste dumps, **5:**119, 124
*Toxic Wastes and Race in the
    United States,* **5:**119
Trachoma, **1:**48
Trade/industry associations and
    WCD report, **3:**169–70
    see also Globalization and
        international trade of
        water
Traditional planning
    approaches, **1:**5
    see also Projections, review of
        global water resources;
        Twentieth-century
        water-resources devel-
        opment
Transfers, water, **1:**27–28, 74–75,
    3:39–40
    see also Dams
Transpiration: loss of water into
    atmosphere, **1:**141, **2:**83,
    4:159–60
Treaties, *see* Law/legal instru-
    ments/regulatory bodies;
    United Nations
Trichuriasis, **1:**48
Trinidad, **2:**108, **5:**72
Trout Unlimited, **2:**118, 123, 128
Tuna/shrimp disputes, **3:**49, 50
Tunisia, **2:**142, **5:**33
Turkey:
    bag technology, water, **1:**202–5
    conflict/cooperation con-
        cerning freshwater,
        1:110, 118, **5:**8
    globalization and interna-
        tional trade of water,
        3:45–47
    terrorism, **5:**22
Turkish Antiquity Service,
    3:183
    see also Southeastern
        Anatolia Project

Turtle/dolphin disputes, **3:**49, 50
Tuvalu, **5:**136
Tuxedo, The, **5:**14
Twentieth-century water-
    resources development:
    Army Corps of Engineers and
        Bureau of Reclama-
        tion, U.S., **1:**7–8
    benefits of, **3:**2
    capital investment, **1:**6–7
    drivers of, three major, **1:**6
    end of
        alternatives to new infra-
            structure, **1:**17–18
        demand, changing nature
            of, **1:**10–14
        economics of water
            projects, **1:**16–17
        environmental movement,
            1:12, 15–16
        opposition to projects
            financed by inter-
            national organiza-
            tions, **1:**17
        overview, **1:**9–10
        shift in paradigm of
            human water use,
            1:5–6
    government, reliance on,
        1:7–8
    limitations to, **1:**8–9, **3:**2–3
    problems/disturbing charac-
        teristics of current
        situation, **1:**1–2
    summary/conclusions, **1:**32
    supply-side solutions, **1:**6
Twenty-first century water-
    resources development:
    agriculture, **1:**23–24
    alternative supplies, **1:**28
    desalination, **1:**29–32
    efficient use of water, **1:**19–20
    industrial water use, **1:**20–21
    overview, **1:**18–19
    Pacific Island developing
        countries, **3:**121–23
    pricing, water, **1:**24–28
    reclaimed water, **1:**28–29
    residential water use, **1:**21–23
    shift in the paradigm of
        human water use,
        1:5–6
    summary/conclusions,
        1:32–33
    see also Soft path for meeting
        water-related needs; Sus-
        tainable vision for the
        earth's freshwater
Typhoid, **1:**48
Typhus, **1:**48

U.S. Filter Company, **3:**63
U.S. National Water Assessment,
    5:112
Uganda, **1:**55, **4:**211
Ultrafiltration and desalination,
    5:55
Unaccounted for water, **3:**305,
    307, **4:**59

Undiminished principle and the
    human right to water, **5:**37
Unilever, **5:**149
United Arab Emirates (UAE),
    5:68–69
United Kingdom, *see* England
United Nations:
    Agenda 21, **1:**18, 44, **3:**90
    arsenic in groundwater,
        2:167, 172
    Children's Fund (UNICEF),
        1:52, 55, **2:**167, 172,
        173
    Commission on Human
        Rights, **2:**5
    Commission on Sustainable
        Development, **2:**10,
        3:90
    Committee on Economic,
        Social, and Cultural
        Rights, **5:**117, 137
    Comprehensive Assessment
        of the Freshwater
        Resources of the World
        (1997), **1:**42–43
    Conference on International
        Organization (1945),
        2:5
    conflict/cooperation con-
        cerning freshwater,
        1:107, 114, 118–19,
        124, 210–30, **2:**36
    Convention on the Non-Nav-
        igational Uses of Inter-
        national Watercourses
        (1997), **1:**107, 114, 124,
        210–30, **2:**10, 36,
        41–42, **3:**191
    data, strict access to water,
        2:41–42
    Declaration on the Right to
        Development (1986),
        2:8–10
    Development Programme
        (UNDP), **1:**52, 82, 171,
        2:172, 173, **3:**90, **4:**7,
        5:100
    diseases, water-related, **5:**117
    dracunculiasis, **1:**52, 55
    drinking water, **1:**40, 251
    droughts, **5:**100
    Earth Summit (1992), **3:**38,
        88, 101
    Environment Programme
        (UNEP), **1:**137, **3:**127,
        164
    environmental justice,
        5:137–38
    Food and Agriculture Organi-
        zation (FAO), **2:**64, 67
    food needs for current/future
        populations, **2:**64, 66,
        67
    Framework Convention on
        Climate Change
        (UNFCC), **3:**126
    Global Water Partnership,
        1:165, 166, 171, 175
    greenhouse gases, **3:**126
    groundwater, **4:**80–81

Human Poverty Index, **3:**87,
        89, 90, 109–11
    human right to water, **2:**3, 5–9,
        14, **4:**208, 214, **5:**117
    International Conference on
        Nutrition (1992), **14**
    public participation and sus-
        tainable water
        planning, **1:**82
    Summit for Children (1990),
        1:52, **2:**14
    Universal Declaration of
        Human Rights, **2:**4–10,
        4:208
    well-being, measuring water
        scarcity and, **3:**90, 96,
        109–11
    World Water Council, **1:**172,
        173, 175
    see also International
        Covenant on Economic,
        Social & Cultural Rights;
        Law/legal instruments/
        regulatory bodies; Millen-
        nium Development Goals
United States:
    availability, water, **2:**217
    bottled water
        brands, leading, **4:**22
        consumption figures,
            4:288–91, **5:**170,
            281, 283
        growth rates, **5:**163
        imports, **3:**44
        pricing, **4:**22
        Pure Life *vs.,* **4:**40
        quality, **4:**25–26
        regulations, federal,
            4:27–34
        sales figures, **3:**43, **4:**292
        tap water and bottled
            water, comparing
            standards for,
            4:36–37
    business/industry, water
        risks, **5:**162–63
    cholera, **1:**56, 266, 270, 271
    climate change, **1:**148
    conflict/cooperation con-
        cerning freshwater,
        1:110, 111, **5:**6–9,
        11–12, 24
    dams, **1:**69–70, **3:**293
    desalination, **2:**94–95, 97,
        5:58–63
    diseases, water-related,
        4:308–12
    dracunculiasis, **1:**52
    drinking water, **1:**253, **3:**257,
        280–88, **5:**242
    droughts, **5:**93
    economic productivity of
        water, **4:**321–24
    environmental flow, **5:**34,
        36–37
    environmental justice,
        5:119–20, 122–23
    floods, **4:**305–7
    food needs for current/future
        populations, **2:**68–69

General Agreement on Tariffs and Trade, **3:**50

groundwater, **3:**2, **4:**82, 86, 96, **5:**125

human right to water, **4:**213

hydroelectric production, **1:**71, 278

irrigation, **1:**299, **2:**265, **3:**289, **4:**296, **5:**299

meat consumption, **2:**79–80

needs for water, basic human, **1:**42–43

North American Free Trade Agreement, **3:**51–54

pesticides, **5:**305–7

population data/issues, **1:**248, **2:**214

precipitation changes, **1:**146–47

privatization, **3:**58–60

renewable freshwater supply, **1:**238, **2:**200, 217, **3:**240, **4:**264, **5:**224

reservoirs, **2:**270, 272, 274

runoff, **2:**23

salinization, **2:**268

sanitation services, **1:**258, **3:**266, 272, **5:**251

terrorism, **5:**21–23, 25

threatened/at risk species, **1:**293, **4:**313–16

usage estimates, **1:**245

water industry revenue/ growth, **5:**303–4

well-being, measuring water scarcity and, **3:**92

withdrawals, water, **1:**243, **2:**207, **3:**247, 308–12, **4:**271, 317–20, **5:**232

see also California; Colorado River; Dams, removing/ decommissioning; Urban water conservation *listings*

United Utilities, **3:**63

United Water Resources, **3:**61, 63

United Water Services Atlanta, **3:**62, **4:**46

Universidad de San Augustin, **2:**179

University of California at Santa Barbara (UCSB), **3:**20–22

University of Kassel, **2:**56

University of Michigan, **3:**183

Upper Atmosphere Research Satellite (UARS), **1:**196

Urban areas:
    droughts, **5:**98
    floods, **5:**104
    municipal water, **4:**29
    pricing, water, **1:**25–27
    privatization, **3:**76
    reclaimed water, **1:**25, **2:**146–49
    soft path for meeting water-related needs, **3:**20–22

Urbanization, **5:**98

Urban water conservation: commercial/industrial water use in California:

background to CII water use, **4:**132–33

calculating water conservation potential, methods for, **4:**143–47

current water use in CII sectors, **4:**133–38

data challenges, **4:**139–40, 148–50, 152–53

defining CII water conservation, **4:**132

evolution of conservation technologies, **4:**149

overview, **4:**131–32

potential savings, **4:**140–43

recommendations for CII water conservation, **4:**150–53

summary/conclusions, **4:**153–54

water use by end use, **4:**138–39

Urban water conservation: residential water use in California:

abbreviations and acronyms, **4:**126–27

agricultural water use, **4:**107

current water use, **4:**105–6

data and information gaps, **4:**108–9

debate over California's water, **4:**102–3

defining conservation and efficiency, **4:**103–5

economics of water savings, **4:**107–8

indoor water use
    dishwashers, **4:**116
    end uses of water, **4:**112–13
    faucets, **4:**117
    leaks, **4:**117–18
    overview, **4:**109
    potential savings by end use, **4:**111–12
    showers and baths, **4:**114
    summary/conclusions, **4:**118
    toilets, **4:**113–14
    total use without conservation efforts, **4:**110
    washing machines, **4:**114–16

outdoor water use
    current use, **4:**119–20
    existing efforts/ approaches, **4:**120–21
    hardware improvements, **4:**122–23
    landscape design, **4:**122–23
    management practices, **4:**121–22
    overview, **4:**118–19
    rate structures, **4:**124–25
    summary/conclusions, **4:**125–26

overview, **4:**101–2

Urfa, **3:**185

Urlama, **5:**5

Usage estimates, **1:**46; **1:**245
    see also Withdrawals, water

Use, defining water, **1:**12

User fees and environmental flows, **5:**41–42

Utilities and risks that face business/industry, large/small, **5:**162–63

Uzbekistan, **1:**52, **4:**40

V is for Vendetta, **5:**14

van Ardenne, Agnes, **4:**196

Vapor compression distillation desalination process, **2:**100–101

Variations in regional water resources, **2:**218

Varieties of Environmentalism (Guha & Martinez-Alier), **5:**123

Velocity, measuring, **2:**305, **3:**323, **4:**330

Veolia Environnement, **5:**162

Vermont Natural Resources Council, **2:**128

Very long baseline interferometry (VLBI), **3:**221

Vibrio cholerae, **1:**56. 1:57, 58

Vietnam, **4:**18, **5:**134, 163

Viking, **3:**214, **5:**177

Virgin Islands, U.S., **3:**46

Virginia, **2:**152

Visalia (CA), **1:**29

Vision 21 process, **2:**3

Vivendi, **3:**61–64, 70, **4:**47

Voith and Siemens, **1:**85

Volume, measuring, **2:**303–4, **3:**321–22, **4:**328–29
    see also Stocks and flows of freshwater

Volumetric allocation systems, **4:**95–97

Waggoner, Paul, **1:**149

Waimiri-Atroari people, **5:**134

Wall Street Journal, **1:**89

Warfare, environmental, **5:**4–5
    see also Conflict/cooperation concerning freshwater; Terrorism

Warming, global, **1:**138
    see also Climate change *listings*

Washing machines, **1:**23, **4:**114–16, **5:**219, 220

Wastewater treatment, **1:**6, **5:**153, 159–60
    see also Reclaimed water; Sanitation services

Water & Process Technologies, **5:**159

Water (Vizcaino, Zolla, Sagan, Margulis, Molina, Bras, Cousteau, Raven, Ceballos, Wolff & Gleick), **5:**219

Water in Crisis: A Guide to the World's Fresh Water Resources (Gleick), **2:**300, **3:**318
Water industry, *see* Business/industry, water risks; Economy/economic issues
Water on Earth, origins of, **1:**193–98, **3:**209–12
Water Policy for the American People, A, **3:**16, **4:**103
Water quality:
    bottled water, **4:**17, 25–26, 31–32, 37–40
    business/industry, water risks that face, **5:**146–49
    climate change, **4:**167–68
    desalination, **5:**74–75
    droughts, **5:**98, 102
    environmental justice, **5:**127–29
    floods, **5:**109
    groundwater, **4:**83, 87
    Guidelines for Drinking-Water Quality, **4:**26–27, 31
    privatization, **3:**78
    see also Desalination; Drinking water, access to clean; Environmental flow; Salinization
Water Resources Policy Committee, **3:**16
Water Supply and Sanitation Collaborative Council (WSSCC), **2:**3, 13–14
Water units/data conversions/constants, **2:**300–309; **3:**318–327; **4:**325–334; **5:**319–328
Water use, defining, **1:**12
Water-use efficiency, **1:**19–20, **3:**77–78, **4:**xvi–xvii, 58–60, **5:**153, 157
    see also Soft path for meeting water-related needs; Sustainable vision for the Earth's freshwater; Twenty-first century water-resources development; Urban water *listings*
WaterAid, **5:**131
Waterborne diseases, **1:**47–49, 274–75, **4:**8
Waynilad Water, **4:**46
WCD, *see* World Commission on Dams
Weather Underground group, **5:**20
Web-sites, water-related, **1:**231–34, **2:**192–96, **3:**225–35
Weight of water, measuring, **2:**309, **3:**327, **4:**334
Well water, **4:**30
    see also Groundwater
Well-being, measuring water scarcity and:

Falkenmark Water Stress Index, **3:**98–100
indicators, quality-of-life
    current, **3:**90–92
    development and use of, **3:**89
    limitations to, **3:**92–96
    what are indicators, **3:**88–89
multifactor indicators
    Human Poverty Index, **3:**87, 89, 90, 109–11
    Index of Human Insecurity, **3:**107, 109
    International Water Management Institute, **3:**197, **108**
    overview, **3:**101
    vulnerability of water systems, **3:**101–4
    Water Poverty Index, **3:**110–11
    Water Resources Vulnerability Index, **3:**105–6
overview, **3:**xviii–xix, 87–88
single-factor measures
    basic water requirement, **3:**101–3
    sanitation services and drinking water, **3:**96–98
summary/conclusions, **3:**111
Wetlands, **1:**6, **3:**141–43, **5:**111
Wetlands, specific:
    Ciénega de Santa Clara, **3:**141–43
    El Doctor, **3:**141
    El Indio, **3:**141, 142
    Rio Hardy, **3:**141
Wheat, **2:**75, **4:**89
White, Gilbert, **3:**16
Williams, Ted, **2:**119
Wind energy and desalination, **2:**105
Wintu people, **5:**123
Wisconsin, **2:**117–18
Withdrawals, water:
    by country and sector, **1:**241–44, **2:**203–11, **3:**243–51, **4:**267–75, **5:**228–36
    conflict/cooperation concerning freshwater, **1:**112
    consumption in U.S. compared to, **1:**12, 13
    defining terms, **1:**12
    gross national product
        China, **3:**316–17
        Hong Kong, **3:**313–15
        United States, **3:**310–12
    population in the U.S., **1:**10, 12, 13
    soft path for meeting water-related needs, **3:**23–24
    threatened/at risk species, **3:**3

total/per-capital, **1:**10, 11
United States, **1:**10, 11, **3:**308–12, **4:**317–20
see also monitoring/management problems *under* Groundwater; Projections, review of global water resources
Wolf, Aaron, **2:**28
Wolff, Gary, **3:**xiv
Women and environmental justice, **5:**126, 134
World Bank:
    arsenic in groundwater, **2:**172–73
    business/industry, water risks that face, **5:**147
    dams, **1:**82–83
    diseases, water-related, **4:**9
    displaced people, dams and, **1:**78
    dracunculiasis, **1:**52
    Global Water Partnership, **1:**165, 171, **5:**183
    human right to water, **2:**10–11
    Lesotho Highlands project, **1:**96, 99
    needs, basic human, **1:**44, 47
    opposition to projects financed by, **1:**17
    overruns, water-supply projects, **3:**13
    privatization, **3:**59, 70, **4:**46
    sanitation services, **5:**273
    self-review of dams funded by, **1:**175–76
    Southeastern Anatolia Project, **3:**190–91
    Three Gorges Dam project, **1:**85, 88
    World Water Council, **1:**173
World Climate Conference (1991), **1:**149
World Commission on Dams (WCD):
    data and feedback from five major sources, **3:**150
    environmental flow, **5:**30
    environmental justice, **5:**134, 135
    findings and recommendations, **3:**151–53
    goals, **3:**150–51
    organizational structure, **3:**149–50
    origins of, **1:**83, 177–79
    overview, **3:**xix
    priorities/criteria/guidelines, **3:**153–58
    reaction to the report
        conventions, international, **3:**166
        development organizations, international, **3:**164–65
        funding organizations, **3:**158, 162–64, 167–69
        governments, **3:**170–71

industry/trade associations, international, **3:**169–70
national responses, **3:**159–62
nongovernmental organizations, **3:**157
overview, **3:**155–56
private sector, **3:**166
regional groups, **3:**165
rights and risk assessment, **3:**153
Southeastern Anatolia Project, **3:**191
summary/conclusions, **3:**171–72
WCD Forum, **3:**150
World Conservation Union (IUCN), **1:**82–83, 121, 177, **3:**164
World Council on Sustainable Development, **5:**158
World Court, **1:**109, 120
World Food Council, **2:**14
World Fund for Water (proposed), **1:**174–75
World Health Assembly, **1:**52
World Health Organization (WHO):
  arsenic in groundwater, **2:**166, 167, 172
  bottled water, **4:**26–27
  cholera, **1:**61, 271

desalination, **5:**75
diseases, water-related, **4:**9, **5:**117
dracunculiasis, **1:**52, 55
drinking water, **3:**2, **4:**2, 208
human right to water, **2:**10–11
needs, basic human, **1:**44
reclaimed water, **2:**154, 155
sanitation services, **1:**256, **3:**2, **4:**2, 208
unaccounted for water, **3:**305
well-being, measuring water scarcity and, **3:**90, 91
World Health Reports, **4:**8, 9
World Meteorological Organization (WMO), **1:**137, **5:**100
World Resources Institute, **2:**27–28, 49–50
World Trade Organization (WTO), **3:**48–50
World Water Council (WWC), **1:**172–76, **3:**165, **4:**192–93, **5:**183
World Water Forum (2000), **3:**xviii, 58, 59, 90, 173
World Water Forum (2003):
  background to, **4:**192–94
  Camdessus Report, **4:**195–96, 206
  efficiency and privatization, lack of attention given to, **4:**192

focus of, **4:**191
human right to water, **4:**212
Millennium Development Goals, **4:**6, 7
Ministerial Statement, **4:**194–95, 200–204
NGO Statement, **4:**192, 198, 205–6
overview, **4:**xv
successes of, **4:**191–92
Summary Forum Statement, **4:**196–97
value of future forums, **4:**192
World Water Forum (2006), **5:**186–88
World Wildlife Fund International, **3:**157
Worldwatch Institute, **2:**28

Xeriscaping, **1:**23, **4:**123–24

Yangtze! Yangtze!, **1:**92
Yeates, Clayne, **1:**194–95
Yemen, **1:**53, 55

Zag meteorite, **3:**210–12
Zambia, **1:**63, **4:**211, **5:**9, 32
Zeugma archaeological site, **3:**185–87
Zimbabwe, **1:**107, **5:**7, 140
Ziyaret Tepe archaeological site, **3:**189
Zuari Agro-Chemical, **1:**21